INFRARED AND MILLIMETER WAVES

VOLUME 5 COHERENT SOURCES
AND APPLICATIONS, PART I

CONTRIBUTORS

E. Affolter

L. R. Barnett

V. L. Granatstein

J. O. Henningsen

M. Inguscio

F. K. Kneubühl

Benjamin Lax

Jun-ichi Nishizawa

M. E. Read

F. Strumia

INFRARED AND MILLIMETER WAVES

VOLUME 5 COHERENT SOURCES AND APPLICATIONS, PART I

Edited by **KENNETH J. BUTTON**

NATIONAL MAGNET LABORATORY
MASSACHUSETTS INSTITUTE OF TECHNOLOGY
CAMBRIDGE, MASSACHUSETTS

1982

ACADEMIC PRESS

A Subsidiary of Harcourt Brace Jovanovich, Publishers

New York   London
Paris   San Diego   San Francisco   São Paulo   Sydney   Tokyo   Toronto

ACADEMIC PRESS, INC.
111 Fifth Avenue, New York, New York 10003

United Kingdom Edition published by
ACADEMIC PRESS, INC. (LONDON) LTD.
24/28 Oval Road, London NW1 7DX

Library of Congress Cataloging in Publication Data
Main entry under title:

Infrared and millimeter waves.

    Includes bibliographies and indexes.
    CONTENTS: v. 1. Sources of radiation--
v. 5. Coherent sources and applications.
    1. Infra-red apparatus and appliances. 2. Millimeter
wave device. I. Button, Kenneth J.
TA1570.I52   621.36'2   79-6949
ISBN 0-12-147705-3 (v. 5)   AACR1

PRINTED IN THE UNITED STATES OF AMERICA

82 83 84 85   9 8 7 6 5 4 3 2 1

CONTENTS

LIST OF CONTRIBUTORS   vii
PREFACE   ix
CONTENTS OF OTHER VOLUMES   xi

Chapter 1   **Coherent Sources and Scientific Applications**
       Benjamin Lax

I.	Introduction	1
II.	Coherent Sources	2
III.	Scientific Applications	9
IV.	Conclusion	25
	References	25

Chapter 2   **Molecular Spectroscopy by Far-Infrared Laser Emission**
       J. O. Henningsen

I.	The Tools	29
II.	CH_3OH—A Case Story	57
	References	124

Chapter 3   **Stark Spectroscopy and Frequency Tuning
in Optically Pumped Far-Infrared
Molecular Lasers**
       F. Strumia and M. Inguscio

I.	Introduction	130
II.	Stark Effect in Polar Molecules	133
III.	Laser Gain Curves in the Presence of an Electric Field	139
IV.	General Performances of Optically Pumped FIR Lasers	155
V.	Experimental Results—dc Electric Fields	176
VI.	Experimental Results—ac Electric Fields	204
VII.	Conclusion	208
	References	210

Chapter 4   **The GaAs TUNNETT Diodes**
            Jun-ichi Nishizawa
     I.   Introduction   215
    II.   TUNNETT Diode   238
   III.   Future of TUNNETT   260
    IV.   Conclusion   264
          References   265

Chapter 5   **Measured Performance of Gyrotron Oscillators
            and Amplifiers**
            V. L. Granatstein, M. E. Read, and L. R. Barnett
     I.   Introduction   267
    II.   Gyrotron Traveling-Wave Tube Amplifiers   272
   III.   Gyromonotron Oscillators   281
          Appendix. Measurement Methods   298
          References   301

Chapter 6   **Distributed-Feedback Gas Lasers**
            F. K. Kneubühl and E. Affolter
     I.   Introduction   305
    II.   Dispersion Relations Based on Hill's Equation   308
   III.   Resonance Conditions   311
    IV.   DFB in Periodic Metal Guides   314
     V.   DFB Gas Laser in Operation   322
    VI.   Future Developments   334
          References   335

INDEX   339

LIST OF CONTRIBUTORS

Numbers in parentheses indicate the pages on which the authors' contributions begin.

E. Affolter * (305), *Physics Department, Eidgenössiche Technische Hochschule, Zürich, Switzerland*

L. R. Barnett (267), *Naval Research Laboratory, Washington, D. C. 20375*

V. L. Granatstein (267), *Naval Research Laboratory, Washington, D. C. 20375*

J. O. Henningsen (29), *Physics Laboratory I, H. C. Ørsted Institute, University of Copenhagen, Copenhagen, Denmark*

M. Inguscio (129), *Istituto di Fisica dell'Università, I-56100 Pisa, Italy*

F. K. Kneubühl (305), *Physics Department, Eidgenössiche Technische Hochschule, Zürich, Switzerland*

Benjamin Lax (1), *Francis Bitter National Magnet Laboratory and Department of Physics, Massachusetts Institute of Technology, Cambridge, Massachusetts 00139*

Jun-ichi Nishizawa (215), *Research Institute of Electrical Communication, Tohoku University, Sendai, Japan*

M. E. Read (267), *Naval Research Laboratory, Washington, D. C. 20375*

F. Strumia (129), *Istituto di Fisica dell'Università, I-56100 Pisa, Italy*

* Present address: Siemens-Albis, AG, CH-8047 Zürich, Switzerland.

vii

PREFACE

The subject of sources of radiation will become a subset of volumes, consisting so far of Volume 1 (Sources of Radiation), Volume 5 (Coherent Sources and Applications, Part I), and Volume 7 (Coherent Sources and Applications, Part II). Volume 1 dealt with gyrotrons, free electron lasers, IMPATT devices, pulsed optically pumped lasers, backward wave oscillators, and the ledatron. This was far from complete, of course, so we have tried to arrange for complete coverage as quickly as possible. Professor Kneubühl and Professor Sturzenegger gave us "Electrically Excited Submillimeter-Wave Lasers" in time to be published as Chapter 5 in Volume 3. Now Professor Kneubühl and Professor Affolter have given us "Distributed-Feedback Gas Lasers" as Chapter 6 in this book. In Volume 4 (Millimeter Systems), N. Bruce Kramer gave us full coverage of traveling-wave tubes, IMPATT and Gunn diodes. We are still waiting for someone to write up the very important extended interaction oscillator and amplifier. We do have the GaAs TUNNETT diodes in this book (Chapter 4), written by Professor Nishizawa, who originally developed this device.

As we continue to develop this subset on sources of millimeter and far-infrared radiation, we have the space to include a broader treatment by dealing with *performance* and *applications*. The keynote of this book, Volume 5, is the opening chapter on coherent sources and their scientific applications by Professor Benjamin Lax. This is followed by a very comprehensive treatment of spectroscopy of molecules by far-infrared laser emission prepared by Professor J. O. Henningsen. A similarly comprehensive treatment of Stark spectroscopy and frequency tuning in optically pumped far-infrared molecular lasers is given by Professor F. Strumia and Professor M. Inguscio. Then we were quite fortunate to get a follow-up on gyrotron oscillators and amplifiers in the form of "Measured Performance of Gyrotron Oscillators and Amplifiers" (Chapter 5) by some of the leaders of the outstanding group at Naval Research Laboratory, Dr. V. L. Granatstein, Dr. M. E. Read, and Dr. L. R. Barnett.

Volume 7 (Coherent Sources and Applications, Part II), will open with "CW Optically Pumped Lasers" by Professor Thomas A. DeTemple and Dr. E. Danielewicz. Professor DeTemple had already published his contribution on pulsed optically pumped far-infrared lasers in Volume 1. This

ix

will be followed by a chapter on mid-infrared optically pumped molecular lasers by Robert G. Harrison and Pradeep K. Gupta and a short chapter on the optimization of optically pumped far-infrared lasers by Dr. Konrad Walzer.

Dr. Philip Sprangle, who is also one of the leaders of the outstanding Plasma Physics Division of the Naval Research Laboratory, has contributed a chapter on free-electron lasers to Volume 7 as a complement to his theoretical chapter on the subject that appeared in Volume 1. Following this is a treatment of the oratron, originally known as the Smith–Purcell free-electron laser, which is a promising version of the FEL for generating coherent millimeter- and submillimeter-wave radiation. A definitive chapter describing this device has been prepared by Dr. Donald E. Wortman and Dr. Richard P. Leavitt to appear in Volume 7 (Coherent Sources and Applications, Part II). In this third book of the subset on sources, we shall have space to expand again into the applications with a chapter on high-frequency gyrotrons and applications to tokamak plasma heating by Kenneth E. Kreischer and Richard J. Temkin of the Plasma Fusion Center at M.I.T. This chapter was commissioned for the forthcoming volume on tokamak plasma diagnostics but its originality and the importance of its concepts earned it early publication in Coherent Sources and Applications, Part II. The subject of nonlinear frequency conversion in bulk crystals and semiconductor diodes will be developed by Dr. Dane D. Bicanic in "Generation of Tunable Laser Sidebands in the Terahertz Region by Frequency Mixing of the HCN Laser and a Microwave Source in a Metal–Semiconductor Diode". The basic subject of metal–semiconductor junctions as frequency converters has been covered by Dr. Martin V. Schneider in Chapter 4 of Volume 6.

CONTENTS OF OTHER VOLUMES

Volume 1:   Sources of Radiation

J. L. Hirshfield, Gyrotrons

H. J. Kuno, IMPATT Devices for Generation of Millimeter Waves

Thomas A. DeTemple, Pulsed Optically Pumped Far Infrared Lasers

G. Kantorowicz and P. Palluel, Backward Wave Oscillators

K. Mizuno and S. Ono, The Ledatron

F. K. Kneubühl and E. Affolter, Infrared and Submillimeter-Wave Waveguides

P. Sprangle, Robert A. Smith, and V. L. Granatstein, Free Electron Lasers and Stimulated Scattering from Relativistic Electron Beams

Volume 2:   Instrumentation

N. C. Luhmann, Jr., Instrumentation and Techniques for Plasma Diagnostics: An Overview

D. Véron, Submillimeter Interferometry of High-Density Plasmas

J. R. Birch and T. J. Parker, Dispersive Fourier Transform Spectroscopy

B. L. Bean and S. Perkowitz, Far Infrared Submillimeter Spectroscopy with an Optically Pumped Laser

Wallace M. Manheimer, Electron Cyclotron Heating of Tokamaks

Volume 3:   Submillimeter Techniques

T. G. Blaney, Detection Techniques at Short Millimeter and Submillimeter Wavelengths: An Overview

W. M. Kelley and G. T. Wrixon, Optimization of Schottky-Barrier Diodes for Low-Noise, Low-Conversion Loss Operation at Near-Millimeter Wavelengths

A. Hadni, Pyroelectricity and Pyroelectric Detectors

A. F. Gibson and M. F. Kimmitt, Photon Drag Detection

F. W. Kneubühl and Ch. Sturzenegger, Electrically Excited Submillimeter-Wave Lasers

xi

Michael von Ortenberg. Submillimeter Magnetospectroscopy of Charge Carriers in Semiconductors by Use of the Strip-Line Technique
Eizo Otsuka, Cyclotron Resonance and Related Studies of Semiconductors in Off-Thermal Equilibrium

Volume 4:   Millimeter Systems

James C. Wiltse, Introduction and Overview of Millimeter Waves
Edward K. Reedy and George W. Ewell, Millimeter Radar
Charles R. Seashore, Missile Guidance
N. Bruce Kramer, Sources of Millimeter-Wave Radiation: Traveling-Wave Tube and Solid-State Sources
Tatsuo Itoh, Dielectric Waveguide-Type Millimeter-Wave Integrated Circuits
M. Tsuji, H. Shigesawa, and K. Takiyama, Submillimeter Guided Wave Experiments with Dielectric Waveguides
Gary A. Gordon, Richard L. Hartman, and Paul W. Kruse, Imaging-Mode Operation of Active NMMW Systems

Volume 6:   Systems and Components

J. E. Harries, Infrared and Submillimeter Spectroscopy of the Atmosphere
D. H. Martin, Polarizing (Martin–Puplett) Interferometric Spectrometers for the Near- and Submillimeter Spectra
P. L. Richards and L. T. Greenberg, Infrared Detectors for Low Background Astronomy
M. V. Schneider, Metal–Semiconductor Junctions as Frequency Converters
Paul F. Goldsmith, Quasi-Optical Techniques at Millimeter and Submillimeter Wavelengths
G. D. Holah, Far-Infrared and Submillimeter Wavelength Filters

Volume 7:   Coherent Sources and Applications, Part II

Thomas A. DeTemple and E. Danielewicz, CW Optically Pumped Lasers
Robert G. Harrison and Pradeep K. Gupta, Mid-Infrared Optically Pumped Molecular Gas Lasers
Konrad Walzer, On the Optimization of Optically Pumped Far-Infrared Lasers
J. P. Pichamuthu, Submillimeter Lasers with Electrical, Chemical, and Incoherent Optical Excitation.

Dane D. Bicanic, Generation of Tunable Laser Sidebands in the Terahertz Region by Frequency Mixing of the HCN Laser and a Microwave Source in a Metal–Semiconductor Diode

Philip Sprangle, Quasi-Optical Laser

Donald E. Wortman and Richard P. Leavitt, The Oratron

Kenneth E. Kreischer and Richard J. Temkin, High-Frequency Gyrotrons and Applications to Tokamak Plasma Heating

INFRARED AND MILLIMETER WAVES

VOLUME 5 COHERENT SOURCES
AND APPLICATIONS, PART I

CHAPTER 1

Coherent Sources and Scientific Applications

Benjamin Lax

Francis Bitter National Magnet Laboratory*
and Department of Physics
Massachusetts Institute of Technology
Cambridge, Massachusetts

I.	INTRODUCTION	1
II.	COHERENT SOURCES	2
	A. *Harmonic Generation and Nonlinear Mixing*	2
	B. *Optical Pumping*	5
	C. *Electronic Devices*	6
	D. *The Free-Electron Laser*	8
III.	SCIENTIFIC APPLICATIONS	9
	A. *Cyclotron Resonance*	9
	B. *Magneto-Optical Effects and the Spin-Flip Laser*	14
	C. *Raman Transitions in Submillimeter Lasers*	15
	D. *Plasma Diagnostics*	18
	E. *Motional Stark Spectroscopy*	21
IV.	CONCLUSION	25
	REFERENCES	25

I. Introduction

Coherent sources for resonance spectroscopy now extend from the microwave to the ultraviolet region. Pioneers in this area of research, such as Professor Gordy, began with microwave sources that were developed during World War II for radar. The challenges of spectroscopy demanded coherent sources at shorter wavelengths. One of the first techniques for this purpose was to use harmonic generation of a microwave source and a crystal multiplier to obtain 5-mm radiation (Beringer, 1946). Professor Gordy and his students adapted this technique to make tunable millimeter sources in 1948 to study a variety of molecules (Gordy, 1948). During subsequent years they made numerous such measurements of gaseous mole-

* Supported by the National Science Foundation.

1

cules in the region of 2–5 mm (Gilliam *et al.*, 1950). They continued to extend these harmonic techniques to 1 mm and beyond into the submillimeter region (King and Gordy, 1953).

Spectroscopy of solids soon followed that of molecular spectroscopy using microwave techniques. The first experiments were those of paramagnetic resonance (Zavoisky, 1945) in 1945, and later ferromagnetic (Griffiths, 1946) and antiferromagnetic resonance (Poulis *et al.*, 1951). Cyclotron resonance at microwaves was first observed in plasmas (Lax *et al.*, 1950) but really reached prominence in semiconductors (Dresselhaus *et al.*, 1953; Lax *et al.*, 1954b) and metals (Galt *et al.*, 1955; Dexter and Lax, 1955). Cyclotron resonance and related magneto-optical studies in semiconductors and semimetals provided strong motivation for seeking resonance studies in the far-infrared region of the spectrum (Lax, 1979). These studies realized their full potential with the advent of lasers into this field of spectroscopy.

It is not surprising that it was molecular spectroscopy that gave rise to the concept of the maser by Gordon *et al.*, (1955; Basov and Prokhorov, 1955). Similarly, the invention of the first solid-state maser by Bloemberger (1956) resulted from basic studies of paramagnetic resonance. Ultimately, the maser led to the proposal of the laser by Schawlow and Townes (1958) and was realized by Maiman (1960). In succession, the He –Ne laser was developed by Javan and co-workers (1961), the CO_2 laser by Patel (1964a). It is the latter that gave rise to the optically pumped lasers in the submillimeter (Chang and Bridges, 1970) and tunable nonlinear sources from the IR to the millimeter regions (Aggarwal *et al.*, 1973; Lax *et al.*, 1973). These new lasers are now a source of interesting research as tools for resonance spectroscopy in solids and plasmas, atoms and molecules. The object of this paper is to present a short review of the status of a variety of coherent sources and their application to science.

II. Coherent Sources

A. Harmonic Generation and Nonlinear Mixing

It was indicated in Section I that harmonic generation was the first technique for extending the coherent sources of the electromagnetic spectrum to higher frequencies. Today this approach is still being pursued. The use of "cat's whisker" as a nonlinear element is being replaced by such devices as Schottky barrier diodes, varactors, and avalanche diodes. The latter may be the most important development for future millimeter and submillimeter spectroscopy. The Schottky barrier diode has been used by Fetterman and co-workers at Lincoln Laboratory both as a harmonic gen-

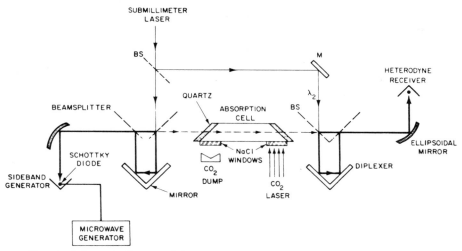

FIG. 1   Schematic diagram of the double resonance spectrometer. The intensity of the sideband beams transmitted through the sample cell was 40 dB above the receiver noise level. (After Blumberg *et al.*, 1979.)

erator and as a tunable sideband mixer with a submillimeter optically pumped laser (Blumberg *et al.*, 1979). Such a spectrometer is shown in Fig. 1 and has been used as a high-resolution instrument to measure the rotational transition in D_2O at 692 245 $\pm$ 1 MHz and $J = 5 \rightarrow 6$ rotational transition in CO at 691 473 $\pm$ 1 MHz. The latter is in good agreement with Helminger *et al.*, (1970).

The varactor diode is a fairly efficient device for generating millimeter waves at higher harmonics using standard microwave sources such as magnetrons, traveling wave tubes, etc. However, the avalanche diode may be actually superior in that its nonlinearity originates in the avalanche process itself, as well as from the nonlinear nature of the Read equation. The avalanche diode exhibits one of the strongest nonlinear processes per volume in any device known. Therefore, it is not surprising that harmonics as high as the 30th have been observed. Such devices have potential to operate as a submillimeter source and provide a wide-band tunable source for spectroscopy. The strength of the nonlinearity was demonstrated recently by the Raytheon group (Adlerstein, 1980), where they obtained tens of milliwatts at 100 GHZ. Harmonic generation has also been observed with ferrites, plasma, and electron beams. However, at present no practical devices are being developed using these techniques.

Perhaps the most interesting devices using nonlinear properties of materials have been demonstrated with lasers involving second- and third-order nonlinearities. The first such demonstration was that by Franken and co-workers in the optical region of the spectrum (Franken *et al.*, 1961). More recently with the development of new lasers and better materials, harmonic generation and nonlinear mixing has produced tunable radiation from the infrared well into the millimeter region. One such scheme was used by Shen and co-workers (Yang *et al.*, 1973), who mixed the output of two dye lasers in $LiNbO_3$ to obtain tunable radiation in the far-infrared and the submillimeter regions. The output and linewidth of this source provided limited possibilities for spectroscopy. Another scheme that was developed by Aggarwal and co-workers (1973) involved two CO_2 lasers which were step tunable with a grating and were mixed noncollinearly in GaAs. This technique was successively improved to operate both in the pulsed (Lee *et al.*, 1974) and continuous modes (Aggarwal *et al.*, 1974). The former produced as much as 10 kW of power and

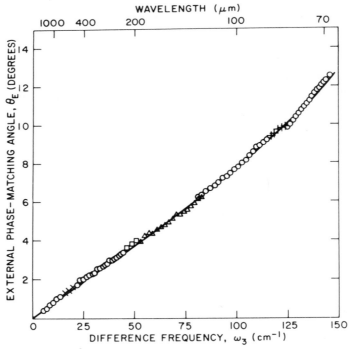

FIG. 2   Tuning curve of the noncollinear phase matched difference frequency of two CO_2 lasers mixing in GaAs crystal. $\bigcirc$, $\omega_2 = 944.2\,cm^{-1}$; $+$, $\omega_1 = 1074.6\,cm^{-1+}$; $\triangle$, $\omega_2 = 974.6$ cm^{-1}; $\square$, $\omega_1 = 979.7\,cm^{-1}$; $\times$, $\omega_1 = 952.9\,cm^{-1}$. (After Lax *et al.*, 1973.)

the latter about a microwatt. As many as 3000 lines were available with this source from 70 μm, well into the millimeter region. A plot of the spectral range containing a large number of these lines is shown in Fig. 2. With the recent development of multiatmosphere pulsed CO_2 lasers, this spectral region can now be covered on a continuously tunable basis.

This technique of noncollinear mixing using a cascade arrangement was also able to generate tunable radiation in the 12-μm region of the spectrum. However, this was surpassed by an equivalent four-photon mixing technique in germanium, which utilizes the third-order nonlinearity of this material (Lee *et al.*, 1977). Two pulsed TEA (Transversely Excited Atmospheric) CO_2 lasers are mixed noncollinearly to generate frequencies such as $2\omega_1 - \omega_2$, which produced a large number of discrete frequencies in the 12–8-μm range. The latter was improved to the stage of producing 600 kW of pulsed power sufficient for laser isotope separation of molecules.

More standard techniques of collinear mixing and parametric generation in the infrared region has received a great deal of attention in recent years. This has been motivated by the need for tunable sources for photochemistry and isotope separation by coherent sources. One was of generating a large number of lines in the middle infrared region is to mix two infrared lasers such as CO_2 or CO in different combinations using high-quality crystals of the chalcopyrites such as GeGaAs. This has been achieved by Mooradian and co-workers (Menyuk *et al.*, 1976) to generate many frequencies in the 2.5–6-μm regions. Operation on a pulse mode, harmonic generating using this material, is very efficient for remote sensing of atmospheric molecules. The Lincoln group, namely Kildal and Brueck (1977) and Brueck and Kildal (1978), have used the third-order nonlinearity in liquefied gases to generate high-power third harmonic of CO_2 lasers in the 3-μm region. Finally, one should mention the work of the Los Alamos group, using CdSe and CO_2 lasers to generate tunable radiation at 16 μm with a parametric oscillator for their laser isotope program (Rockwood, 1976).

B. OPTICAL PUMPING

One of the most effective ways of generating a large number of discrete coherent frequencies in the far-infrared and submillimeter regions of the spectrum is to pump one molecular laser by another. This has been particularly successful in the submillimeter region. It was first shown by Chang and Bridges (1970) that one can pump such molecular gases as CH_3F and others with a CO_2 laser to obtain radiation at submillimeter wavelengths. The principle involves a coincidence of one of the grating tunable lines of the CO_2 and a vibrational transition between a particular set of J values of the pumped molecule. The selection rules usually involve $\Delta J = 0, \pm 1$. Using this principle and a CO_2 laser, hundreds of lines have been produced from the far-infrared into the millimeter region (Rosenbluh *et al.*, 1976). The spectral density of this scheme is illustrated in a bar diagram of Fig. 3. Other analogous optical pumping techniques have also generated lines in the infrared from 4–30 μm. Thus almost three decades of discrete lines are now available by this method.

What makes optical pumping such a fascinating occupation is that it involves a knowledge of molecular spectroscopy, optical techniques, and the physics of quantum electronics. Later, in Section III, a detailed discussion of the physics of two such optically pumped molecular lasers will be presented. Recently Biron *et al.*, (1979a) have shown that the Raman mode of pumping CH_3F at higher pump intensities greater than 10 MW/cm^2 of the CO_2 laser can excite as many as 25 different transitions. This permits step tuning over a spectral range of 150–500 μm.

C. Electronic Devices

To extend the coherent sources to shorter wavelengths, one of the best approaches was the carcinotron or the backward wave oscillator, although klystrons and magnetrons have operated in the millimeter region. The carcinotron, however, has reached the submillimeter region and has been used very effectively for solid-state spectroscopy. Rudashevsky at the Lebedev Physical Institute has studied antiferromagnetic resonance using such sources (Rudashevsky *et al.*, 1974). Cyclotron resonance has been observed by the French group in semiconductors and metals (Couder and Goy, 1970).

Another electron beam approach to generate higher frequencies has

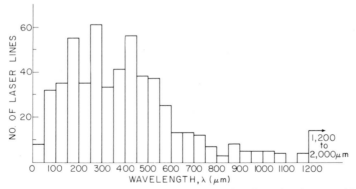

FIG. 3   Spectral density of optically pumped lasers lines in molecular gas with grating tunable CO_2 laser as the pump source. (After Rosenbluh *et al.*, 1976.)

been the cyclotron resonance maser (Bott, 1964), which relies on the relativistic properties of an electron beam in high-magnetic fields. This was first developed by Hirschfield and Wachtel (1966). This device has operated at both millimeter and submillimeter frequencies. Its extension by the Russian group (Gapanov *et al.*, 1975) to a high-intensity beam in which the longitudinal energy is transformed into transverse motion by the mirror action as the beam enters the magnetic field. The gyrotron, as the device is called, then emits radiation at the fundamental and harmonic cyclotron frequencies. With a 50-kG superconducting magnet, they have generated 22 kW of continuous energy at 2 mm and 1.5 kW at 0.9 mm with efficiencies of 22% and 6%, respectively (Zaytsev *et al.*, 1974). This radiation is tunable with variation of the magnetic field, although the cavity structure may limit this frequency tuning. Varian Associates has developed an oscillator which delivers 212 kW of cw power at 11 mm with an efficiency of 50% (Jory *et al.*, 1980). A diagram of the oscillator is shown in Fig. 4.

The extension of the cyclotron resonance masers (CRM) using intense relativistic electron beams to produce coherent pulse sources in the microwave and submillimeter region has been demonstrated by the Naval Research Laboratory group, of Granatstein and co-workers (1975). Excellent reviews of the work on the CRM and gyrotrons have been given by Hirshfield and Granatstein (1977) and by Andronov *et al.* (1978).

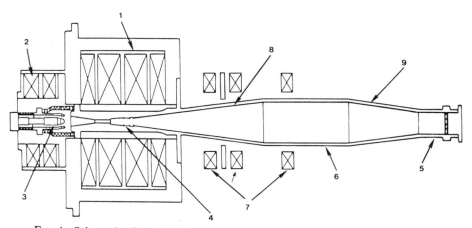

FIG. 4 Schematic of the 28-GHz cw gyrotron with tapered output guide illustrating the detailed construction of the electronic and magnetic components. 1, main magnet coils; 2, gun magnet coil; 3, electron gun; 4, cavity; 5, output waveguide and window; 6, beam-collector area; 7, collector magnet coils; 8, output guide up-taper; g, output guide down-taper. (After Jory *et al.* 1977.)

D. The Free-Electron Laser

The last device that we shall discuss in this presentation is called the free-electron laser. Its origin can be traced back to the undulator, which was started about thirty years ago (Motz, 1951). A related device which is again under consideration uses the Smith Purcell effect (Gover and Yariv, 1978). Both of these devices are seriously being developed as coherent sources from the infrared to the millimeter region. The free-electron laser is essentially analogous to inverse Compton scattering. If an electromagnetic wave of frequency ν_0 is scattered backward from a relativistic beam $\gamma \gg 1$, and hence $\beta \approx 1$, then in the rest frame of the electron beam, the frequency of the incoming wave is $\nu' = \nu_0 (1 - \beta \cos \theta) \approx 2\gamma\nu_0$. This is Thomson scattered in this rest frame. When the wave is again transformed into the laboratory frame, the light is scattered in a relatively narrow angle in the forward direction of the beam with the frequency $\nu = 4\gamma^2\nu_0$ with a certain fractional frequency spread. The difference between this and the free-electron laser built by Madey and co-workers (Elias et al., 1976) is that they produced an alternating magnetic field using two counterwound superconducting helical coils. As the beam passed through this coil in its rest frame, it saw an alternating electric field whose wavelength was the Lorentz contracted $\lambda = l/\gamma$, where l is the period of the magnetic structure. The electron beam scatters the "wave" with the Thomson cross section, which in the forward direction appears in the laboratory frame as an electromagnetic wave of $\nu' = 2\gamma^2 c/l$. Deacon et al. (1977) have shown that there is gain at infrared wavelength and there is narrowing of the line, hence, lasing when a Fabry–Perot cavity is used. The other difference between inverse Compton scattering and the free-electron laser is that the intensity of the oscillating electromagnetic field that the electron sees is much larger in the latter case; hence the number of photons scattered is much greater. However, the gain is not very large in any event. Consequently, several schemes have been devised that can increase the cross section and hence the gain. One of these involves stimulated Raman scattering of the plasma mode of the electrons in the beam (Sprangle and Granatstein, 1974). This is most effective for high current densities and short pulses. The other method is to take advantage of the cyclotron resonance in the rest frame of the electron beam (Sprangle and Granatstein, 1978). This implies that there is a longitudinal field about the electron beam which satisfies the relation

$$\omega' = 2\pi\gamma c/l = \omega_c = eB/m$$

or for an incident microwave field, $\omega' = 2\gamma\omega_0 = \omega_c$. The result is that the cross section and hence the gain are increased by the factor

$$\sigma' = [\sigma_0 \omega'^2 / (\omega' - \omega_c)^2] + \nu^2$$

where σ_0 is the Thomson cross section and ν is the radiation loss, which can be substantial. This factor can be very large and can increase the efficiency of the free-electron laser and is particularly suitable for the submillimeter and millimeter regions. However, in order to maintain resonance along the beam, even in the Compton regime, one must design the periodic field to compensate for the large radiation loss. The problem of even the single electron motion in a wiggler tailored to resonance is highly nonlinear. So far the solution has been treated primarily in the small signal limit by Friedland and Hirshfield (1980) and others. The self-consistent solution, which maintains resonance in the strong signal limit and takes advantage of the enhancement represented by the above equation, has not yet been properly formulated. The optimum result must include a prescription of the tailored wiggler, a large radiation loss factor that is also resonant. Hence the problem in the longitudinal field differs from that considered by Kroll et al. (1980) and others for resonant-enhanced free-electron lasers. Such a device would be very useful in the submillimeter and could use a superconducting magnet of the order of 1–2 T (10 kG is 1 T). In the IR and visible regions magnet technologies exist to achieve the resonant condition. The submillimeter would be very useful for cyclotron resonant heating of magnetically confined fusion plasmas and for nonlinear spectroscopy. The IR and visible are of interest for high-power pulsed lasers for inertial confined pellet fusion and military applications.

III. Scientific Applications

A. Cyclotron Resonance

The availability of microwave sources after World War II afforded a variety of opportunities for studies of resonance spectroscopy. Among these was the study of cyclotron resonance of electrons in gaseous plasmas (Lax et al., 1950). In 1948 I initiated a project on the effect of cyclotron resonance on the breakdown of gases in a magnetic field. These were the first observations of microwave cyclotron resonance and showed the dramatic lowering of the breakdown threshold at low pressures as the resonance was approached. The effect is shown graphically in Fig. 5 as a function of magnetic field for a number of pressures. The phenomenon was well accounted for by a classical theory using the Boltzmann equation or a single-electron treatment, including diffusion of electrons in a magnetic field. This experiment has recently been extended to higher frequencies with a highpower submillimeter laser (Hacker et al., 1976). The phenomenon was more complicated than in the microwave case because the elec-

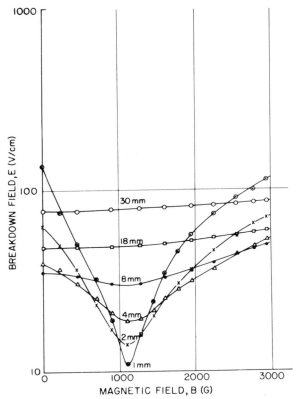

FIG. 5 Cyclotron resonance effect on the breakdown of helium in transverse electric and magnetic fields at microwaves as a function of magnetic field for different pressures: diameter, 7.32 cm; height 4.60 cm. (After Lax *et al.*, 1950.)

tric field of the laser was inhomogeneous. A modification of the theory to take this into account showed good agreement with the experiment (Biron *et al.* 1979b) as shown in Fig. 6.

Perhaps the most significant application of microwave cyclotron resonance was to that of semiconductors (Dresselhaus *et al.*, 1953; Lax *et al.*, 1954b). These experiments were the first to clarify the nature of the band structure of holes and electrons in germanium and silicon. With the use of light excitation, the group at Lincoln Laboratory was able to observe the anisotropy of both carriers in these two materials and determine the mass parameters of both electrons and holes for the first time (Lax *et al.*, 1954b). A typical multiresonant trace observed in germanium is shown in Fig. 7. The extension of these experiments to the infrared (Burstein *et al.*, 1956) and higher fields (Keyes *et al.*, 1956), and later to the millimeter per-

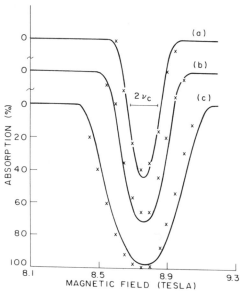

FIG. 6 Laser absorption as magnetic field for He at 5.5 torr and laser energy of (a) 0.035 mJ, (b) 0.07 mJ, (c) 0.20 mJ. The experimental points are indicated by (x). The theoretical fit (solid curve) is for $N_0 = 5 \times 10^9$ (initial number of electrons). The laser source was 1.22-mm optically pumped $^{13}CH_3F$ gas with 9.63-μm R(32) line of a TEA CO_2 laser. (After Biron et al., 1979.)

mitted the quantitative study of a few other new materials (Rauch, 1961, 1962), diamond for example, as shown in Fig. 8. The optimum system for cyclotron resonance studies came with the introduction of submillimeter molecular lasers (Button et al., 1966). These combined with high fields opened the door for the measurement of other materials and the study of

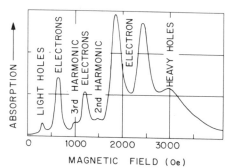

FIG. 7 Cyclotron resonance trace in germanium at 4 K excited by light at 23 GHz. (After Dexter et al., 1956.)

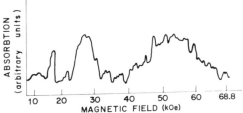

F<small>IG</small>. 8   Cyclotron resonance in *p*-type diamond at 70 GHz and 1.55 K using tungsten light for excitation of holes and carrier modulation. (After Rauch, 1962.)

new phenomena. The power of this combination of new techniques is illustrated in the comparison of the millimeter and submillimeter resonance of holes in InSb (Button, 1976) as shown in Fig. 9. The quantum effects of the transitions between the lower-lying Landau levels is clearly resolved at submillimeter and barely seen at millimeter wavelength. Another important accomplishment of submillimeter cyclotron resonance was the study of the polaron problem in CdTe (Waldman *et al.*, 1967). Larsen showed (1964) that owing to the coupling of electrons to the optical phonon, the mass of the electron would change with magnetic field. It required a number of lines of different molecular lasers to probe the necessary range of frequencies and magnetic field to demonstrate this phenomenon. The variation of mass with frequency (and field) is shown in Fig. 10. This experiment not only verified the theory of the polaron quantitatively, but it also measured the electron–phonon coupling directly (Litton *et al.*, 1976).

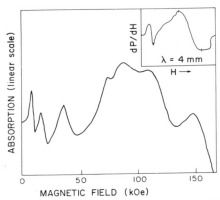

F<small>IG</small>. 9   Cyclotron resonance of holes in InSb at 337 μm of an HCN laser: $p - \text{InSb}(10°)$, $T = 35°\text{K}$, and $\lambda = 337$ μm. Quantum effects of light and heavy holes are observed. Inset is millimeter resonance. (After Button, 1976.)

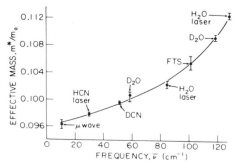

FIG. 10 Variation of polaron effective mass in CdTe as measured with different laser sources and magnetic fields. ⊥, experiment; ———, theory; $\alpha = 0.40$. (After Litton *et al.*, 1976.)

Another submillimeter experiment was the study of cyclotron resonance in CdS at low temperatures (Cohn *et al.*, 1971). It was observed that the extrapolation of the resonance at zero field had an intercept on the energy scale. We were unable to account for this by any impurity structure because the binding energy was too small. We postulated that it was due to the existence of an H^- ion. Recently this was theoretically analyzed and quantitatively explained by Larsen (1979), as shown in Fig. 11. He also demonstrated that at fairly high-magnetic fields even the triplet H^- state would be bound. With the existence of many more lines with the optically pumped submillimeter lasers, the studies of the H^- ion, the po-

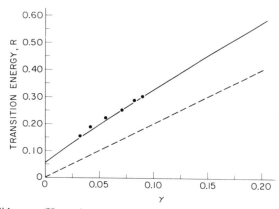

FIG. 11 Solid curve (H^- cyclotron resonance) shows theoretical value of the energy required to excite an electron from H^- ground state to first Landau state as a function of dimensionless magnetic field γ. The free electron cyclotron resonance is shown as a dashed line. The dots are experimental points. (After Larsen, 1979.)

laron, and other new materials is now possible. In fact, spectrometers using superconducting magnets and optically pumped submillimeter lasers are being built throughout Western Europe, Japan, and the U.S. for resonance spectroscopy.

B. MAGNETO-OPTICAL EFFECTS AND THE SPIN-FLIP LASER

The observation of the quantum effects in semiconductors naturally led to the discovery of the analagous phenomenon of the interband quantum effect in a magnetic field. These were observed and studied in semiconductors and semimetals with great success. It extended our knowledge of the band structure of these materials and supplemented the results of cyclotron resonance (Lax and Mavroides, 1967). One of the important discoveries made with this technique by the group at Lincoln Laboratory was the existence of the anomolous g factor in InSb (Zwerdling et al., 1957). They interpreted the experiments and found a value of $g = -50$. This was then explained on the basis of the band structure by Roth et al. (1959). The extension of this work by Pidgeon and Brown (1966) at higher fields established a thorough understanding of the wave functions and eigenvalues of the bands in a magnetic field. This discovery in InSb formed the basis for the theoretical development by Wolff (1966) and Yafet (1966) of the magneto-Raman scattering and spin-flip scattering effects in InSb. These phenomena were observed by Patel and co-workers (Slusher et al., 1967) by spontaneous Raman scattering, in which both the cyclotron and spin-flip transitions were excited in InSb with a pulsed CO_2 laser. It became evident from this work that the spin-flip transition would be more favorable for the stimulated emission. Indeed, this proved to be correct, and the tunable spin-flip Raman laser was demonstrated by Patel and Shaw. (1970) Using a high-power TEA laser, it was shown by Aggarwal et al. (1971) that considerable tuning range could be achieved with high magnetic fields when the Stokes, second Stokes, and antistokes transitions were also excited. The tuning curve is shown in Fig. 12. The range of tuning varies from 9 to 14 μ. Mooradian and co-workers (1970), using a CO laser and utilizing the resonant enhancement of the virtual interband transition, were able to obtain cw operation of the tunable spin-flip laser in the $5-6$-μm range. Since then a number of other materials such as $Hg_xCd_{1-x}Te$ (Weber et al., 1975) and CdS (Romestain et al., 1974) have been operated as spin-flip lasers.

A very interesting effect that was observed in the high-field regime of operation by Favrot et al. (1970) was the anomalous behavior of the intensity of the laser output as a function of magnetic field. This was attributed to the complicated cyclotron resonance absorption of both the pump and the Stokes radiation in a large crystal. Combination resonances of phonon

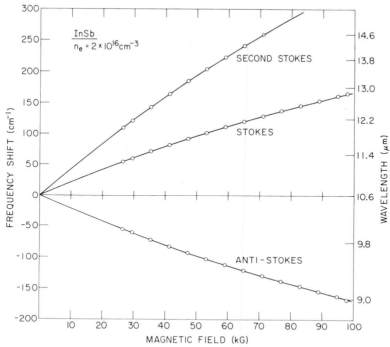

FIG. 12   Frequency shift of a spin-flip Raman laser as a function of magnetic field. The pump was a CO_2 TEA laser. (After Aggarwal *et al.*, 1971.)

and spin-flip assisted transitions of the fundamental and its harmonics played important roles in explaining the phenomenon. To demonstrate this, a magneto-absorption study in the 10-μm region was used to demonstrate the presence of these combination cyclotron resonance phenomena as shown in Fig. 13. These experiments showed that there was an anisotropy associated with the resonance absorptions. Weiler (1979), using the appropriate band model, was able to account for the resonances theoretically. The partial results of these experiments, together with other data, obtained by McCombe and Wagner (1971) and Lee (1976) using submillimeter laser resonance techniques to observe spin resonance, combination resonance, and harmonics, are compared with the theory as shown in Fig. 14.

C.   RAMAN TRANSITIONS IN SUBMILLIMETER LASERS

The transitions involved in an optically pumped laser that is pumped with another high-intensity laser requires a rigorous treatment of the density matrix equations of a three-level system in which two strong oscillat-

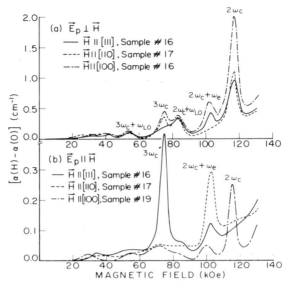

FIG. 13   Magnetoabsorption as a function of magnetic field in InSb at 10.6 μm showing anisotropy of harmonics and combination resonance. (After Favrot *et al.*, 1970.)

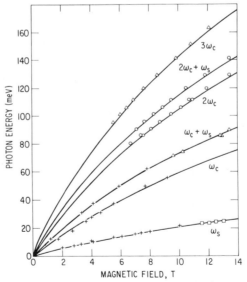

FIG. 14   Harmonic and combination resonances of electrons in InSb showing spin and spin-flip cyclotron resonance. $\bigcirc$, E$\perp$H and $\triangle$, E$\parallel$H: Lee; $\square$: Favrot; $+$: McCombe. Solid lines are theoretical. (After Weiler, 1979.)

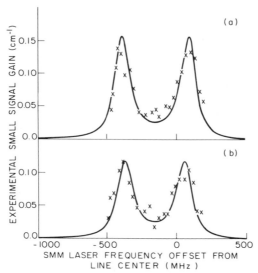

FIG. 15 Comparison of theoretical and experimental data of optically pumped CH_3F laser with high power CO_2 TEA laser. The lower frequency peak is the Raman shifted transition and upper peak is the Stark shifted resonant peak. (After Drozdowicz, Temkin, and Lax, 1979. *IEEE Journal of Quantum Electronics* **15**, 865. © 1979 IEEE.)

ing fields are present. The solution of the five coupled equations was treated by Panock and Temkin (1977). The results of the solution showed that in addition to the Autler–Townes effect, i.e., the ac Stark effect, there is a Raman transition present as well as normal resonant transition. Both of these, of course, are shifted by the ac Stark effect. To demonstrate the existence of the dual set of modes, Drozdowicz *et al.* (1979) used a single-mode and hybrid CO_2 TEA laser to pump a submillimeter D_2O laser in a short cavity. The single-mode laser cavity was tunable over a GHz range. The gain measurement of the laser was then measured as a function of frequency. This was then compared with the theory. The results agree quite well with the experiment showing the existence of the Raman mode as well as the normal resonant mode. This is indicated in Fig. 15.

The importance of the Raman mode was recently demonstrated by Biron *et al.* (1979a) in CH_3F. They obtained as many as 14 new transitions in this molecule by using a pump intensity of a single-mode laser as high as 20 MW/cm². The arrangement is shown in Fig. 16. The laser intensity was above the threshold required to pump the Raman transition for many values of J as high as 35. This allowed the excitation of laser transitions as far from resonance pumping as 0.8 cm⁻¹. In this way a large number of

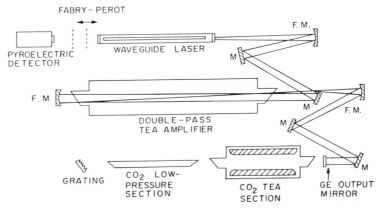

FIG. 16 Experimental arrangement of optically pumped $^{12}CH_3F$ waveguide laser using single mode tunable CO_2 TEA laser and amplifier to observe large number of Raman-shifted rotational transitions. (After Biron, 1980.)

frequencies from 200 cm^{-1} to over 500 cm^{-1} were excited. This suggests that a multiatmosphere TEA laser which is continuously tunable can pump a significant portion of the submillimeter frequency range in a quasi-continuous mode. The tuning curve using the existing grating tunable CO_2 pump shows the range of frequencies that are available. This is indicated in Fig. 17, which shows only one set of optically pumped laser lines of the R branch of CH_3F molecule when excited with the 9-μm branch of the CO_2 laser.

D. PLASMA DIAGNOSTICS

As the plasma densities of fusion machines have increased, the use of microwave interferometers became impractical because the plasma cut-off frequency exceeded those of even millimeter sources. With the advent of submillimeter molecular lasers, it became possible to overcome this difficulty. At MIT, in conjunction with the Alcator program, we have developed a very reliable and stable interferometer using an optically pumped cw methyl alcohol laser which operates at 119 μm (Wolfe *et al.*, 1976). The output of a CO_2 laser is beam split to pump two submillimeter lasers, one of which is offset in frequency by 1 MHz. A fraction of the output of these two lasers is fed into a cooled Ge detector and the output amplified to produce a MHz rectangular signal. The other portion of one of the lasers is passed through the plasma and then is also combined with its beat frequency to produce a rectangular signal. The latter is phase shifted relative to the first due to the change in refractive index of the plasma. In this way a direct reading of the plasma density as a function of

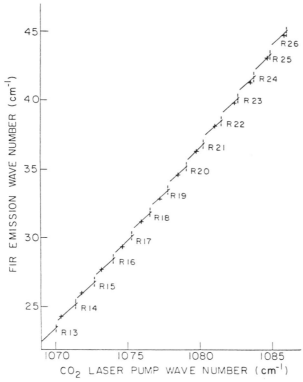

FIG. 17 $^{12}CH_3F$ molecular absorption and theoretical tuning curve of $^{12}CH_3F$ laser for
R-branch transitions. Laser emission is shown by crosses for all available CO_2 TEA laser
lines with grating tuning: +, experimental data, ¦, line center absorption. (After Biron, 1980.)

time during the plasma duration is obtained. This submillimeter interferometer has become the best such instrument for modern plasma machines.

Another important parameter of fusion plasmas is the temperatures of
electrons and ions. These, of course, may be different. Thomson scattering by a ruby laser is scattered by individual free electrons since $\lambda \ll \lambda_D$,
where λ_D is the Debye length of the plasma. The scattered radiation is
Doppler shifted by the velocity distribution. The resultant spectrum is
usually a gaussian which represents the Maxwell–Boltzmann distribution
of the electrons. From this, the electron temperature is determined. In
order to *solve* the analogous problem for the ions, it is necessary to go to
the other limit, namely, $\lambda > \lambda_D$. This means that the scattering is by col-
lective effects or fluctuations. In effect, the submillimeter wavelength that

satisfies this condition is essentially scattered by the Debye sphere of electrons that surround the ion. The Debye sphere follows the motion of the ion, and hence the spectral distribution reflects the ion temperature. The shape of the spectral distribution is more complicated and depends on the relative temperature of the ions and electrons. Since these line shapes have been theoretically determined, once the electron temperature is known from the ruby scattering, curve fitting gives the ion temperature.

To accomplish the objective of doing ion Thomson scattering at submillimeter wavelengths, it was necessary to develop a laser system and a sensitive heterodyne receiver in this region of the spectrum. The laser had to be high power of the order of 1 MW, single mode with a linewidth of the order of 50 MHz. It was a rather difficult task to accomplish. We developed a D_2O laser which is optically pumped by a single-mode CO_2 oscillator–amplifier chain of 30 J and ~ 150 nsec pulse width (Woskoboinikow *et al.*, 1974). The oscillator is tunable with a Fabry–Perot étalon of CdSe, and the lasers amplifiers are isolated by the appropriate molecular gas isolators. The single-mode tunable arrangement has not only allowed greater efficiency, as shown in Fig. 18, by tuning to the optimum frequency of the Raman transition, but has also reduced the superradiant background of the D_2O laser. In order to make the latter operate on a single mode, we have used a Fox–Smith oscillator arrangement. One of the mirrors in the Fox–Smith cavity was replaced by a grating which suppresses any cas-

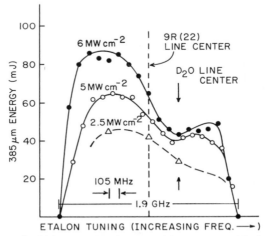

FIG. 18 Output of a high-power D_2O laser at 385 μm, versus CO_2 laser frequency for three pump intensities, showing enhanced power at the Raman transition below the resonant transition. This is made possible by etalon tuning of the pump CO_2 oscillator. (After Woskoboinikow *et al.*, 1974.)

cade transitions of the D_2O molecule. This combination of single-mode pump and oscillator has increased the efficiency and power output by a factor of 3 to 4 to yield about 750 kW at 385 pm. The spectrum of output has also been considerably improved with about a 30-MHz linewidth, and the superradient background highly diminished. The measurement of the far-infrared laser pulses was performed with a real time spectrum analyzer utilizing a special SAW (Surface Acoustic Wave) dispersive delay line developed by the Lincoln group (Fetterman *et al.*, (1979). The output of this laser is to be directed into the Alcator tokamak, and the scattered signal is to be detected by a heterodyne receiver. The mixer is the sensitive Schottky diode developed by the Lincoln group (Fetterman *et al.*, 1978) with a sensitivity of $\sim 10^{-19}$ W/Hz. The local oscillator is a formaldehyde optically pumped cw submillimeter laser.

Another approach that we have recently proposed for measuring the ion temperature of plasmas is the coherent four-photon scattering which depends on the third-order nonlinearity of the plasma. We have shown theoretically by solving the Vlasov equation to higher order that the scattering is orders of magnitude more efficient than the ordinary Thomson scattering (Praddaude *et al.*, 1979). The latter scatters in 4π solid angle, while the coherent scattering is limited to about 1 mrad. The resonant enhancement normally expected by this analog of the CARS (Coherent Antistokes Raman Scattering) technique is limited by the large Landau damping of the ions. Another advantage of this system is that by using two beams tuned to the proper band width of the fluctuation parallel to the magnetic field, we obtain a simple spectrum. The geometry of this arrangement is much better than that of the present Thomson scattering. The latter depends on the thermal fluctuation, whereas the coherent technique uses two frequencies ($\Delta\lambda > \lambda_D$) whose difference excites a coherent fluctuation which then scatters the photons of one of the beams. Since this is a highly nonlinear process at high laser powers, the efficiency of this process exceeds that of the Thomson scattering. Furthermore, the theory which shows that the coherent scattering is inversely proportional to the sixth power of the frequency immediately suggests that a submillimeter laser system is preferable to that using pulsed dye lasers. The spectra are similar to those of submillimeter Thomson scattering as shown in Fig. 19.

E. MOTIONAL STARK SPECTROSCOPY

During the past few years, we have pursued a new technique for investigating the properties of Rydberg states in He in high-magnetic fields with great precision, using laser spectroscopy (Rosenbluh *et al.*, 1977). We excite an He atom with an electron beam parallel to the magnetic field and monitor one of the Balmer lines, for example, the 7^1S to 2^1P transition.

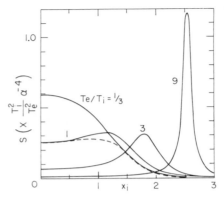

FIG. 19   Spectral function for four-wave scattering versus $x_i = \omega_s/(k_s v_{Ti})$ for hydrogen plasma ions with $\alpha = 10$ and various values of T_e/T_i. The dashed line is a Thomson scattering spectrum for $\alpha = 10$ and $T_e/T_i = 1$ normalized at $x_i = 0$. (After Praddaude et al., 1979.)

We then use a CO_2 laser that is grating tuned to excite the electron from the 7^1S to the 9^1P or other L values of the $n = 9$ manifold. We then sweep the magnetic field and monitor the intensity of the 7^1S-2^1P transition with a photomultiplier. The resultant line shape is highly distorted from the usual Doppler profile and depends also on the particular transition induced by the CO_2 laser, as well as the intensity of the laser.

The fundamental phenomenon we use is the result of the motional Stark effect (MSE) on the line shape. As the atom crosses the magnetic field, it sees in its moving frame an effective electric field $\vec{E} = (\vec{v} \times \vec{B})/c$. This in turn produces a large quadratic Stark effect whose energy must be added

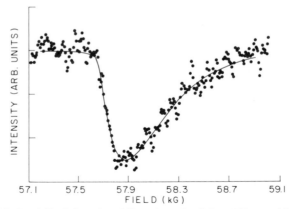

FIG. 20a   Motional Stark broadened line shape for $7'S$ to $9'D$ transition in He as observed by resonant transitions with tunable CO_2 laser. (After Panock et al., 1980.)

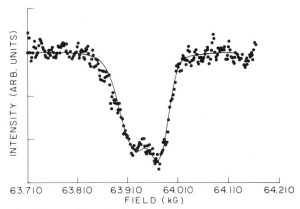

FIG. 20b   Motional Stark effect of a forbidden transition 7^1S to 9^1F in He excited by CO_2 laser. (After Panock *et al.*, 1980.)

to the Doppler term. The result of the theory for the allowed transition 7^1S-9^1P is to produce a sharp edge which is Doppler free and a longtail which is broadened by the MSE. The line shape fits the theory very well and allows us to carry out high-resolution spectroscopy (Rosenbluh *et al.*, 1978).

The MSE also allows transitions to those normally forbidden, such as

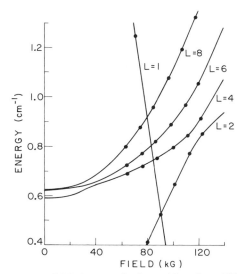

FIG. 21   Zeeman spectra of CO_2 laser excited transitions from 7^1S to $n = 9$ even states. The solid lines are theoretical; the dots are experimental. (After Panock *et al.*, 1980.)

the 9'D, because it couples this state to the 9'P (Panock *et al.,* 1979, 1980). The diamagnetic term is the Hamiltonian coupled states of $\Delta L = \pm 2$, and hence we can see both the even and odd manifold of $n = 9$. However, in order to observe those weaker transitions by the flourescence, we must use higher laser power. This tends to saturate such transitions, and hence we must include velocity terms in the matrix elements in the numerator and denominator of the line shape. These more complicated lines shapes shown in Fig. 20 are fitted to the theory. From the accurate fit we determine line center. We use these values to study the complicated Zeeman pattern of these Rydberg states and also to determine the zero field energies and hence the quantum defect corrections for these states. The comparison of the Zeeman energies for both even and odd states is made with theory as shown in Figs. 21 and 22, respectively. Using a submillimeter laser at 119 pm, we were able to observe transi-

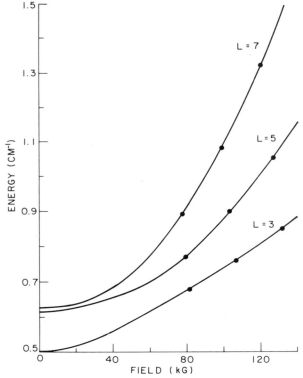

FIG. 22   Zeeman spectra of CO_2 laser excited transitions from 7'S to $n = 9$ odd states. The solid lines are theoretical; the dots are experimental. (After Panock *et al.,* 1980.)

tions between $7'S-7'P$ and $8'S-8'P$ as well (Rosenbluh *et al.*, 1981) and extend the utility of the MSE technique to investigate additional Rydberg states.

IV. Conclusion

It is apparent from this incomplete review that the development of new coherent sources—lasers, solid state, and electron beam devices such as the avalanche diodes, gyrotrons, and free-electron lasers, which will be improved and become available—will provide tunable sources for the future. These will have a tremendous impact on spectroscopy of atoms, molecules, solids, and plasmas from the millimeter into the infrared regions of the spectrum, as well as applications for commercial and military purposes. I selected only a few examples from my own first-hand knowledge to illustrate the historical evolution of the impetus that basic research has made on the development of new sources. The converse in which these developments allowed the extension of our knowledge of the basic properties of matter and the discovery of new phenomena, which in turn provided a feedback to additional inventions of new types of sources, was also indicated. The interdependence of science and technology is enriched in an environment in which scientist and engineers collaborate to enhance the base of our knowledge and the development of practical devices. The above story of coherent sources and their scientific applications is an eloquent testimony of this fundamental principle that is not always appreciated or practiced under adverse circumstances.

ACKNOWLEDGMENTS

I would like to thank Drs. H. Fetterman, R. Temkin, and R. Aggarwal for their help and suggestions. Many thanks to Dr. Howard Jory for allowing the use of the gyrotron drawing, and to Mrs. Ferriabough for her conscientious work in typing and assembling this manuscript.

REFERENCES

Adlerstein, M., private communication, 1980.
Aggarwal, R. L., Lax, B., Chase, C. E., Pidgeon, C. R., Limbert, D., and Brown, F. (1971). *Appl. Phys. Lett.* **18**, 383.
Aggarwal, R. L., Lax, B., and Favrot, G. (1973). *Appl. Phys. Lett.* **22**, 329.
Aggarwal, R. L., Lax, B., Fetterman, H. R., Tannenwald, P. E., and Clifton, B. J. (1974). *J. Appl. Phys.* **45**, 3972.
Andronov, A. A. *et al.* (1978). *Infrared Phys.* **18**, 385.
Basov, N. G., and Prokhorov, A. M. (1955). *Sov. JETP* **1**, 184.
Beringer, R. (1946). *Phys. Rev.* **70**, 53.
Biron, D. G. (1980). PhD Thesis.

Biron, D. G., Temkin, R. J., Lax, B., and Danly (1979a). *Opt. Lett.* **4**, 381.
Biron, D. G., Temkin, R. J., and Lax, B. (1979b). *J. Magn. Magn. Mater.* **11**, 47.
Bloembergen, N. (1956). *Phys. Rev.* **104**, 324.
Blumberg, W. A. M., Fetterman, H., Peck, D. D., and Goldsmith, P. F. (1979). *Appl. Phys. Lett.* **35**, 582.
Bott, I. B. (1964). *Proc. IEEE* **52**, 330.
Brueck, S. R. J., and Kildal, H. (1978). *Opt. Lett.* **2**, 33.
Burstein, E., Picus, G. S., and Gebbie, J. A. (1956). *Phys. Rev.* **103**, 825.
Button, K. J. (1976). *Proc. Int. Conf. Appl. High Magn. Fields Semicond. Phys., Wurzburg* p. 176.
Button. K. J., Gebbie, H. A., and Lax, B. (1966). *IEEE J. Quantum Electron.* **2**, 202.
Chang, T. Y., and Bridges, T. J. (1970). *Opt. Commun.* **1**, 423.
Cohn, D. R., Lax, B., Button, K. J., and Dreybrodt, W. (1971). *Solid State Commun.* **9**, 441.
Couder, Y., and Goy, P. (1970). *Proc. Symp. Submm. Waves* p. 417. Polytechnic Press.
Deacon, D. A. G., Elias, L. R., Madey, J. M. J., Ramian, G. J., Schwettman, H. A., and Smith, T. J. (1977). *Phys. Rev. Lett.* **38**, 892.
Dexter, R. N., and Lax, B. (1955). *Phys. Rev.* **100**, 1216.
Dexter, R. N., Zeiger, H. J., and Lax, B. (1956). *Phys. Rev.* **104**, 637.
Dresselhaus, G., Kip, A. F., and Kittel, C. (1953). *Phys. Rev.* **92**, 827.
Drozdowicz, Z., Temkin, R. J., and Lax, B. (1979). *IEEE J Quantum Electron.* **15**, 865.
Elias, L. R., Fairbank, W. M., Madey, J. M. J., Schwettman, H. A., and Smith T. J. (1976). *Phys. Rev. Lett.* **36**, 717.
Favrot, G., Aggarwal, R. L., and Lax, B. (1970). *Solid State Commun.* **18**, 577.
Fetterman, H. R., Tannenwald, P. E., Clifton, B. J., Parker, C. D., Fitzgerald, W. D., and Ericson, N. R. (1978). *Appl. Phys. Lett.* **33**, 151.
Fetterman, H. R. *et al.* (1979). *Appl. Phys. Lett.* **34**, 123.
Franken, P. A., Hill, A. G., Peters, C. W., and Weinreich, G. (1961). *Phys. Rev. Lett.* **7**, 118.
Friedland, L., and Hirshfield, J. L. (1980). *Phys. Rev. Lett.* **44**, 1456.
Galt, I. K., Yager, W. A., Merritt, F. R., Cetlin, B. B., and Dail, H. W. Jr. (1955). *Phys. Rev.* **100**, 748.
Gaponov, A. V., Gol'denberg, A. L., Grigor'ev, D. P., Pankratova, T. B., Petelin, M. I., and Flyagin, V. A. (1975). *Izv. VUZ. Radiofiz.* **18**, 280.
Gilliam, O. R., Johnson, C. M., and Gordy, W. (1950). *Phys. Rev.* **78**, 140.
Gordon, J. P., Zeiger, H. J., and Townes, C. H. (1955). *Phys. Rev.* **99**, 1264.
Gordy, W. (1948). *Rev. Mod. Phys.* **20**, 668.
Gover, A., and Yariv, A. (1978). *Phys. Quantum Electron.* **5**, 197.
Granatstein, V. L., Parker, R. K., and Sprangle, P. (1975). *Proc. Int. Topical Conf. Electron Beam Res. Technol.* 401–423. Document SAND 76-5122.
Griffiths, I. H. E. (1946). *Nature (London)* **158**, 670.
Hacker, M. P., Temkin, R. J., and Lax. B. (1976). *Appl. Phys. Lett.* **29**, 146.
Helminger, P., DeLucia, F. C., and Gordy, W. (1970). *Phys. Rev. Lett.* **25**, 1397.
Hirshfield, J. L., and Granatstein, V. L. (1977). *IEEE Trans. Microwave Theory Tech.* **MTT-25** (6), 522.
Hirshfield, J. L., and Wachtel, J. M. (1966). *Phys. Rev. Lett.* **17**, 348.
Javan, A., Bennett, W. B., Jr., and Herriott, D. R. (1961). *Phys. Rev. Lett.* **6**, 106.
Jory, H. R. *et al.* (1980). *Proc. Int. Electron Devices Meeting,* Washington, D. C. This device was developed by Varian Associates under subcontract to Oak Ridge National Laboratory operated by Union Carbide Corporation for the Department of Energy.
Keyes, R. J., Zwerdling, S., Foner, S., Kolm, H. H. and Lax, B. (1956). *Phys. Rev.* **104**, 1804.

Kildal, H., and Brueck, S. R. J. (1977). *Phys. Rev. Lett.* **38**, 347.

King, W. C., and Gordy, W. (1953). *Phys. Rev.* **90**, 319.

Kroll, N. M., Morton, P., and Rosenbluth, M. N. (1980). *Phys. Quantum Electron.* **7**, 113.

Larsen, D. M. (1964). *Phys. Rev.* **135A**, 419.

Larsen, D. M. (1979). *Phys. Rev. Lett.* **42**, 742.

Lax, B. (1979). *J. Magn. Magn. Mater.* **11**, 1.

Lax, B., and Mavroides, J. G. (1967). Interband magneto effects, "Physics of III-V Semiconductors," pp. 321–401. Academic Press, New York.

Lax, B., Allis, W. P., and Brown, S. C. (1950). *J. Appl. Phys.* **21**, 1297.

Lax, B., Zeiger, H. J., Dexter, R. N., and Rosenblum E. S. (1954a). *Phys. Rev.* **93**, 1418.

Lax, B., Zeiger, H. J., and Dexter, R. N. (1954b). *Physica* **20**, 818.

Lax, B., Aggarwal, R. L., and Favrot (1973). *Appl. Phys. Lett.* **23**, 679.

Lee, K. (1976). B. S. Thesis.

Lee, N., Aggarwal, R. L., Lax, B., and Chase, C. E. (1974). *Opt. Commun.* **11**, 339.

Lee, N., Aggarwal, R. L., and Lax, B. (1977). *J. Appl. Phys.* **48**, 2470.

Litton, C. W., Button, K. J., Waldman, J., Cohn, D. R., and Lax, B. (1976). *Phys. Rev.* **13B**, 5392.

Maiman, T. H., (1960). *Nature (London)* **187**, 493.

McCombe, B. D., and Wagner, R. J., (1971). *Phys. Rev.* **134**, 1285.

Menyuk, N., Iseler, G. W., and Mooradian, A. (1976). *Appl. Phys. Lett.* **29**, 492.

Mooradian, A., Brueck, S. R. J., and Blum, F. A. (1970). *Appl. Phys. Lett.* **17**, 481.

Motz, H. (1951). *J. Appl. Phys.* **22**, 527.

Panock, R. L., and Temkin, R. J. (1977). *IEEE J. Quantum Electron.* **13**, 425.

Panock, R., Rosenbluh, M., and Lax, B. (1979). *Phys. Rev. Lett.* **42**, 172.

Panock, R., Rosenbluh, M., Miller, T. A., and Lax, B. (1980). *Phys. Rev.* **A22**, 1050.

Patel, C. K. N. (1964a). *Phys. Rev. Lett.* **12**, 588.

Patel, C. K. N. (1964b). *Phys. Rev.* **136A**, 1187.

Patel, C. K. N., and Shaw, E. D. (1970). *Phys. Rev. Lett.* **24**, 451.

Pidgeon, C. R., and Brown, R. N. (1966). *Phys. Rev.* **146**, 575.

Poulis, N. J., van der Handel, J., Nabink, J., Poulis, J. A., and Gordon, C. J. (1951). *Phys. Rev.* **82**, 552.

Praddaude, H. C., Scudder, D. W., and Lax, B. (1979). *Appl. Phys. Lett.* **35**, 766.

Rauch, C. J. (1961). *Phys. Rev. Lett.* **7**, 83.

Rauch, C. J. (1962). *Proc. Int. Conf. Phys. Semicond., Exeter* p. 276.

Rockwood, S. (1976). "Tunable Lasers and Applications." (A. Mooradian *et al.*, eds.), p. 140. Springer-Verlag, Berlin and New York.

Romestain, R., Geschwind, S., Devlin, G. S., and Wolff, P. A. (1974). *Phys. Rev. Lett.* **33**, 10.

Rosenbluh, M., Temkin, R. J., and Button, K. J. (1976). *Appl. Opt.* **15**, 2635.

Rosenbluh, M., Le, H., Lax, B., Panock, R., and Miller, T. A. (1981). *Opt. Lett.* **6**, 2.

Rosenbluh, M., Miller, T. A., Larsen, D. M., and Lax, B. (1977). *Phys. Rev. Lett.* **39**, 874.

Rosenbluh, M., Panock, R., Lax, B., and Miller, T. A. (1978). *Phys. Rev.* **A18**, 1103.

Roth, L. M., Lax, B., and Zwerdling, S. (1959). *Phys. Rev.* **114**, 90.

Rudashevsky, E. S., Prokhorov, A. S., and Valikov, L. V. (1974). *Proc. Int. Conf. Submm. Waves Their Appl.* p. 129.

Schawlow, A. L., and Townes, C. H. (1958). *Phys. Rev.* **112**, 1440.

Slusher, R. E., Patel, C. K. N., and Fleury, P. A. (1967). *Phys. Rev. Lett.* **18**, 77.

Sprangle, P., and Granatstein, V. L. (1974). *Appl. Phys. Lett.* **25**, 377.

Sprangle, P., and Granatstein, V. L. (1978). *Phys. Quantum Electron.* **5**, 296.

Waldman, J., Larsen, D. M., Tannenwald, P. E., Bradley, C. C., Cohn, D. R., and Lax, B. (1967). *Phys. Rev.* **106**, 51.

Weber, B. A., Sattler, J. P., and Nemarich, H. (1975). *Appl. Phys. Lett.* **27,** 93.

Weiler, M. H. (1979). *J. Magn. Magn. Mater.* **11,** 131.

Wolfe, S. M., Button, K. J., Waldman, J., and Cohn, D. R. (1976). *Appl. Opt.* **15,** 2645.

Wolff, P. A. (1966). *Phys. Rev. Lett.* **16,** 225.

Woskoboinikow, P., Praddaude, H. C., Mulligan, W. J., Cohn, D. R., and Lax, B. (1979). *J. Appl. Phys.* **50,** 1125.

Yafet, Y. (1966). *Phys. Rev.* **152,** 858.

Yang, K. H., Morris, J. R., Richards, P. L., and Shen, Y. R. (1973). *Appl. Phys. Lett.* **23,** 669.

Zavoisky, E. (1945). *J. Phys. USSR* **9,** 211–328.

Zaytsev, N. I., Pankratova, T. B., Petelin, M. I., and Flyagin, V. A. (1974). *Radio Eng. Electron. Phys.* **19,** 103.

Zwerdling, S., Lax, B., and Roth, L. M. (1957). *Phys. Rev.* **108,** 1402.

CHAPTER 2

Molecular Spectroscopy by Far-Infrared Laser Emission*

J. O. Henningsen

Physics Laboratory I
H. C. Ørsted Institute
University of Copenhagen
Copenhagen, Denmark

I.	THE TOOLS	29
	A. *Introduction*	29
	B. *Instrumentation*	32
	C. *Diagnostic Techniques*	44
II.	CH₃OH—A CASE STORY	57
	A. *A Priori Information*	58
	B. *Added Information*	70
	C. *Conclusion*	93
	Appendix A The Data Base	94
	Appendix B Assigned Lines	101
	Appendix C The C–O Stretch Torsional Ground-State Band	101
	REFERENCES	124

I. The Tools

A. INTRODUCTION

The development of lasers has opened in a dramatic way for a detailed study of excited molecular and atomic states. Their spectral purity makes it possible to record absorption spectra with a resolution depending on the state of the system under investigation rather than on the instrumentation. In addition, the power available from many lasers is sufficient to significantly change the population of levels involved in a transition; this is the basis for an entirely new class of spectroscopic experiments.

Vibrationally excited molecules can be investigated with lasers oscillating in the mid-infrared range. Tunable lead salt diode lasers provide

* Supported by the Danish Natural Science Research Council under Grants Nos. 511-1787, 511-10161, and 511-15476.

absorption spectra with a Doppler-limited resolution of typically 0.002 cm^{-1}, an improvement by more than an order of magnitude compared with the conventional techniques employing wide-band sources in conjunction with grating monochromators. A further increase in resolution is obtained by using saturated absorption techniques in which the molecular transition is saturated by one monochromatic beam and probed with a second, counter propagating beam. For such experiments, however, sufficient laser power is at present only available from gas lasers, which have limited tunability. This means that only transitions that are accidentally in near coincidence with a suitable laser line can be studied.

In 1970 Chang and Bridges (1970), while working with CH_3F as saturable absorber for a CO_2 laser, discovered that the $P_9(20)$ line was in near coincidence with the $Q(1,12)$ and $Q(2,12)$ absorption lines to the C–F stretch and that population inversion could be obtained between the $J = 12$ and 11 levels of the vibrationally excited molecule (Fig. 1). By letting the absorption take place in a properly tuned resonator, they were able to observe stimulated emission on the pure rotational transitions and hence operate the first optically pumped far-infrared (FIR) laser. In fact, pumping was so efficient that they were able to obtain lasing on two additional transitions for each K, a cascade transition in the excited C–F stretch state, and a refilling transition in the vibrational ground state. This

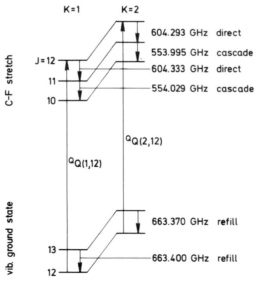

FIG. 1   First optically pumped FIR laser: CH_3F pumped by $P_9(20)$ of a CO_2 laser.

initial work on CH_3F also contains the first spectroscopic application of the technique since by measuring the emission frequencies to an accuracy of ± 3 MHz more accurate values could be derived for the rotational constant B and the centrifugal stretching coefficients D_J and D_{JK} of the C–F stretch. However, being a simple symmetric top molecule, CH_3F was already fairly well studied. The truly exciting advance was that a new source of monochromatic radiation had become available in an otherwise very difficult region of the electromagnetic spectrum, and little attention was paid to the potential spectroscopic applications of the technique.

Over the past decade far-infrared laser action has been obtained on more than 1350 individual lines in more than 60 different molecules, essentially all being pumped by a CO_2 laser (Yamanaka, 1976a; Knight, 1981). Most of these molecules are considerably more complicated than CH_3F and their excited vibrational states have not at all been studied in detail. Consequently, only a very small minority of the observed laser lines have ever been assigned to specific transitions, and since such assignments become increasingly important as the theoretical models for the lasers grow in sophistication, this in itself provides motivation for reconsidering the spectroscopic aspects.

The method can only be applied to molecules which have a permanent dipole moment, and it suffers from the same limitation as the saturated absorption techniques, since it relies on accidental near coincidences with available strong pump lines. Its prime advantage is that frequently several different far infrared lines can be generated by a given pump line. This already follows from Fig. 1 and applies to an even larger extent for more complicated molecules, where additional degrees of freedom may be involved and selection rules for electric dipole transitions may be more liberal owing to reduced symmetry. Under such circumstances, the same experiment provides information about an entire family consisting of one IR pump line and a number of related FIR emission lines. This contrasts with saturated absorption experiments, which contain information about the IR transition only. The technique may be considered equivalent to fluorescence spectroscopy that is widely used at shorter wavelengths. Upon moving into the infrared and far infrared, there is a large reduction in fluorescent yield owing to the ν^{-3} dependence of the radiative lifetime. In the present case an enhancement in signal by many orders of magnitude is obtained by observing stimulated rather than spontaneous fluorescence.

The relatively strong signal obtained in this way has several beneficial consequences. Detection of the signal is usually no problem. Its wavelength can be measured with a relative accuracy of about 10^{-4} with a scanning Fabry–Perot interferometer, or the frequency can be measured to

10^{-7} by using heterodyne techniques. The signal strength, frequency and polarization can be studied under various conditions for the lasing gas, or under the influence of external fields. By careful analysis of such experiments, information can be extracted which can hardly be obtained by other techniques at present.

B. INSTRUMENTATION

The general layout of an optically pumped far-infrared laser system is shown in Fig. 2, and we shall in this section go through its various elements, focusing on those aspects that are of particular relevance to a spectroscopic application of the technique.

1. *The Pump Laser*

As any table of FIR laser lines will immediately reveal, the CO_2 laser has a total dominance as pump radiation source although other lasers such as HF and N_2O have occasionally been used. In this section we shall discuss the various types of CO_2 lasers in some detail, keeping in mind that any powerful infrared laser is potentially useful as pump.

a. *Low-Pressure CO_2 Lasers.*   Under the most favorable conditions, FIR laser emission can be observed with a pump power level of less than 100 mW (Henningsen and Jensen, 1975). For practical purposes, 10 W in the strongest transitions is a more realistic requirement, and 50 to 100 W represents the fully adequate level. A widely favored configuration employs a 2–3 m long, wide bore ($\simeq 20$ mm) flowing gas tube, which, by using external mode limiting apertures, offers about 30–50 W cw in the strongest transitions, with good TEM_{00} beam quality. Alternatively, a shorter ($\simeq 1$ m), sealed off, narrow bore tube ($\simeq 8$ mm) can provide 10–15

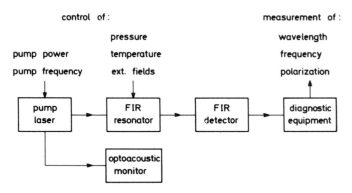

FIG. 2   Schematic diagram of FIR laser emission spectrometer.

W in a mode that borders a waveguide mode. The power level can then be boosted to about 50-W peaks by pulsing the power supply at a repetition rate of about 300 Hz. This scheme has the advantage of providing a larger tunability, due to the increased free spectral range, and at the same time the stability is improved by the absence of flow and by the reduced thermal load on the optical components. Since chopping is always introduced to enhance the signal-to-noise ratio, the pulsing, with the associated duty cycle of a few percent, represents no disadvantage as long as the FIR detectors are reasonably sensitive and not too slow.

While a rigid, thermally stable construction is always desirable, it is not for our purpose mandatory to introduce active stabilization of the frequency. Even relatively weak FIR lines will usually oscillate over a pump frequency range of at least 5–10 MHz, and for a well-designed free-running CO_2 laser, the frequency excursions will stay well within this range for hours. The requirements are illustrated in Fig. 3, which shows some results obtained when CH_3OH is pumped by the $P_9(16)$ line of a CO_2 laser. In Fig. 3a, we have shown the CO_2 output over the approximately

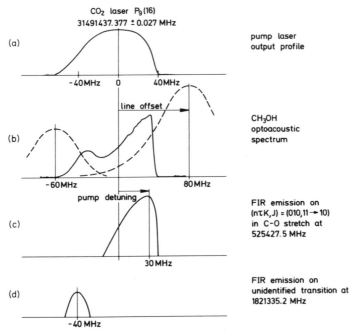

FIG. 3   Pumping and emission in CH_3OH. Line offsets given by Inguscio et al. (1980b) are -56 ± 8 and 64 ± 8 MHz.

± 40 MHz tuning range of the laser. The profile is asymmetric since the narrow bore tube does not allow a clean TEM_{00} mode at the edges of the tuning range. Figure 3b shows the absorption of pump radiation as measured in an optoacoustic cell. It is found that $P_9(16)$ can pump two different absorption lines. The strongest one is the $R(010,10)$ line located approximately 80 MHz above the line center (Sattler *et al.*, 1979), while the other is a considerably weaker, unidentified line with a negative offset of about -60 MHz. Figures 3c and 3d show the FIR output as a function of pump frequency for the strongest lines pumped by the two transitions. In the following we shall use the term "line offset" for denoting absorption line center frequency minus pump line center frequency, and the term "pump detuning" for denoting true pump frequency minus pump line center frequency.

For a more thorough study of single lines, where it is important to keep the pumping conditions strictly unchanged for a long time, active stabilization may be desirable. For stronger absorption lines, the optoacoustic technique may be used for locking to the peak of the absorption versus frequency curve (Busse *et al.*, 1977) or to the center of the transferred Lamb dip (Inguscio *et al.*, 1979a). For lines originating from weak pump transitions, the absorption may be too small to produce an acceptable optoacoustic signal-to-noise ratio. In such cases a variety of alternative schemes may be used. The simplest involves locking to a specific point on the output versus frequency curve of the CO_2 laser (Lund *et al.*, 1979). However, this scheme does not work well around the line center where the tuning curve is flat; it also has the disadvantage that any spurious increase in power output of the CO_2 laser will be compensated by a frequency shift to keep the level constant. A second method involves locking to the maximum of the CO_2 output versus frequency curve by using an adjustable large amplitude square wave dither to produce the required offset (Weiss and Heppner, 1979). Finally, the most sophisticated approach involves saturated fluorescence stabilization of an auxiliary reference CO_2 laser to the CO_2 line center (Freed and Javan, 1970) and subsequent frequency offset locking of the pump laser to the reference laser (Weiss and Kramer, 1976).

b. *TEA Lasers.* The restrictions imposed by the limited tunability of low-pressure gas lasers can be relaxed in various ways. One possibility is to use TEA lasers which can be tuned over a wider frequency range owing to their higher operating pressure, and which can also pump further away from the absorption line center by virtue of their high peak power. TEA lasers are widely used for maximizing the FIR output at selected lines, and have also been used for more systematic studies of various molecules, such as NH_3 (Gullberg *et al.*, 1973), D_2O (Keilmann *et al.*, 1975), PH_3 (Malk *et al.*, 1978), and CF_4 (McDowell *et al.*, 1979). Recently a continuously tunable TEA laser operating at pressures of up to 15 atm has been used for generating hundreds of new far-infrared emission lines in CH_3OH (Feld, 1979). Although this represents a significant advance, these results can not yet be used in the context of high-resolution spectroscopy owing to the large spectral width of the pump laser and the rather large uncertainty of both pump and emission frequencies. To obtain the required precision, it is desirable that the pump laser operates on a single longitudinal and transverse mode, and for such a system a tunability of ± 1.5 GHz has been demonstrated (McDowell *et al.*, 1979).

While there seems to be no fundamental obstacles in improving the spectral purity of the high-pressure TEA laser (Izatt and Mathieu, 1978), two additional complications should not be overlooked. As the pump power level increases, single-photon resonances are shifted by the dynamic Stark effect. At the same time the gain for stimulated Raman scattering becomes significant (Chang, 1977; Panock and Temkin, 1977; Petuchowsky *et al.*, 1977) (Fig. 4), and recent experiments on CH_3F have produced Raman emission with pump offsets of up to 20 GHz (Biron *et al.*, 1979). A TEA laser tuned to a specific frequency is therefore potentially capable of pumping over a frequency range which, in the case of a dense spectrum, may contain hundreds of absorption lines, and competition effects may restrict FIR emission to a few strong ones. Full benefit from a continuously tunable TEA laser may therefore require a drastic reduction of the peak power, but with a pulse length of 50 nsec and a repetition rate of 10 Hz, the necessary attentuation may imply nontrivial detection problems.

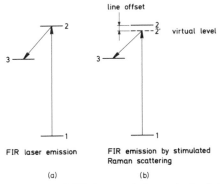

FIG. 4   Level schemes for normal FIR laser emission and stimulated Raman scattering.

c. *Waveguide lasers.* A second way of partly overcoming the coincidence limitation is to use waveguide CO_2 lasers which operate cw at pressures up to 300 torr, are tunable over more than 1 GHz (Abrams, 1974), and deliver continuous output of up to 40 W (Carter and Marcus, 1979). The first FIR emission generated by pumping with waveguide lasers was reported for CDF_3 (Tobin and Koepf, 1980) and for CH_3OH (Ioli *et al.*, 1980; Inguscio *et al.*, 1981). To obtain sufficient power and line discrimination in the pump lasers used in these experiments, some tunability had to be sacrificed. However, even with a modest increase by a factor of 3 compared with conventional CO_2 lasers, significant new results were obtained, and waveguide lasers now offer a very attractive alternative as pump sources.

d. *Two-Photon Pumping.* A fundamentally different approach to pumping involves the simultaneous application of two monochromatic fields for inducing coherent two-photon transitions. NH_3 was two-photon pumped by two TEA lasers (Jacobs *et al.*, 1976), and later the same scheme was used for CH_3F (Prosnitz *et al.*, 1978). In both cases molecules were pumped from the ground state to an overtone band, and subsequent lasing was observed on transitions to the fundamental. To enhance the transition probability, an intermediate level is required (Fig. 5), and the separation between this level and the virtual level was of the order of 6 GHz in both cases. Willenberg *et al.* (1980) have demonstrated cw lasing in NH_3 following two-photon pumping with a 12-W CO_2 laser and a 300-mW, 24-GHz klystron. In this case the sum frequency is fully tunable, but an intermediate level is still required and the limiting factor is the frequency mismatch between this level and the virtual level. In their experiment, the detuning was up to 2 GHz or 40 Doppler half-widths, and a max-

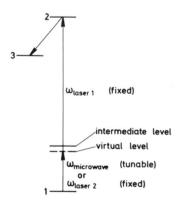

FIG. 5 Level scheme for two-photon pumping.

imum allowable value of about 10 GHz is predicted if the laser and microwave power is increased to the readily available levels of 50 and 3 W, respectively.

With conventional microwave sources the two-photon technique relies on the existence of coupled microwave–infrared transitions, and although such transitions are abundantly present in NH_3 owing to the inversion splitting, they are in general only accidentally present, unless J is very low. If, on the other hand, the technique can be used with carcinotrons, the number of possible candidates for two-photon pumping is vastly increased.

   e. *Sequence and Isotopic Lasers.* Having exhausted the possibilities offered by the normally used bands of CO_2, the attention may be turned to sequence band lasers (Reid and Siemsen, 1976; Weiss *et al.*, 1977). With an intracavity cell containing pure CO_2, lasing can be suppressed on the $[00^01] \rightarrow [10^00,02^00]_{I,II}$ transitions, and up to 10-W cw can be generated on the $[00^02] \rightarrow [10^01,02^01]_{I,II}$ hot-band transitions (Berger *et al.*, 1977). Finally, as a last step one may use CO_2 molecules with the isotopes ^{13}C and ^{18}O both in the normal (Petersen *et al.*, 1980) and in the sequence laser configuration. Figure 6 shows the coverage that is obtained of the 10-μm region with references to frequency tabulations.

2. The FIR Resonator

The FIR resonator serves a dual purpose by acting as a resonator for the FIR laser signal, while confining the pump radiation so that an optimum spatial overlap with the FIR mode is obtained. Weiss (1976) has measured the absorption coefficient of a number of known pump lines and found values down to 10^{-4} cm^{-1} torr^{-1}. When pumping such a line in a 1-m resonator at a typical operating pressure of 0.5 torr, more than 200 double passes are required before 90% of the pump radiation is absorbed. This shows that minimization of pump radiation losses at the mirrors and at

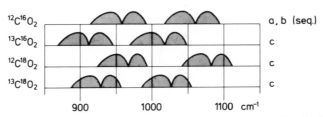

FIG. 6 Spectral coverage with CO_2 lasers: (a) Petersen *et al.* (1974); (b) Siemsen and Whitford (1977); and (c) Freed *et al.* (1974).

confining walls is of major concern, and high reflectivity coatings must be used.

The pump radiation is usually injected through a hole in one of the mirrors. This hole should be kept as small as possible since it also acts as a leak for the pump radiation, and quite apart from thus causing an undesirable loss, any signal reflected from the FIR resonator may be fed back into the pump laser and cause instability. Such feedback can in principle be eliminated by passing the linearly polarized pump radiation through a polarizer and subsequently converting it to circular polarization in a $\lambda/4$ plate. Any feedback radiation will be reconverted to a 90° rotated linearly polarized signal by the $\lambda/4$ plate and hence be rejected by the polarizer. It should be noted, however, that the polarization characteristics of the emitted FIR radiation contains important diagnostic information which is lost if pumping is effected with circularly polarized radiation. Another way of eliminating feedback is to use a traveling wave FIR resonator with unidirectional pumping (Heppner and Weiss, 1978), but this implies some loss of efficiency, which may be undesirable for weak lines.

By making one mirror movable the resonator length can be adjusted to the desired FIR resonance, and the wavelength of the emitted radiation can be determined by using the resonator itself as a scanning Fabry–Perot interferometer. Figure 7 shows the mode pattern for three different types of resonators, and in the following sections we shall discuss their relative merits.

a.   *Open Resonators.*   Stable open resonators with spherical mirrors were used in the initial work of Chang and Bridges (1970), and although somewhat bulky, they still remain attractive for a variety of reasons. To keep diffraction losses at a tolerable level at wavelengths up to $\lambda = 2$ mm, a resonator of length $d = 1$ m requires mirror diameters of about $2a = 10$ cm, corresponding to a Fresnel number of $N = a^2/\lambda d \simeq 1.25$. To facilitate analysis of spectra for shorter FIR wavelengths where a large number of higher order transverse modes may oscillate, the resonator can be provided with an adjustable mode-limiting aperture. The beam diameter of the Gaussian pump mode is much smaller than that of the FIR mode, and this would lead to a poor spatial overlap if the pump beam were mode matched to the resonator in the usual sense. Instead, the beam divergence should be chosen large enough to distribute the pump radiation over a suitable volume or, alternatively, injected at an angle relative to the optical axis (Koepf and McAvoy, 1977). This means that a large number of pump modes are excited, and their different phase shifts lead to a spatial averaging, so that troublesome pump resonances are avoided.

The FIR radiation may be coupled out through a hole, or by inserting

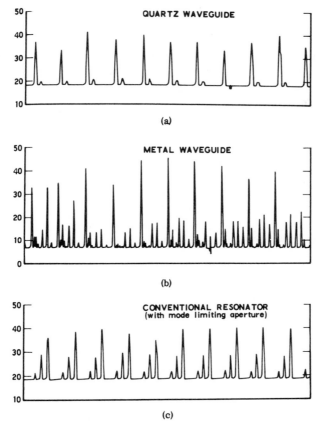

FIG.7 Resonator scans for three types of FIR resonators, oscillating on the same line (Hodges and Hartwick, 1973).

into the resonator a small 45° mirror whose position may be optimized to the desired mode (Petersen *et al.*, 1980). Interferometric couplers have also been used (Weiss, 1977; Tobin, 1979), but since the wire grids that are used as partial reflectors for the FIR radiation are rather hard on the CO_2 radiation, this technique is not suitable for low loss systems. If non-adjustable hole coupling is used, it should be chosen as weak as possible to optimize conditions for the low gain lines.

The normal modes of an open cylindrically symmetric resonator with spherical mirrors are products of a Gaussian and a generalized Laguerre polynomial L_p^l, where l and p are integers (Kogelnik and Li, 1966). The phase of the electromagnetic field on the axis is given by

$$\phi(z) = kz - (2p + l + 1) \, \text{arctg}(2z/kw_0^2), \tag{1}$$

where the beam waist radius w_0 is determined by the mirror configuration, and where z is measured from the beam waist. The free-space wavelength $\lambda_0 = 2\pi/k$, and the apparent wavelength λ as determined by scanning one of the mirrors, are related through

$$\lambda = \lambda_0[1 + (2p + l + 1)\lambda_0^2/4\pi^2 w_0^2]. \tag{2}$$

Since the diffraction correction in Eq. (2) depends on p and l, it is in principle necessary to analyze the mode pattern in order to evaluate λ, and for an open resonator this is usually relatively easy to do (Yamanaka *et al.*, 1974). However, when inserting reasonable numbers, it is found that the relative contribution of the correction term only at the longest wavelengths exceeds 10^{-4}, and since it is in any event difficult to push the experimental uncertainty beyond this level, the correction can usually be neglected.

b. *Metallic Waveguide Resonators.* Metal waveguides were introduced as FIR resonators by Hodges and Hartwick (1973), and their propagation and loss characteristics have been discussed by Yamanaka (1976b) and by Gamble and Danielewicz (1981). For the FIR radiation they act as the equivalent of an oversized microwave cavity, and the mode volume is now defined by the geometry of the guide, which is typically a cylindrical tube with a length of 1–2 m and a diameter of 10–30 mm. For long FIR wavelengths it may be difficult to pump the entire mode volume of an open resonator, whereas in the waveguide resonator the pump radiation is more or less uniformly distributed owing to reflections from the wall. If the gas is a good absorber, a waveguide resonator will therefore usually be more efficient at long wavelengths. For weakly absorbing gases at shorter wavelengths, the additional pump radiation losses caused by absorption in the walls may more than outweigh the benefits of more uniform pumping, and here open resonators remain the best choice.

The normal modes for a cylindrical waveguide of radius a are Bessel functions, and for an oversized guide, the phase of the field is

$$\phi(z) \simeq kz[1 - \tfrac{1}{2}(u_{nm}/ka)^2], \tag{3}$$

where u_{nm} is the mth root of $J_n'(u) = 0$ for TE or of $J_n(u) = 0$ for TM modes. The apparent wavelength in this case relates to the free-space wavelength through

$$\lambda \simeq \lambda_0(1 + u_{nm}^2\lambda_0^2/8\pi^2 a^2). \tag{4}$$

The losses remain small up to very high m and n, and this usually results in a complicated and rather unpredictable mode pattern, which can generally not be analyzed (Fig. 7b). Since, moreover, the correction term of Eq. (4) remains significant even at short wavelengths, it is not possible to evaluate λ with the required accuracy from a resonator scan.

The large spectral density of low loss modes is useful, however, when searching for cascade lines. The upper level of a cascade transition is populated by lasing transitions from a directly pumped level, and the resonator must therefore be simultaneously resonant at two different frequencies. This is readily achieved in a metal waveguide where strong lines are frequently found to be lasing for all resonator settings (Henningsen et al., 1979).

c. *Hollow Dielectric Waveguide Resonators.*   Dielectric waveguides were introduced for optically pumped FIR lasers by Hodges and Hartwick (1973), and their use as FIR resonators has been discussed by Hodges et al. (1977). The theory for passive resonators was developed by Marcatili and Schmeltzer (1964) and Steffen and Kneubühl (1968), and a comprehensive review has been given by Degnan (1976). For cylindrical symmetry, the phase is

$$\phi(z) \simeq kz\left[1 - \frac{1}{2}\left(\frac{u_{nm}}{ka}\right)^2\left(1 - \frac{i2\nu_n}{ka}\right)\right], \tag{5}$$

where u_{nm} is the mth root of $J_{n-1}(u) = 0$. ν_n depends on the complex dielectric constant ε of the bounding material through

$$\nu_n = \begin{cases} \dfrac{1}{(\nu^2 - 1)^{1/2}} & \text{for TE}_{om} \text{ modes } (n = 0) \\[3mm] \dfrac{\nu^2}{(\nu^2 - 1)^{1/2}} & \text{for TM}_{om} \text{ modes } (n = 0) \\[3mm] \dfrac{\nu^2 + 1}{2(\nu^2 - 1)^{1/2}} & \text{for EH}_{nm} \text{ modes } (n \neq 0) \end{cases} \tag{6}$$

with $\nu = \sqrt{\varepsilon/\varepsilon_0}$. If the material is lossless, ν_n is real. The guide wavelength is determined by the real part of $\phi(z)$, and its relation to the free-space wavelength is then given by

$$\lambda \simeq \lambda_0[1 + u_{nm}^2\lambda_0^2/8\pi^2a^2]. \tag{7}$$

The presence of an imaginary contribution to the phase means that even for a lossless dielectric, radiation is lost to the surroundings by leakage through the walls. The attenuation constant is given by

$$\gamma = u_{nm}^2 \nu_n \lambda^2 / 4\pi^2 a^3, \tag{8}$$

and its dependence on n and m means that higher order modes suffer progressively increasing losses. Contrary to the metallic waveguide, the hollow dielectric waveguide therefore supports only a few low-loss modes, and the interferogram that results from scanning the resonator length is much easier to analyze (Fig. 7a).

For waveguide resonators, the same output coupling schemes can be used as for open resonators. In addition, a variety of mirrors have been developed which provide close to 100% reflectivity for the infrared pump radiation and a specified reflectivity for the far-infrared signal. These include interferometric couplers using uncoated semiconductors as partial reflectors (Evenson et al., 1977) and hybrid mirrors where the IR reflectivity is provided by an FIR transparent dielectric multilayer, while the FIR reflectivity is taken care of by a metal mesh (Danielewicz et al., 1975) or by a large hole in a metal film (Hodges et al., 1976). With such mirrors a diffraction limited output beam is obtained, which can very efficiently be coupled into a point contact diode, for example. However, since the FIR reflectivity is frequency dependent, these mirrors are less suitable if a search is to be conducted for weak emission lines throughout the entire FIR range.

d. *Hybrid Resonators.* Hybrid resonators combine different features of the resonators described in the preceding sections. The hollow rectangular hybrid waveguide is particularly useful for Stark effect investigations where a static electric field is applied to the lasing gas. Two of the resonator walls are plane parallel metal plates which at the same time serve as Stark electrodes. Lateral confinement is provided either by dielectric walls (Tobin and Jensen, 1976; Bionducci et al., 1979; Koo and Claspy, 1979) or by using curved mirrors for optical stability (Henningsen, 1980a). The theory of hybrid resonators has been reviewed by Degnan (1976), and a thorough discussion of the design considerations dictated by the presence of the static electric field is given in Chapter 3. As for the all-metal waveguide, many low loss modes exist, all of which are polarized parallel to the metal plates.

3. *Detection of the FIR Emission*

The far infrared is a rather difficult region of the electromagnetic spectrum as far as detectors are concerned. Quantum detectors which provide ultimate sensitivity at optical frequencies, are difficult to realize in the FIR because the light quanta are so small and liquid Helium operating temperatures are required (Lengfellner and Renk, 1977). Extrapolation of microwave techniques into the THz range is difficult due to problems with spreading resistance and shunt capacitance, and require detecting elements of submicron size. However, for our purpose, sensitivity is usually not a major problem, and a wide variety of different detectors will perform adequately. A review of available FIR detectors has been given by Kimmitt (1970), and here we shall restrict the attention to a few types that are particularly widely used.

The standard detector for wide-band FIR spectroscopy is the Golay cell. It operates at room temperature and its main virtue is the flat frequency response, which is essential in Fourier spectroscopy, but less so in the present context. So many parameters can influence the intensity of a given FIR line that a precise quantitative comparison with other lines is not very revealing. In addition, the rather slow response is a disadvantage, particularly if an internally chopped pump laser is used. A popular alternative is the pyroelectric TGS (Tri Glycine Sulphate) detector which also operates at room temperature. At low chopping frequencies (20 Hz) its performance is comparable to that of the Golay cell, but it remains useful to frequencies beyond 100 kHz (Hadni *et al.*, 1978).

Cryogenic detectors are more sensitive and may be advantageous when dealing with very weak lines. The Ge:Ga detector has a particularly interesting feature. At wavelengths below 300 μm it has a fast photoconductive response, but in addition it has a slow bolometric response that extends to the millimeter range so that it gives complete converage of the far infrared. The photoconductive response diminishes as the cutoff wavelength is approached, and this can be used for providing a coarse indication of the FIR frequency, either by watching on an oscilloscope the response to square wave chopping (Fig. 8) or by chopping simultaneously at two different frequencies and monitoring the ratio of the two signals.

The last detection scheme which will be mentioned in this section employs the superheterodyne principle. A local oscillator signal is mixed with the FIR laser radiation to produce a beat note that can subsequently

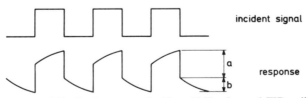

FIG. 8   Response of Ge:Ga detector to incident 16-Hz chopped FIR radiation: (a) fast photoconductive; (b) slow bolometric.

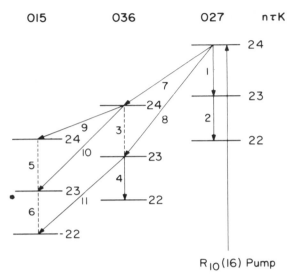

FIG. 9. Level scheme for $^{13}CH_3OH$ showing cascade transitions (2 and 4) identified by heterodyne measurements (Henningsen *et al.*, 1979).

be handled by conventional electronics. This gives a very large improvement in sensitivity over video detection (Putley, 1966), but two other features of the technique are more important. First, if the frequency of the local oscillator is known and the beat frequency is counted, the frequency of the FIR signal can be measured with an accuracy limited only by the accuracy with which the laser can be tuned to the center of the gain profile. This aspect will be further discussed in Section I.C.1. Second, since the bandwidth can be made arbitrarily small, the capability of discriminating between signals of closely spaced frequencies is limited only by their spectral widths. This is particularly important when searching for cascade lines which always appear on a background of the much stronger parent lines. An example is provided by the family of lines observed when $^{13}CH_3OH$ is pumped by $R_{10}(16)$ of the CO_2 laser (Henningsen *et al.*, 1979). In Fig. 9, the cascade lines 2 and 4 have frequencies of 1 069 853.4 and 1 069 771.4 MHz, respectively, and would have been difficult to resolve in the presence of the strong line 1 at 1 116 245.5 MHz by any other technique.

C. DIAGNOSTIC TECHNIQUES

1. *Wavelength and Frequency Measurements*

a. *Absorption Lines.* For a quick survey of an unknown spectrum, the optoacoustic technique has proved very useful (Busse *et al.*, 1977). A microphone mounted either in the FIR resonator itself, or in a separate cell, responds to chopped pump radiation at a rate depending on the absorption coefficient of the gas. This immediately identifies the potential pump laser lines as well as the optimum pump detuning (Fig. 10).

Since the frequencies of most potential pump lasers are known to better than a few MHz, any uncertainty in pump transition frequency derives from a lack of precise knowledge of the line offset (see Fig. 2). However, if emission is observed with cw or quasi-cw pump lasers, the absorption line center is not likely to be more than about one Doppler width away from the pump frequency, and its position can usually be estimated to within ± 30 MHz.

If the absorption line center is located within the tuning range of the CO_2 line, the precise location may be indicated by a saturation dip in the absorption or by a transferred dip in the FIR emission (Inguscio *et al.*, 1979a). If the absorption line center is outside the CO_2 tuning range, saturation techniques can still be used if a dc electric field can Stark tune the individual M components into resonance and if the permanent dipole moment is sufficiently well known. If the saturation dip is monitored on the infrared absorption, some care should be exercised in the interpretation. Several transitions may contribute, and those giving the most prominent

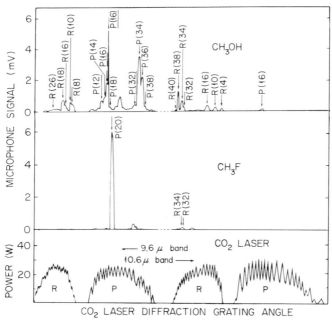

FIG. 10 Optoacoustic response of CH_3OH and CH_3F pumped by a CO_2 laser (Busse and Renk, 1978).

Stark signals are not necessarily those which are also most efficient in producing FIR laser emission (Bedwell *et al.*, 1978). This is apparent in Fig. 11, which shows the laser Stark spectrum recorded with $P_9(34)$ pumping of CH_3OH. The unresolved structure near zero field [$Q(125,9)$] gives rise to 15 FIR lines. The absorption line labeled $E_2(6,6)$ [$Q(026,6)$] produces two FIR lines if a Stark field of more than 1 kV/cm is applied (the assignment $A(6,6)$ [$Q(016,6)$] is suggested for this pump line by Henningsen, (1981)). None of the remaining lines produces any emission at all.

Under strong pumping conditions the molecular levels may be shifted by the dynamic Stark effect, so that the measured frequencies do not correspond to level separations for the unperturbed molecule. For cw or quasi cw pumping these shifts, though measurable, are not a matter of concern at the present level of accuracy (Heppner *et al.*, 1980). For TEA laser pumping, however, they constitute a major problem in the analysis (see Section I.B.1).

b. *Emission Lines.* The simplest way of measuring the frequency of the emitted FIR radiation is to determine the wavelength from a resonator scan and then convert to frequency $\nu = c/\lambda$ with $c = 2.997\ 924\ 58$ m/sec. For metal waveguide lasers the accuracy is limited to about 1% since the large waveguide correction to the wavelength [Eq. (4)] cannot be reliably evaluated. For shorter wavelengths such an uncertainty amounts to several cm^{-1}, and this is too much in a spectroscopic context. For hollow dielectric waveguides the correction is of the same order of magnitude [Eq.

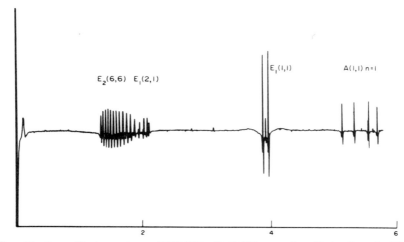

FIG. 11   Laser Stark spectrum of CH_3OH with $P_9(34)$ pump line (Bedwell *et al.*, 1978).

(7)] but can in principle be taken into account because the modes can be identified. In both cases, however, a better accuracy is obtained by using an external scanning Fabry–Perot interferometer. The problems are avoided if an open structure resonator is used, since in this case the corrections are of the order of 10^{-4} [Eq. (2)]. A comparison of measured wavelengths for 22 lines in $^{13}CH_3OH$ (Henningsen and Petersen, 1978) with subsequent frequency measurements shows an rms deviation of about 2×10^{-4}, and the same accuracy holds for measurements of Chang et al. (1970), which were performed with an external scanning Fabry–Perot. Unless more sophisticated techniques are used, such as interferometric monitoring of the mirror travel and fast scanning to eliminate thermal drift during the course of the measurement, this seems to be a realistic limit on the accuracy.

Although this is already superior to what can be obtained by conventional spectroscopy, a further improvement by several orders of magnitude is possible by directly measuring the emission frequencies, using heterodyne techniques. The uncertainty can then be reduced to the point where it is determined by the accuracy with which the FIR laser can be set at the maximum of its gain curve, i.e., to less than 1 MHz. The technique was first applied by Chang and Bridges (1970), who determined the six CH_3F frequencies (Fig. 1) with an accuracy of ± 3 MHz by beating the FIR laser signal with harmonics of a stable, tunable microwave source in a point contact diode. The results were used for deriving improved band parameters for the C–F stretch, and by incorporating measurements of additional lines, as well as information from other sources, Arimondo and Inguscio (1979) have carried this even further.

Nb–Nb Josephson point contacts allow very high-order mixing thanks to their extreme nonlinearity. McDonald et al. (1972) have reported operation at 3.8 THz, using the 401st harmonic of an X-band klystron, and Bleaney et al. (1978) have performed frequency measurements up to 2.06 THz, using the 102nd harmonic of a 20-GHz klystron. The advantage of the Josephson point contact is that it minimizes the number of steps required to make contact with a cesium clock, which is the fundamental frequency standard (9.192 631 770 GHz), but it is mechanically fragile, and the liquid helium operating temperature is inconvenient.

Schottky diodes are moving upwards steadily into the far infrared. Fetterman et al. (1974) have reported 33rd harmonic mixing at 2.52 THz with a 74-GHz klystron, and Schottky diodes have been used for frequency measurements up to 1.29 THz (Radford et al., 1977). They are rugged, practical room-temperature devices, and show very reliable performance.

Ni–W point contacts are used in the extensive and systematic frequency measurements carried out at the National Bureau of Standards,

Boulder (Radford *et al.*, 1977; Petersen *et al.*, 1980). Like the Josephson point contacts, they are susceptible to mechanical disturbances, but they operate at room temperature and perform reliably under normal laboratory conditions. Their response extends into the near visible (Jennings *et al.*, 1979), and they appear to be superior to Schottky diodes at the highest FIR frequencies.

Frequency measurements are routinely performed at many laboratories, using stable microwave sources in a harmonic mixing scheme. However, Petersen *et al.* (1975) have introduced a different technique, which avoids high-order mixing by synthesizing an appropriate local oscillator signal from the difference frequency of two Lamb dip fluorescence stabilized CO_2 lasers. More than 100 lines for each of the $^{12}C^{16}O_2$, $^{13}C^{16}O_2$, $^{12}C^{18}O_2$, and $^{13}C^{18}O_2$ lasers are known to an accuracy of between 30 kHz and 3 MHz (Section I.B.1), and the possible difference frequencies provide quasi-continuous coverage of the entire FIR range. In the scheme shown in Fig. 12, a microwave oscillator has been added, since this allows full coverage to be obtained with just one CO_2 isotope.

Although the frequency measurement technique in accuracy is far superior to wavelength measurements, the latter are not superfluous. For surveying an unknown spectrum, the bandwidth of the superheterodyne scheme is too low to be practical, and the most efficient procedure is to

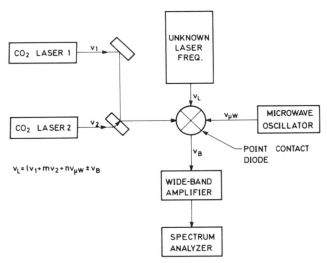

FIG. 12 Schematic diagram of FIR frequency measurement by CO_2 laser frequency synthesis (Petersen *et al.*, 1975. *IEEE J. Quantum Electron.* **QE-11**, 838–843. © 1975 IEEE).

determine in advance the frequencies to an accuracy which guarantees that the beat note will be within the bandwidth of the spectrum analyzer for a particular choice of local oscillator frequency. For the 1.5-GHz width of the widely used HP8558, this implies a relative a priori accuracy of between 10^{-3} and 10^{-4}, which is attainable with wavelength measurements (Henningsen *et al.*, 1979; Petersen *et al.*, 1980).

2. *Polarization Effects*

If the pump radiation is linearly polarized, the emitted FIR radiation will also be linearly polarized, either parallel or orthogonal to the pump radiation (Chang and Bridges, 1970). If an open resonator is employed, this is solely a consequence of the nonisotropic interaction between the pump radiation and the molecules, and a polarization study then provides information about the transitions involved. Following Chang and Bridges, this is intuitively clear if reference is made to a classical picture of a symmetric top. Consider the interaction between the pump field and an ensemble of molecules with transition dipole moment μ and isotropic angular momentum distribution (Fig. 13). If the angular momentum is oriented along $\bar{E}_{\text{pump}}$, the field will see a dipole oscillating at the pure vibrational frequency ν_{vib} and hence be able to induce mainly Q transitions, which

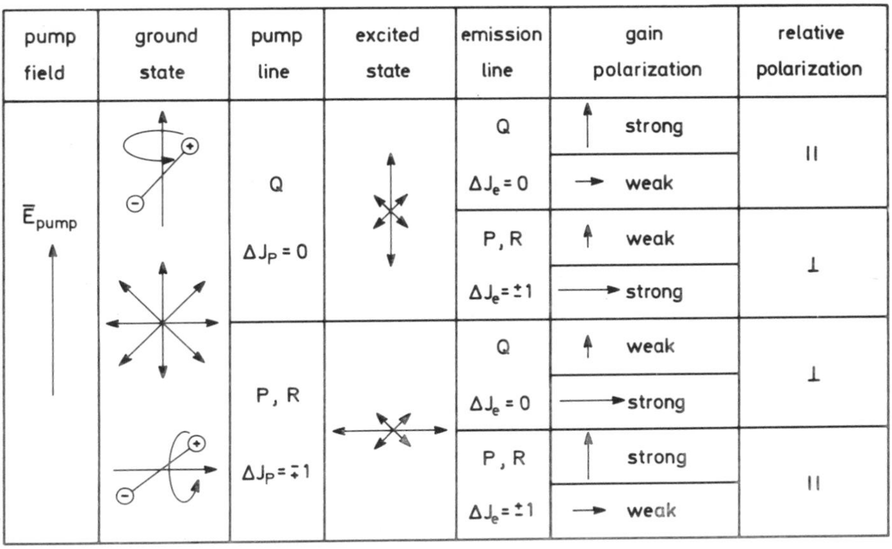

pump field	ground state	pump line	excited state	emission line	gain polarization	relative polarization
$\bar{E}_{\text{pump}}$		Q $\Delta J_P = 0$		Q strong		∥
				$\Delta J_e = 0$ weak		
				P, R weak		⊥
				$\Delta J_e = \pm 1$ strong		
		P, R $\Delta J_P = \mp 1$		Q weak		⊥
				$\Delta J_e = 0$ strong		
				P, R strong		∥
				$\Delta J_e = \pm 1$ weak		

FIG. 13   Angular momentum distribution in ground and excited state and its effect on the gain polarization.

leave the angular momentum unchanged. Conversely, if the angular momentum is orthogonal to $\bar{E}_{pump}$, the field will see a dipole oscillating at $\nu_{vib} \pm \nu_{rot}$, and the field will induce mainly P or R transitions, which change the magnitude of the angular momentum. A similar argument relates the polarization of the emitted radiation to the orientation of the emitting molecule, and all of this condenses into a simple rule for the relative polarization of pump- and emission field (Henningsen, 1977)

$$\Delta J_p + \Delta J_e = \begin{cases} \text{even} \Rightarrow \parallel \\ \text{odd} \ \Rightarrow \perp, \end{cases} \tag{9}$$

where ΔJ_p and ΔJ_e denote the change in J induced by the pump- and the emission transitions.

This rule holds for asymmetric as well as for symmetric top molecules. For asymmetric tops the motion of the molecule is difficult to visualize classically since the instantaneous rotation axis does not coincide with the angular momentum vector. In quantum-mechanical language the anisotropy of the interaction between field and molecule is expressed through the M dependence of the squared transition dipole matrix element. Table I shows this quantity for allowed transitions from a symmetric top state (JKM) as given by Table 4–4 in Townes and Schawlow (1955). For asymmetric tops K is not a good quantum number, and the wave function for a state (JM) is a linear combination of the limiting symmetric top wave functions with same (JM) and different K. The validity of the polarization

TABLE I

OVERALL ROTATIONAL CONTRIBUTION TO THE OSCILLATOR STRENGTH FOR
ALLOWED TRANSITIONS FROM (J,K) AS A FUNCTION OF M

Transition	$S_R(J,K,M)$	Optimum M
$J,K \rightarrow J,K$	$\mu_{\parallel}^2 \dfrac{K^2 M^2}{J^2 (J + 1)^2}$	J
$J,K \rightarrow J + 1,K$	$\mu_{\parallel}^2 \dfrac{[(J + 1)^2 - K^2][(J + 1)^2 - M^2]}{(J + 1)^2 (2J + 1)(2J + 3)}$	0
$J,K \rightarrow J - 1,K$	$\mu_{\parallel}^2 \dfrac{(J^2 - K^2)(J^2 - M^2)}{J^2 (4J^2 - 1)}$	0
$J,K \rightarrow J,K \pm 1$	$\mu_{\perp}^2 \dfrac{M^2 (J \mp K)(J \pm K + 1)}{4J^2 (J + 1)^2}$	J
$J,K \rightarrow J + 1,K \pm 1$	$\mu_{\perp}^2 \dfrac{(J \pm K + 1)(J \pm K + 2)[(J + 1)^2 - M^2]}{4(J + 1)^2 (2J + 1)(2J + 3)}$	0
$J,K \rightarrow J - 1,K \pm 1$	$\mu_{\perp}^2 \dfrac{(J \mp K)(J \mp K - 1)(J^2 - M^2)}{4J^2 (4J^2 - 1)}$	0

rule then follows from the factorization of the matrix elements into two functions which do not mix the K and the M dependence.

Drozdowicz *et al.* (1979) have derived expressions for the gain anisotropy for all 18 combinations of pump transition, emission transition, and relative field polarization. Choosing the direction of the pump $\bar{E}$ field as a common axis of quantization, they sum the matrix element product over M, using for the pump transition $\Delta M_p = 0$, and for the emission transition $\Delta M_e = 0$ in the parallel case and $\Delta M_e = \pm 1$ in the orthogonal case. In this way they arrive at the absolute strength of the combined transition expressed in terms of a function $F(J)$, which multiplies the product of the M-averaged strengths of the individual transitions. All of the gain anisotropy is thus contained in $F(J)$, and in the limit of large J, the 18 cases reduce to three, which are essentially different:

(a) $\Delta J = \pm 1$ for both transitions $\Rightarrow F_{\parallel}/F_{\perp} = 4/3,$ (10)

(b) $\Delta J = \pm 1$ for one transition
        $\Delta J = 0$ for the other transition $\Big\} \Rightarrow F_{\parallel}/F_{\perp} = 1/2,$ (11)

(c) $\Delta J = 0$ for both transitions $\Rightarrow F_{\parallel}/F_{\perp} = 3/1.$ (12)

This result is the quantitative verification of Eq. (9).

In deriving these polarization rules, it is tacitly assumed that no M-changing collisions take place while the molecules are in the upper laser level. This assumption is reasonable since M-changing collisions are not dipole allowed and therefore in general will occur much less frequently than J-changing collisions, which eventually limit the lifetime of the molecule in a given state. An experimental support for the assumption is provided by Stark-effect measurements on the CH_3F laser (Inguscio *et al.*, 1979b). The individual M transitions were resolved, and the observed relative intensity of the emission lines clearly showed that the molecules remember the pump transition, although a residual effect from M collisions was discernible for the highest and lowest M.

Before concluding this section, it should be emphasized that the measured ratio of parallel to orthogonal polarization of the emission signal is related to, but not necessarily equal to the gain anisotropy. In an open resonator, where there are no external constraints on the field, it depends on the configuration of the FIR-resonator mode, and—in the case of hole coupling—on where the mode is sampled. In a waveguide resonator the polarization of the output is dominated by the constraints imposed by the boundary conditions (Yamanaka *et al.*, 1975). This is illustrated in Fig. 14, which shows two lines observed when CH_3OH is pumped by $P_9(38)$ (Henningsen, 1980a). The resonator is a hybrid, consisting of two plane-parallel metal plates with lateral confinement being provided by a stable optical mirror configuration. The modes with low losses are those that have the E field polarized parallel to the plates, and all modes on both

CH₃OH
P₉ (38)
193.2 μm
Ē_P
292.2 μm
Ē_P

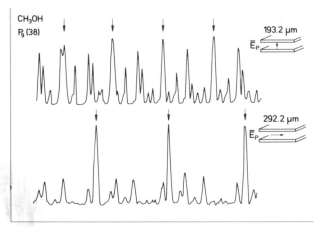

FIG. 14 Far-infrared emission for pump field polarized orthogonal and parallel to the plates of a plane parallel metal waveguide. CH_3OH pumped by $P_9(38)$.

traces are strongly polarized accordingly, despite the predictions of Eq. (9). The situation is analyzed in Fig. 15. In this case the information about the transitions is not directly contained in the polarization of the output, but it can still be recovered from the relative intensities observed for the two mutually orthogonal directions of pump polarization, where the gain polarization works either with or against the loss polarization.

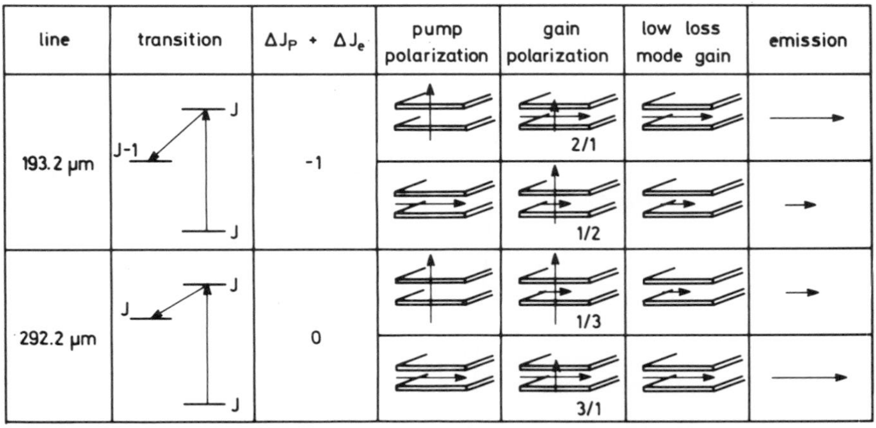

line
transition
ΔJ_P + ΔJ_e
pump polarization
gain polarization
low loss mode gain
emission
193.2 μm
J
J-1
J
-1
2/1
1/2
292.2 μm
J
J
J
0
1/3
3/1

FIG. 15 Analysis of polarization effect in plane parallel waveguide resonator. CH_3OH pumped by $P_9(38)$.

3. *Stark Effect*

Just as in conventional spectroscopy, the application of a static electric field provides additional information about the energy levels by removing the M degeneracy. This general topic is discussed in detail in Chapter 3 and here we shall only consider those aspects that are of particular relevance to Section II of this chapter.

The electric field splits the gain curve, and in the ideal case, lasing is expected on each of the individual M components by appropriate tuning of the FIR resonator. Information on J is then obtained simply by counting. However, real life is not that easy. The gain of the various components is strongly M dependent, and only the strongest transitions will be above threshold for lasing. In addition, the requirement of having enough molecules for sustaining FIR laser action means that the pressure and the associated homogeneous linewidth cannot be reduced below a certain minimum, and this minimum pressure tends to increase with the electric field owing to the broadening of the gain curve. Unfortunately, this frequently forces the laser to operate around the minimum of the Paschen curve, where even a modest electric field will cause breakdown of the gas. Increasing the attainable electric field by narrowing the Stark plate separation has the detrimental effect of increasing both pump losses and FIR losses. The lasing transitions usually involve $J \gtrsim 10$, and the combination of relatively high pressure ($\gtrsim 100$ mtorr) and relatively low electric field ($\lesssim 1$ kV/cm) means that the M-structure is generally unresolved (Fig. 16). Exceptions to this are CH_3F (Inguscio *et al.*, 1979b) and NH_3 (Redon *et al.*, 1979), which both can operate at very low pressure. The same holds for some lines of CH_3OH (Gastaud *et al.*, 1980), but for general applica-

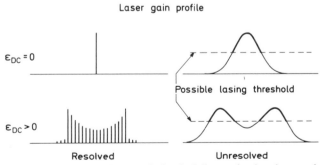

FIG. 16   Stark effect on FIR laser emission in fully resolved and unresolved case.

tion of the Stark effect it is necessary to focus on the unresolved limit (Henningsen, 1980a).

The theory of molecules in a static electric field is presented in Chapter 10 of Townes and Schawlow (1955). For $\pm K$-degenerate symmetric rotor states the Stark effect is linear, and the energy is given by

$$W_{\text{Stark}} = -\mu_a \varepsilon K M / J(J + 1), \tag{13}$$

where μ_a is the permanent dipole moment and ε the electric field. For nondegenerate asymmetric rotor states the Stark effect is quadratic in ε, while an intermediate case occurs in, for example, CH_3OH, where states of internal rotation A symmetry are split for low K by the slight asymmetry of the molecule. Here the Stark effect is quadratic until the Stark shifts become comparable to the asymmetry splitting, and with further increasing electric field, it gradually turns linear (Ivash and Dennison, 1953). When evaluating Stark shifts on the basis of Eq. (13), it should be recalled that the permanent dipole moment is not necessarily the same in the upper and the lower laser level, and that even a small difference can modify the spectrum considerably.

The dominant contribution $\Delta \nu_e(M)$ to the net Stark shift is caused by the Stark effect on the emission transition as calculated from Eq. (13) or from analogous expressions. However, an additional effect, which may arise from a possible Stark splitting of the pump transition, can contribute. Pumping is effected by a monochromatic signal, and in zero electric field the molecules that are pumped are those which are Doppler shifted into resonance. If the pump is at its line center, these molecules have a velocity $v_z = -\delta c / \nu_p$ in the propagation direction of the pump beam, where δ is the line offset as defined in Fig. 3, c is the velocity of light, and ν_p is the pump frequency. This leads to a Doppler shift of the emitted frequency ν_e of $\Delta_D \nu_e = \mp \delta \nu_e / \nu_p$, where the two signs correspond to counter- and copropagating emission. A Stark shift $\Delta \nu_p(M)$ of the pump transition leads to an M dependence of v_z, and the resulting emission frequencies are then

$$\nu_e + \Delta \nu_e(M) \mp [\delta + \Delta \nu_p(M)] \nu_e / \nu_p. \tag{14}$$

In general, there is an asymmetry between the counter- and copropagating gain, but in the usual resonator configuration with simultaneous pumping by a forward and backward running pump beam, symmetry is restored, and two identical peaks are expected for each M. The effect is counteracted by velocity cross relaxation which tends to restore a Maxwellian velocity distribution in the upper laser level, and if the pressure is not kept very low, it is more likely to act as an additional line broadening mechanism. The effect of δ is well known, and for numerous FIR lines in

CH₃OH, for example, it leads to double-peaked resonances which merge into a single line as the pressure is increased.

Having determined the net Stark shifts of the various M components, the resulting gain profile is calculated by summing over M with intensities evaluated from Table 4–4 of Townes and Schawlow (1955). Since the Stark plates are usually an integral part of a hybrid waveguide laser, the polarization of the emitted radiation is constrained to be parallel to the plates (see Section I.C.2 and Fig. 15), and hence only $\Delta M_e = \pm 1$ transitions need to be considered. For an intermediate $J(= 10)$ and different $K(= 1, 5,$ and $10)$, Henningsen (1980a) has evaluated the gain profile for the three transitions $\Delta(J, K) = (0, -1), (-1, 0),$ and $(-1, -1)$, assuming $P, Q,$ or R pumping, different ratios α of lower to upper state dipole moment $(= 1.1, 1.0,$ and $0.9)$, and pump polarization parallel or orthogonal to the static electric field. With increasing electric field the profile may show a broadening or a symmetric splitting into two or more peaks. Many of the individual signatures are quite similar, but a comparison between the profiles observed with the two different modes of pump polarization frequently screens out a large number of possibilities (Fig. 17). Nevertheless, Stark data usually must be supplemented by other data for complete assignment (see Section II.B.5).

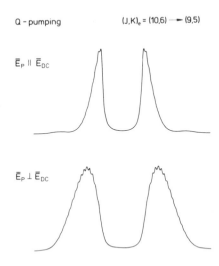

FIG. 17 Calculated FIR gain profile for two different modes of pump polarization relative to static electric field. Parameters correspond to Fig. 40.

4. *Double Resonance Techniques*

The concept of double resonance is based on two radiation fields that are simultaneously resonant on transitions sharing a common level, or being tightly coupled by nonradiative transitions. The presence of one resonance is detected on the other through its effect on the level populations. Double resonance thus involves two separate, but coupled, one-photon transitions and should not be confused with coherent two-photon transitions.

In optically pumped FIR lasers the technique has been used for both verifying and disproving assignments. Lees *et al.* (1979) considered CH_3OH pumped by $P_9(16)$, where two alternative assignments had been suggested for the pump transition (Henningsen, 1977; Danielewicz and Coleman, 1977). Figure 18 shows the level scheme for the microwave–infrared double resonance experiment, in which a probe beam tuned to three different transitions in the 20-GHz range was applied to CH_3OH at a pressure of 50 mtorr. In cases a and b, application of the $P_9(16)$ laser beam causes an increase in probe absorption because the pump depletes a level which is collisionally coupled to the upper probe level. Conversely, in

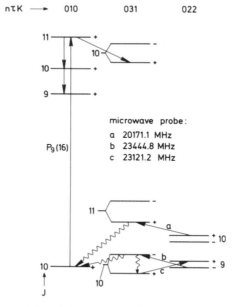

FIG. 18   Microwave–infrared double resonance experiment. (Adapted from Lees *et al.*, 1979.)

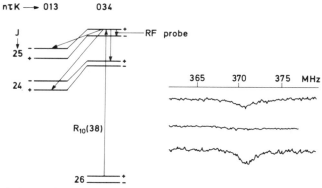

FIG. 19   Scheme of rf-FIR-IR triple resonance experiment. (Adapted from Arimondo *et al.*, 1980.)

case c, a decrease in probe absorption is observed, since the lower pump level and the lower probe level are coupled by two-step dipolar collisional transfer, while at least three steps are needed for coupling the upper probe level to the lower pump level. A similar technique was used for disproving suggested assignments of CH_3OH lines observed at a different offset of the $P_9(16)$ pump line (Lees, unpublished), and for confirming assignments of lines pumped by $R_{10}(38)$ (Young *et al.*, 1979).

The $R_{10}(38)$ assignments have been further verified by a triple resonance technique in which the three resonances share a common level (Arimondo *et al.*, 1980). The scheme is shown in Fig. 19. In this case the FIR laser emission was monitored, and by applying rf radiation at 370.8 MHz, a decrease in emission was observed owing to a depletion of the upper laser level. Apart from proving the assignment, this experiment also provides a very accurate determination of the asymmetry splitting.

II. CH_3OH—A Case Story

Methyl alcohol is particularly well suited for illustrating the potential of the optical pumping technique. It was early recognized as a source of strong FIR lasing (Chang *et al.*, 1970), and following this initial work, hundreds of lines have been generated by the various isotopic modifications of this molecule. A main reason is the excellent overlap that exists between the strongly absorbing C–O stretch band and the CO_2 laser spectrum, but equally important features are the complexity of the rotational spectrum, brought about by the internal rotation degree of freedom, and the fairly large permanent dipole moments both along the quasi-symmetry axis and orthogonal to this axis, which make the electric dipole selection rules less restrictive than for symmetric top molecules.

The same complexity that makes the molecule such a prolific FIR source also makes it a nontrivial spectroscopic object. In Section II.A we shall review what is known from conventional spectroscopy and present the model that has been very successful in accounting for the structure of the vibrational ground state. In Section II.B the C–O stretch data obtained by optical pumping are interpreted with reference to this model, and by combining with an analysis of diode laser spectra (Sattler *et al.*, 1978, 1979) precise band parameters are determined for the C–O stretch. Systematic deviations are found, which indicate the presence of a strong perturbation. Direct contact is established with some of the perturbing levels by optical pumping, and from analysis of the associated FIR emission some characteristics of the perturbing states are deduced.

A. A PRIORI INFORMATION

1. *Infrared Spectrum in the C–O Stretch Region*

a. *Vibrational Modes.*   The equilibrium structure of the CH_3OH molecule is indicated in Fig. 20. It is a slightly asymmetric top, and from general considerations it is expected to have 12 fundamental vibrational modes. Six of these are similar to modes of the methyl halides: the symmetric

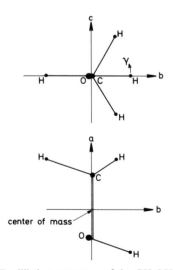

FIG. 20   Equilibrium structure of the CH_3OH molecule.

TABLE II

VIBRATIONAL MODES OF CH_3OH

	Borden and Barker (1938) vapor	Barnes and Hallam (1970) matrix	Serrallach et al. (1974) vapor	matrix	
OH stretching A'	3683	3667.3	3681.5	3667.0	$\nu_1(a')$, ν(OH)
CH_3 asym. stretch A'	2978	3005.3	2999.0	3005.5	$\nu_2(a')$, ν(CH_3)asym.
CH_3 asym. stretch A''	2978	2961.9	2970 ± 4	2961.5	$\nu_9(a'')$, ν(CH_3)asym.
CH_3 sym. stretch A'	2845	2847.9	2844.2	2847.5	$\nu_3(a')$, ν(CH_3)sym.
CH_3 asym. deform. A'	1477	1474.1	{1479.5 / 1477.2}	1473.0	$\nu_4(a')$, δ(CH_3)asym.
CH_3 asym. deform. A''	1477	1465.8	1465 ± 3	1466.0	$\nu_{10}(a'')$, δ(CH_3)asym.
CH_3 sym. deform. A'	1455	1451.4	1454.5	1451.5	$\nu_5(a')$, δ(CH_3)sym.
OH bending A'	1340	1335	{1339.5 / 1332.0}	1334.0	$\nu_6(a')$, δ(COH)
CH_3 rocking A''		1157	1145 ± 4		$\nu_{11}(a'')$, $\gamma_\perp$(CH_3)
CH_3 rocking A'		1076.7	1074.5	1076.5	$\nu_7(a')$, $\gamma_\parallel$(CH_3)
CO stretching A'	1034.2	1034.0	1033.5	1033.5	$\nu_8(a')$, ν(CO)
OH torsion A''		240		271.5	$\nu_{12}(a'')$, τ(OH)

and the asymmetric CH_3 stretch, the symmetric and the asymmetric CH_3 deformation, the CH_3 rocking mode, and the C–O stretch. Of these, the two asymmetric CH_3 modes and the CH_3 rocking mode appear as doublets owing to the slight asymmetry of the molecule. The components of these doublets are designated A' and A'' for even or odd symmetry with respect to reflection in the COH plane. The remaining three modes are associated with the OH group: the O–H stretch, the C–O–H bending, and the O–H torsion.

The fundamental frequencies of the vibrational modes are located between 200 and 4000 cm⁻¹, and the infrared spectrum in this range has been extensively studied since the mid thirties. One of the early investigations which for many years remained unsurpassed, was carried out by Borden and Barker (1938), who resolved seven distinct frequencies (see Table II). Among the most recent works are those of Barnes and Hallam (1970) and Serrallach *et al.*, (1974). Barnes and Hallam studied argon matrix isolated CH_3OH at cryogenic temperatures over a range of M(atrix)/A(bsorber) from 2000 to 20. The increased resolution obtained by thus suppressing the rotational structure enabled them to locate all 12 frequencies. In addition, they observed numerous lines which they associated with dimer and multimer absorptions, and with monomer absorptions from molecules trapped at a different site in the argon matrix.

Serrallach *et al.* observed the fundamental frequencies plus a number of combination and overtone frequencies, both in matrix isolation and in the gas phase. They reproduced the matrix isolated frequencies of Barnes and Hallam, except for the A'' CH_3 rocking mode, which was not observed, and the O–H torsion, which they located at 271.5 cm rather than at 240

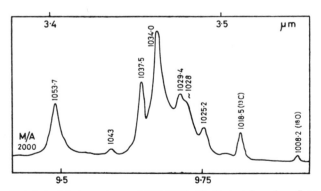

FIG. 21   Matrix isolated spectrum of CH_3OH in C–O stretch region. Suggested assignments: 1053.7 dimer (?), 1043 multimer, 1037.5 2nd site monomer, 1034.0 monomer, 1029.4 dimer, 1028 multimer, 1025.2 dimer (Barnes and Hallam, 1970).

cm^{-1}. In the gas phase their results are in essential agreement with those of Borden and Barker.

Since we shall be concerned mainly with the C–O stretch, we shall summarize in slightly more detail the results obtained in this region. Figure 21 shows the matrix spectrum of Barnes and Hallam together with their suggested assignments. It should be noted that these are not based directly on concentration studies, but rather on a similarity with the structure of the O–H stretch region, where such studies were carried out. Serrallach *et al.* observe essentially the same structure at $M/A \simeq 1000$ (Fig. 22). At $M/A \simeq 3000$ most of the satellites are reduced in intensity relative to the 1034.0 cm^{-1} absorption, but a strong line is still present at a reported frequency of 1027.5 cm^{-1}, its relative intentisy having in fact gone up by a factor of 2. This outrules the interpretation of the 1029.4/1028 structure of Fig. 21 as dimer or multimer absorption, and Serrallach *et al.* suggest instead either monomeric absorption from molecules trapped at a different site or torsional hot band absorption. This could in principle be decided on the basis of gas-phase spectra. However, in the gas phase all of the structure present in Figs. 21 and 22 is completely buried under the rotational structure of the fundamental C–O stretch, and to observe a possible rotational structure associated with any of the satellites requires a resolution which is far better than that offered by a conventional spectrometer.

 b. *Vibration–Rotation Spectrum.* Superimposed on all of the bands of Table II is the structure arising from the rotation of the molecule, and

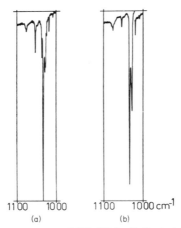

FIG. 22 Matrix isolated spectrum of CH$_3$OH in C–O stretch region (Serrallach *et al.*, 1974): (a) $M/A = 1000$; (b) $M/A = 3000$.

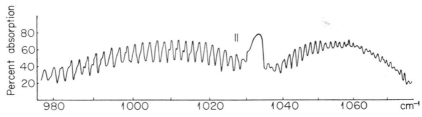

FIG. 23   Rotational structure of C–O stretch band (Borden and Barker, 1938).

the infrared transitions actually occur between the combined vibration–rotation states. On a gross scale, the spectrum in the C–O stretch region is similar to that observed for rigid symmetric tops such as the methyl halides. The rotational state is labeled by the quantum numbers J, measuring the total angular momentum, and K, measuring its projection on the quasi-symmetry axis of the molecule. Since for the C–O stretch the transition dipole moment is directed essentially along the a axis, no torque can be excerted about it, and hence K must remain unchanged. Thus a simple parallel band is expected with a P, Q, and R branch corresponding to the electric dipole selection rules $\Delta J = -1, 0$, and 1 in the absorption; Borden and Barker (1938), in fact, observed this spectrum (Fig. 23).

This remained the ultimate in resolution until Woods (1970) published his thesis, and it was only then that the complexity of the spectrum became apparent (Fig. 24a). What appear as single lines in Fig. 23 are actually multiplets, each covering a range of about 1 cm^{-1}, and the number of components increasing with J. To understand the origin of this, it is necessary to consider the structure of the molecule in more detail.

c. *Internal Rotation.*   The truly complicating feature in CH$_3$OH is the existence of an internal degree of freedom in which the OH group rotates about the axis of the CH$_3$ group, the rotation being opposed by a hindering threefold potential. For an infinitely high barrier, the stationary states would be harmonic oscillator-like torsional levels, labeled $n = 0, 1, 2, \ldots$, with each n being threefold degenerate owing to the three equivalent positions of the OH group. The finite barrier height and the associated probability for tunneling removes the degeneracy, and an additional quantum number $\tau = 1, 2, 3$ is needed to keep track of the split components.

The lowest lying mode of Table II, labeled OH torsion A'', refers to transitions from $n = 0$ to $n = 1$ as observed in matrix isolation, where rotational motions are frozen. In the gas phase this mode is strongly coupled

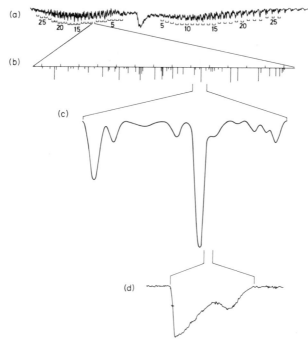

FIG. 24.   A segment of the C–O stretch band viewed with progressively increasing resolution: (a) grating monochromator spectrum (Woods, 1970; reproduced from Danielewicz and Coleman, 1977. *IEEE J. Quantum Electron.* **QE-13**, 485–490. © 1977 IEEE); (b) diode laser spectrum (Sattler *et al.*, 1979); (c) as (b), but direct recording; (d) optoacoustic scan over tuning range of CO_2 laser oscillating in $P_9(16)$.

to the overall rotation about the molecular axis, and it is therefore not possible to treat these degrees of freedom separately. The theoretical framework for the combined problem has been provided mainly by Dennison toghether with a number of co-workers, leading up to Kwan and Dennison (1972), and by Lees and Baker (1968). We shall summarize this theory in Section II.A.2 since it forms the basis for the interpretation of the optical pumping results. At present, we shall just note that four quantum numbers $(n\tau K,J)^v$ are needed to specify a particular level belonging to the vibrational state v. Although only J can change in transitions from the vibrational ground state to the C–O stretch, the transition frequencies will depend parametrically on n, τ, and $\check{K}$ if the energy barrier, moments of inertia, etc. are slightly different in the two states.

Allowing for this, Woods could, without actually resolving the structure of
Fig. 24a, account for the spread of each multiplet by assuming the hindering barrier to be 19 ± 2 cm^{-1} higher in the C–O stretch than in the vibrational ground state.

   d. *Diode Laser Spectra.*   The next big step forward was taken when
tunable diode lasers were introduced in infrared spectroscopy. Sattler *et
al.* (1978, 1979) recorded absorption spectra with Doppler limited resolution, using PbSnSe lasers, and the dramatic improvement is illustrated in
Fig. 24b, where the $R(10)$ multiplet is shown, now fully resolved. No less
than 56 lines are included in the diagram, which only includes lines with
an intensity of more than 5% of the strongest, and from the small segment
of the original spectrum (Fig. 24c) it is seen that many weaker lines are
present.

   Figure 25 shows the diode laser spectrum for the entire R branch of the
C–O stretch according to Sattler *et al.,* (1979). The spectrum is broken up
into segments, each corresponding to an $R(J)$ multiplet, and the positions

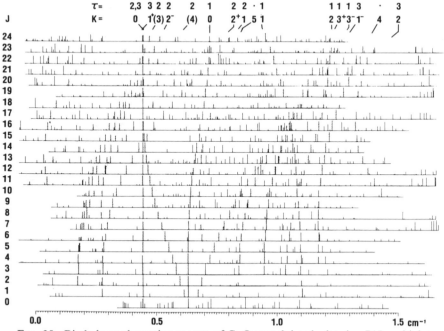

FIG. 25   Diode laser absorption spectra of C–O stretch band, showing $R(J)$ multiplets
up to $J = 24$ (Sattler *et al.,* 1979).

of the segments are normalized such that the lines corresponding to $(n\tau K) = (020)$ are vertically above each other. In such a representation the $R(J)$ series corresponding to other $(n\tau K)$ are expected to follow smooth curves which tend to be vertical at low J and terminate at $J = K$. Using Woods' assumptions for the C–O stretch parameters, it proved possible to identify all series up to $K = 2$, one $K = 3$ series, and to make a partial identification of another four series with K ranging up to 5. However, despite extensive efforts no further progress could be made. It is at this stage that the optically pumped far infrared laser demonstrates its potential, but before proceeding, we shall describe in detail the model on which the interpretation will be based.

2. The Torsion–Rotation Model

a. *Energy Levels.* We shall assume that the energy associated with the combined vibration, overall rotation, and O–H torsion, or internal rotation, can be expressed as an expansion in powers of $J(J + 1)$:

$$E(n\tau K,J)^v = E_{vib}^v + W^v(n\tau K) + B^v(n\tau K)J(J + 1)$$
$$- D^v(n\tau K)J^2(J + 1)^2$$
$$- H^v(n\tau K)J^3(J + 1)^3(+ \text{asymmetry splitting}), \qquad (15)$$

where E_{vib}^v denotes the pure vibrational energy. For the vibrational ground state ($v = 0$), the expansion coefficients can be expressed in terms of molecular parameters and phenomenological interaction constants (Lees and Baker, 1968), and we shall assume that the vibration–rotation interaction can be accounted for by letting these depend on v. In the following, the superscript will be omitted wherever possible.

The expansion coefficients are given by

$$B(n\tau K) = \tfrac{1}{2}(B + C) + F_v\langle 1 - \cos 3\gamma \rangle + G_v\langle P_\gamma^2 \rangle$$
$$+ L_v K\langle P_\gamma \rangle - D_{JK}K^2 + b(n\tau K), \qquad (16)$$

$$D(n\tau K) = D_{JJ} - d(n\tau K), \qquad (17)$$

$$W(n\tau K) = \tfrac{1}{2}V_3\langle 1 - \cos 3\gamma \rangle + F\langle P_\gamma^2 \rangle$$
$$+ [A - \tfrac{1}{2}(B + C)]K^2 + \Delta E(n\tau K), \qquad (18)$$

where γ is the angle of internal rotation, P_γ its canonically conjugate momentum, and the bracketed quantities are internal rotation expectation values.

Choosing a Cartesian coordinate system abc with the a axis through the molecular center of mass, parallel to the symmetry axis of the CH_3 group, and the b and c axes orthogonal to a, with b in the COH plane (Fig. 20), the constants B, C, F, and A are given by the moments of inertia:

$$B = \frac{I_b}{I_b^2 + I_{ab}^2} \frac{\hbar}{4\pi} \simeq \frac{1}{I_b} \frac{\hbar}{4\pi}, \tag{19}$$

$$C = \frac{1}{I_c} \frac{\hbar}{4\pi}, \tag{20}$$

$$F = \frac{I_a I_b - I_{ab}^2}{I_{a1} I_{a2} I_b - I_{a2} I_{ab}^2} \frac{\hbar}{4\pi} \simeq \left(\frac{1}{I_{a1}} + \frac{1}{I_{a2}}\right) \frac{\hbar}{4\pi}, \tag{21}$$

$$A = \left(\frac{I_a + I_b}{I_a I_b - I_{ab}^2} - \frac{I_b}{I_b^2 + I_{ab}^2}\right) \frac{\hbar}{4\pi} \simeq \frac{1}{I_{a1} + I_{a2}} \frac{\hbar}{4\pi}. \tag{22}$$

Here I_{a2} is the moment of inertia of the CH_3 group, and $I_{a1} \equiv I_a - I_{a2}$. The approximate expressions apply if the small cross moment of inertia I_{ab} is neglected. The terms with F_v, G_v, and L_v account for the J-dependent part of the centrifugal distortion effects on the internal rotation. D_{JJ} and D_{JK} correspond to rigid rotor centrifugal stretching constants. The first two terms of $W(n\tau K)$ are the expectation values of the internal rotation potential and kinetic energy, while $\Delta E(n\tau K)$ lumps together all residual J-independent contributions. It is given by

$$\begin{aligned}
\Delta E(n\tau K) = &\tfrac{1}{2}V_6 \langle 1 - \cos 6\gamma \rangle - D_{KK}K^4 + k_1 K^3 \langle P_\gamma \rangle + k_2 K^2 \langle P_\gamma^2 \rangle \\
&+ k_3 K \langle P_\gamma^3 \rangle + k_4 \langle P_\gamma^4 \rangle + k_5 K^2 \langle 1 - \cos 3\gamma \rangle \\
&+ k_6 K \langle P_\gamma (1 - \cos 3\gamma) \rangle + w(n\tau K),
\end{aligned} \tag{23}$$

where V_6 measures the departure of the hindering potential from a perfect sinusoid, D_{KK} corresponds to a rigid rotor centrifugal stretching constant, and the terms involving k_1 to k_6 are the so-called Kirtman terms, which measure the J-independent centrifugal distortion effects on the internal rotation.

Owing to the threefold symmetry of the hindering potential, the internal rotation states may be classified according to their symmetry properties under operations of the point group C_3 as E_1, A, and E_2 states for $\tau + K = 3N$, $3N + 1$, and $3N + 2$, respectively, N being integral. For all states, the small asymmetry of the molecule introduces a shift, which can be evaluated by second-order perturbation theory. This shift is included through the terms $b(n\tau K)$, $d(n\tau K)$, and $w(n\tau K)$. For a symmetric top, the $\pm K$ states are degenerate, but the asymmetry leads for $K \neq 0$ to a splitting of A states into an A^+ and an A^- state. The splitting increases with J and decreases rapidly with increasing K. According to Ivash and Dennison (1953), it can be expressed as

$$S(J,K) = S(K) \frac{(J + K)!}{(J - K)!}, \tag{24}$$

where $S(K)$ can be evaluated by perturbation theory. For any K, E_1, and E_2, states are widely spaced owing to the coupling of external and internal rotation. However, the levels $E_1(\pm K)$ and $E_2(\mp K)$ remain degenerate as long as the threefold symmetry is not broken.

While all of the terms discussed so far have been considered by Lees and Baker (1968), we have introduced, in addition, a third-order coefficient $H(n\tau K)$ to get a better representation of the energy levels at the high J, which are frequently encountered in far-infrared laser emission.

b. *Selection Rules and Line Intensities.* Since CH_3OH has sizable permanent dipole moment components both along the a axis and along the b axis (see Table III), the selection rules for rotational transitions inside a

TABLE III

MOLECULAR PARAMETERS FOR CH_3OH^a

	CH_3OH ground state	CH_3OH C–O stretch	Unit
I_b	34.003 856	34.2828(26)	
I_c	35.306 262	35.6380(26)	
I_{ab}	−0.1079	−0.1079	kg m² × 10⁻⁴⁷
I_{a1}	1.2504	1.2523(8)	
I_{a2}	5.3331	5.3334(8)	
E_{vib}	0	1030.084(3)	
V_3	373.21	392.35(30)	
V_6	−0.52	−0.52	
D_{KK}	0.38 × 10⁻⁴	0.38 × 10⁻⁴	
k_1	−0.48 × 10⁻⁴	−0.48 × 10⁻⁴	
k_2	−18.41 × 10⁻⁴	−18.41 × 10⁻⁴	
k_3	−53.73 × 10⁻⁴	−53.73 × 10⁻⁴	
k_4	−85.50 × 10⁻⁴	−85.50 × 10⁻⁴	
k_5	137.07 × 10⁻⁴	137.07 × 10⁻⁴	cm⁻¹
k_6	67.85 × 10⁻⁴	67.85 × 10⁻⁴	
k_7	0	0	
F_v	−2.389 × 10⁻³	−6.546 × 10⁻³	
G_v	−1.168 × 10⁻⁴	−1.67 × 10⁻⁴	
L_v	−2.26 × 10⁻⁶	−2.26 × 10⁻⁶	
D_{JK}	9.54 × 10⁻⁶	9.54 × 10⁻⁶	
D_{JJ}	1.6345 × 10⁻⁶	1.6345 × 10⁻⁶	
μ_a	2.952	3.055	Cm × 10⁻³⁰
μ_b	4.80	4.80	

^{a} Moments of inertia are converted to amu Å² by dividing the numbers of the table by 1.660 531. Dipole moments are converted to Debye by dividing by 3.335 64.

given vibrational state are less restrictive than for vibration–rotation transitions to the C–O stretch. The selection rules for parallel transitions involving μ_a are

$$\Delta J = 0, \pm 1 \qquad \text{and} \qquad \Delta K = \Delta n = \Delta \tau = 0. \qquad (25)$$

For transitions involving μ_b, which is associated with the OH group, we have $\Delta K = \pm 1$, with Δn arbitrary and $\Delta \tau$ determined such that internal rotation symmetry is preserved. Furthermore, each of these Q-band transitions has associated with it a P- and an R-type transition corresponding to a simultaneous change of $\Delta J = \pm 1$. For Δn even, the split A states obey the rules $\pm \rightarrow \pm$ for $|\Delta J| = 1$ and $\pm \rightarrow \mp$ for $\Delta J = 0$, while the reverse holds for Δn odd.

Assuming pumping has been effected to a state $(n\tau K,J)$, the selection rules in general allow FIR emission on numerous transitions. To identify those that are most likely to lase, we note that the FIR gain for an i $\leftrightarrow$ j transition is proportional to

$$\nu f F(J) |\mu_{ij}|^2, \qquad (26)$$

where ν is the frequency, f is a nuclear spin factor which is 2 for E_1 and E_2 states and 4 for unresolved A doublets (Kwan and Dennison, 1972), while $F(J)$ is a polarization factor that depends on the combination of pump and emission transitions and on the relative polarization of pump field and

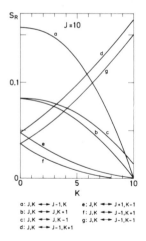

FIG. 26 Overall rotation contribution to the oscillator strength for transitions from (K,J).

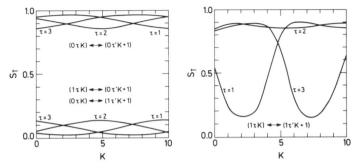

FIG. 27   Internal rotation contribution to the oscillator strength for transitions from $(n\tau K, J)$.

emitted field (Drozdowicz *et al.*, 1979). For emission polarizing along the optimum direction (see Section I.C.2), $F(J)$ in the high-J limit equals 1.20 for all combinations except QQ, where $F(J) = 1.80$. The squared dipole matrix element may be expressed as a product

$$|\mu_{ij}|^2 = \mu^2 S_R S_T, \tag{27}$$

where μ is the appropriate component of the permanent dipole moment, while S_R refers to the overall rotation and S_T to the internal rotation. S_R can be found from Table I by averaging over M, and the result is shown in Fig. 26 for transitions from (K,J) with $J = 10$. S_T is plotted in Fig. 27 for transitions involving $n = 0$ and $n = 1$. Similar curves are given by Burkhard and Dennison (1959), but note that they give the magnitude of the matrix element rather than its square.

c. *Emission Frequencies.* The Q-band frequencies depend on J owing to the $(n\tau K)$ dependence of the expansion coefficients in Eq. (15). For a transition $(n\tau K, J) \rightarrow (n'\tau'K', J)$, the frequencies may be expressed as a power series in $J(J + 1)$:

$$\nu(n\tau K \rightarrow n'\tau'K', J) = q_0 + q_1 J(J + 1) + q_2 J^2(J + 1)^2 \\ + q_3 J^3(J + 1)^3 + \cdots, \tag{28}$$

where the coefficients are given by

$$q_0 = W(n\tau K) - W(n'\tau'K'),$$
$$q_1 = B(n\tau K) - B(n'\tau'K'),$$
$$q_2 = -D(n\tau K) + D(n'\tau'K'),$$
$$q_3 = -H(n\tau K) + H(n'\tau'K'). \tag{29}$$

Vibrational excitation is known to have a large effect on the hindering barrier V_3, and consequently on $W(n\tau K)$. It is therefore a major aim of the analysis to extract the Q-band origin q_0 as reliably as possible from Q-band transitions, which are frequently measured at high J. The information required for evaluating q_1, q_2, and q_3 is in part obtained from observed $\Delta K = 0$ transitions, whose frequencies are well approximated by

$$\nu(n\tau K, J \rightarrow J - 1) = 2JB(n\tau K) - 4J^3D(n\tau K) - 6J^5H(n\tau K). \quad (30)$$

d. *The Vibrational Ground-State Model.* The model described above was developed for the vibrational ground state, and the best set of parameters is given in Table III. Its experimental basis is a series of microwave measurements up to about 200 GHz by Lees and Baker (1968), and a grating spectrometer investigation up to about 300 cm^{-1} with a resolution of 0.025 cm^{-1} performed by Woods (1970) and analyzed by Kwan and Dennison (1972). With these parameters, the model represents torsional ground-state levels ($n = 0$) to an accuracy of about 5×10^{-4} cm^{-1} in the range $K \leq 6$ and $J \leq 10$, and torsionally excited levels to within the experimental accuracy of 0.025 cm^{-1}. Later, Stern *et al.* (1977) have studied internal rotation transitions in the range 80–250 cm^{-1} and thus verified the model up to torsional $n = 5$ states. The analysis of Lees and Baker was not carried beyond second order in $J(J + 1)$, but some information about higher order terms is contained in the analysis of the (012 $\rightarrow$ 021) Q branch by Hughes *et al.* (1951).

e. *The C–O Stretch Model.* By considering transitions that occur within the C–O stretch state, microwave spectroscopy can in principle furnish the same information about this band as about the vibrational ground state. However, the levels are only sparsely populated, and the transitions are difficult to identify. They are masked by the strong spectrum of ground-state transitions and carry no label that associates them with the C–O stretch. Despite this, Lees (1972) succeeded in identifying the two parallel transitions (010,1 $\rightarrow$ 0) and (020,1 $\rightarrow$ 0), and 16 transitions of the (012 $\rightarrow$ 021,J) Q band, all of which are located in the microwave range. His analysis supported Woods concerning the barrier height, but the data did not contain enough information for arriving at a precise value. The parallel transitions are not sensitive to changes in the internal rotation parameters, and at least three different Q bands are needed in order to unambiguously determine V_3, I_{a1}, and I_{a2}.

B. ADDED INFORMATION

1. *The Data*

The first observation of far-infrared laser emission by optical pumping of CH$_3$OH with a CO$_2$ laser was reported by Chang *et al.* (1970). They observed 23 lines and measured their wavelengths to a relative accuracy of about 10^{-4}, as well as their relative polarization. Only one line, the 570.5-μm line pumped by $P_9(16)$ was partially assigned on the basis of Borden and Barkers spectrum as a $J = 11 \rightarrow 10$ emission line pumped by an $R(10)$ C–O stretch absorption line. Subsequently the assignment problem was taken up independently by Danielewicz and Coleman (1977), Henningsen (1977, 1978), and Yano and Kon (1977). Initially the motivation was the need for precise information concerning degeneracies, populations, etc. for the levels involved, a need arising in connection with modeling and optimization of optically pumped far-infrared lasers. However, as the number of observed emission lines kept increasing, and as their characterization became more precise, in particlar through the introduction of frequency measurements (Petersen et al., 1975, 1980), it became obvious that the data might provide the clue to a better understanding of the C–O stretch region of the infrared spectrum. To see this clearly, we may consider Fig. 24d, which shows the optoacoustic absorption over a ± 40-MHz tuning range of the $P_9(16)$ line of the CO_2 laser. This line happens to be in near coincidence with two absorption lines, one on either side, and both can be pumped by choosing an appropriate positive or negative pump offset. Choosing the positive offset, three FIR lines are observed and the measured frequencies lead to the assignment $(n\tau K) = $ (010) (Henningsen, 1977). From Fig. 25 it is seen that the absorption line in question is one of a strong series, and the optical pumping experiment thus confirms the assignment of the entire (010) family. In the negative pump offset, three different far-infrared lines are observed. A tentative assignment was disproved by infrared microwave double resonance experiments (see Section I.C.4), and these lines still remain unassigned.

Figure 28 shows a superposition of the CO_2 laser spectrum and the C–O stretch of CH_3OH. Heavy vertical lines indicate where far-infrared emission has been observed, the length of a line measuring the number of emission lines. All measured frequencies are listed in Appendix A, together with some characteristics of the lines, and Appendix B contains a list of all assignments. In the following, we shall indicate how the lines are assigned and how information about the band parameters is extracted from the data.

2. *Assignment and Data Analysis*

a. Combination Relations among Emission Lines. The interpretation of a spectroscopic experiment will always to some extent be coupled to existing models for the energy levels involved. However, to retain a possibility of learning also about the unexpected, such coupling must be minimized. This can be done by exploiting internal consistencies, and an important aid in this respect is provided by combination relations. They

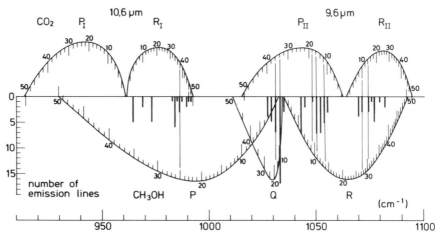

FIG. 28   Superimposed C–O stretch spectrum and CO_2 laser spectrum. Heavy lines indicate measured FIR emission, thin lines that at least some of these lines have been assigned to the C–O stretch. (Adapted from Henningsen, 1977. *IEEE J. Quantum Electron.* **QE-13**, 248–252. © IEEE 1977.)

may be approximate or exact, they may involve measured frequencies only, or they may involve combinations of measured frequencies and frequencies that can be reliably calculated. In the following, we shall give a number of examples to illustrate what can be learned from them.

With pumping of a state $(n\tau K, J)^v$, the possible emission lines are given by the selection and intensity rules of Section II.A.2. Some typical schemes are shown in Fig. 29, but modifications may occur depending on the relative location of the energy levels. In particular, it should be observed that the energy is not necessarily a monotonically increasing function of K^2. According to Fig. 29, a triad of emission lines may be observed, which satisfy an approximate combination relation. Examples of this are provided by the $P_9(36)$, $R_9(14)$, and $R_9(10)$ pump lines as illustrated in Fig. 30. Having identified the lines as belonging to a triad, the polarization of the lines unambiguously specifies their relative location in the scheme. From the general rule of Eq. (9), it follows that the a and the c lines, which both have $\Delta J = -1$, will have the same polarization. The third line is then identified as the Q-band transition with $\Delta J = 0$. As a check, the a and a' frequencies, when divided by a suitable even integer, should lead to a reasonable value for the effective rotational constant $B(n\tau K)$, and this in turn determines J.

The polarization of the Q-band line b, in conjunction with Eq. (9), now provides information about the pump transition. In the parallel case, the

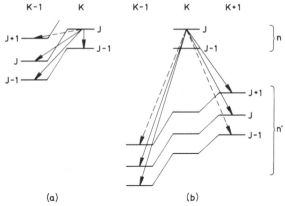

FIG. 29 Dominant emission lines within (a) a torsional state and between (b) different torsional states. Dotted arrows indicate allowed, but weak transitions.

pumping is effected by a Q-type absorption line, and the pump frequency thus should be located in the vicinity of the vibrational frequency as given in Table II. In the perpendicular case, the absorption line is of the P or R type. The case can easily be settled since J is known, and this serves as a further check on the J assignment.

To determine the quantum numbers n, τ, and K, it is usually necessary to resort to comparison with calculations. This step is quite difficult since the transitions are frequently at high J, where in particular second-order

	$R_9 (14)$	$R_9 (10)$	$P_9 (36)$
c	99.199981	103.602901	84.150935
b	51.529607	60.685781	58.624765
a'	47.670374	42.917119	25.526170
a	47.634874	42.929683	25.505733
a'-a	0.035500	- 0.012564	0.020437

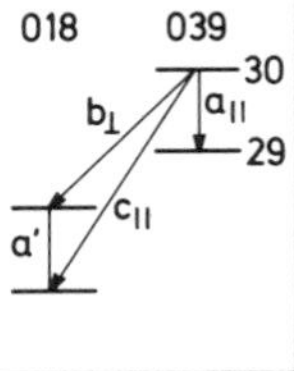

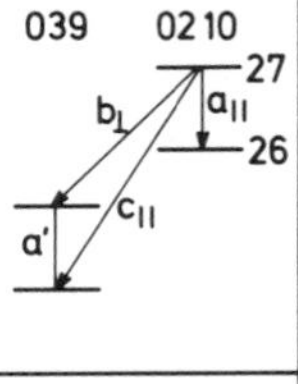

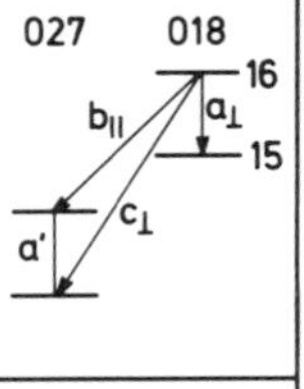

FIG. 30 Combination relations between emission lines in torsional $n = 0$ state (units of cm^{-1}).

asymmetry shifts may add a very significant and strongly K-dependent contribution.

Occasionally, triads of lines are found which seem to conform with the pattern of Fig. 30, despite the fact that $a' - a$ is too large to be explained by the τK dependence of the rotational constants. This may indicate that a and a' are in different torsional states, but a second possibility is that the levels involved are internal rotation A states, which are split owing to the asymmetry of the molecule (Fig. 31). The large $a' - a$ derives from the fact that the splitting is strongly K dependent, and at the same time the $\pm$ selection rule is different for $\Delta J = 0$ and $\Delta J = -1$ transitions (see Section II.A.2).

b. *Combination Relations Involving Pump Transitions.* A different type of combination relations, involving the pump transitions, may provide an additional check on the assignments and also yield information about the expansion coefficients of Eq. (15). The three pump lines of Fig. 30 are pumping E_1 states, which are coupled by allowed transitions, and the combined energy level diagram is shown in Fig. 32. Here single-ended arrows are measured lines, while double-ended arrows are calculated from the ground-state parameters of Table III. The plain lines correspond to frequencies that can be evaluated by combination relations, and since they are expressible in terms of $B^{co}(n\tau K)$, $D^{co}(n\tau K)$, and $H^{co}(n\tau K)$, they constitute a very useful supplement to the information supplied by the emission frequencies.

c. *Data Analysis.* The aim of the data analysis is to obtain reliable values for as many as possible of the parameters in Eqs. (16), (17), and (18). Information concerning $B(n\tau K)$ and $D(n\tau K)$ is contained in the $\Delta K = 0$ frequencies and can be extracted without knowing $W(n\tau K)$. Information about $W(n\tau K)$, on the other hand, is contained in the Q-band frequencies and the pump frequencies, which both may contain a significant J-dependent contribution. To extract $W(n\tau K)$, the measured frequencies must therefore be reduced to $J = 0$.

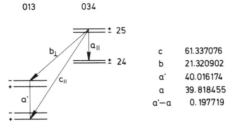

c	61.337076
b	21.320902
a'	40.016174
a	39.818455
a'−a	0.197719

FIG. 31   Combination relations between emission lines involving asymmetry split A states (units of cm^{-1}).

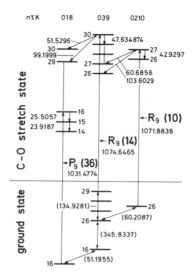

FIG. 32 Combination loops involving pump transitions (units of cm^{-1}) (Henningsen, 1980b).

Reduction of a Q-band frequency requires a knowledge of the coefficients q_1, q_2, and q_3 of Eq. (28), and for complete determination, at least three $\Delta K = 0$ transitions must be known for each of the two adjacent $(n\tau K)$ ladders. This much information is usually not available, and even if it were, there would be a risk that minor perturbations of any of the levels involved could lead to a substantial error in the extrapolated Q-band origin. A different approach is suggested by analyzing the $(012 \rightarrow 021)$ Q band, which has been studied by Lees (1972) for the C–O stretch. Despite the fact that this band is strongly affected by asymmetry contributions, q_2 as calculated from a preliminary model based on Woods' results, agrees very well with experiments. On the other hand, this does not hold for q_1, and the procedure adopted is therefore to use the measured $\Delta K = 0$ frequencies for determining $B^{CO}(n\tau K)$, using calculated $D^{CO}(n\tau K) \simeq D^0(n\tau K)$ and neglecting $H^{CO}(n\tau K)$.

In some cases redundant information is available, so that the procedure can be checked. For $P_9(36)$ we may evaluate $B^{CO}(018)$, $D^{CO}(018)$, and $H^{CO}(018)$ from the $J = 30 \rightarrow 29$, $29 \rightarrow 16$, and $16 \rightarrow 15$ level separations (Fig. 32) with results as given in Table IV. Evaluating the $15 \rightarrow 14$ cascade frequency from these values, we arrive at 23.914 827 cm^{-1}, in mild disagreement with the measured value of 23.918 714 cm^{-1}. A further check is possible since the $R(8)$ to $R(24)$ lines for $(n\tau K) = (018)$ can be

TABLE IV

ENERGY EXPANSION COEFFICIENTS IN CM^{-1}

	$B^{CO}(018)$	$D^{CO}(018)$	$H^{CO}(018)$
FIR laser analysis	0.797893	1.542×10^{-6}	0.252×10^{-9}
Diode laser analysis	0.797997	1.72×10^{-6}	0.23×10^{-9}

identified in the diode laser spectrum (Fig. 25). The coefficients deduced from an analysis of these lines are also listed in Table IV, and the agreement is seen to be quite satisfactory.

Figure 33 shows the deviation between the measured $R(J)$ frequencies and those calculated from an interpolation scheme, and it is seen that $R(14)$ and $R(15)$ are displaced upward by about 0.005 cm^{-1}. This indicates a slight perturbation of the $J = 15$ and $J = 16$ levels of the C–O stretch and explains the discrepancy for the $15 \rightarrow 14$ transition.

Reduction to $J = 0$ of a pump frequency is less demanding since we stay within the same ($n\tau K$) ladder. It is here assumed that all ground-state contributions can be reliably calculated. The reduced frequency is expressed as

$$\nu(n\tau K) = E_{\text{vib}} + W^{CO}(n\tau K) - W^O(n\tau K). \qquad (31)$$

3. *C–O Stretch Model for Torsional Ground State*

By analyzing the optical pumping data along the guidelines described above and combining them with the diode laser spectra, a model has been constructed for the torsional $n = 0$ state of the C–O stretch (Henningsen, 1981). The parameters I_{a1}, I_{a2}, I_b, I_c, V_3, F_v, and G_v were adjusted, while the remaining ones were fixed at their ground-state values. The results are given in Table III. An important feature of the optical pumping data is that they contain information about $W^{CO}(n\tau K)$ up to $K = 10$ and hence make it

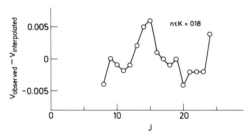

FIG. 33   Diode laser data for $R(018,J)$, indicating minor perturbation of levels $J = 15$ and 16.

possible to determine A, which enters in the K^2 coefficient of Eq. (18). When combined with F, this allows separate determination of I_{a1} and I_{a2}. G_v cannot be uniquely determined and is arbitrarily chosen to -5 MHz. This affects the determination of F_v, but not of the other parameters.

With this model at hand, it is possible to identify an additional number of $R(J)$ series in the diode laser spectrum, ranging up to $K = 10$. From each series we determine the $B^{CO}_{meas}(n\tau K)$ and $W^{CO}_{meas}(n\tau K)$ needed to reproduce the spectrum, and when these measured quantities are compared with the predictions of the model, an interesting pattern is revealed. Plots of $\Delta W = W^{CO}_{meas}(n\tau K) - W^{CO}_{model}(n\tau K)$ and $\Delta B = B^{CO}_{meas}(n\tau K) - B^{CO}_{model}(n\tau K)$ point to the existence of a significant perturbation around $K = 5$ to 7 (Fig. 34). Calculated values for $W(n\tau K)$, $B(n\tau K)$, and $D(n\tau K)$ are given in Tables V and VI, supplemented for the C–O stretch by the empirical corrections ΔW, ΔB, and H. For $(n\tau K) = (025)$ the levels appear split. The splitting $\Delta(025,J)$ is tabulated in Table VII and expressed in terms of $\Delta(025,J)/J(J + 1)$ in Fig. 35. The available information as presented in Section II.A gives no immediate clue as to the source of the perturbation.

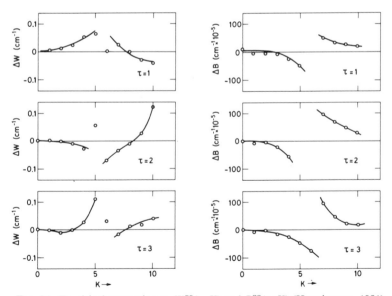

FIG. 34   Empirical correction to $W^{CO}(n\tau K)$ and $B^{CO}(n\tau K)$ (Henningsen, 1981).

TABLE V

ENERGY EXPANSION COEFFICIENTS FOR VIBRATIONAL
GROUND STATE

$n\tau K$	W^0 (cm^{-1})	B^0 (cm^{-1})	D^0 (cm^{-1} × 10^{-6})
010	127.975	0.806772	4.91
011	131.851	0.806768	6.45
012	143.435	0.806761	−3.03
013	162.587	0.806724	1.13
014	189.072	0.806684	1.66
015	222.553	0.806628	1.80
016	262.618	0.806550	1.81
017	308.841	0.806443	1.78
018	360.901	0.806302	1.72
019	418.699	0.806126	1.67
0110	482.396	0.805920	1.64
020	137.097	0.806865	12.86
021	142.602	0.806858	0.69
022	154.183	0.806861	1.67
023	171.559	0.806815	1.61
024	194.697	0.806733	1.51
025	223.820	0.806622	1.45
026	259.319	0.806489	1.44
027	301.615	0.806341	1.47
028	351.077	0.806183	1.53
029	407.975	0.806020	1.63
0210	472.479	0.805855	1.75
030	137.097	0.806865	12.86
031	138.080	0.806831	3.15
032	145.975	0.806763	−2.83
033	161.136	0.806696	−0.15
034	183.818	0.806620	0.77
035	214.174	0.806536	1.25
036	252.264	0.806443	1.54
037	298.057	0.806340	1.72
038	351.430	0.806224	1.82
039	412.166	0.806094	1.87
0310	479.949	0.805946	1.88

We shall return to the problem in Section II.B.5, where optical pumping data are discussed which involve the perturbing levels in a more direct way.

Appendix C contains tables of R, Q, and P transitions to the C–O

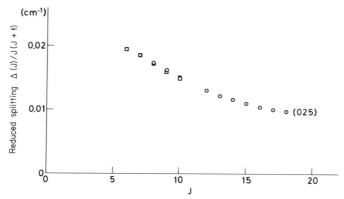

FIG. 35 Measured splitting of (025) levels. Circles are from diode laser spectrum (Sattler *et al.*, 1979), squares from FIR laser emission (Henningsen, 1980a).

stretch for $0 \le J \le 35$ and $0 \le K \le 10$, as evaluated on the basis of Tables V and VI. For $J \le 25$ they are believed to be accurate to within ± 0.005 cm^{-1}, while at higher J, errors are expected from the high-order terms in the energy expansion. For transitions to asymmetry-split A states the splittings are evaluated by using formula (24) and Table VIII. For transitions to the perturbation-split (025) states, the splittings are evaluated for $J \le 20$ by using Table VII.

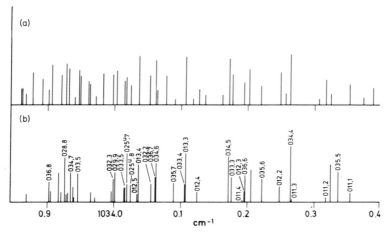

FIG. 36 C–O stretch Q band of CH$_3$OH from (a) diode laser spectroscopy (Sattler *et al.*, unpublished) and (b) calculated.

TABLE VI

ENERGY EXPANSION COEFFICIENTS FOR THE C–O STRETCH STATE[a]

$n_\tau K$	W^{CO}(cm^{-1})	ΔW (cm^{-1})	B^{CO}(cm^{-1})	ΔB (cm^{-1} × 10^{-5})	D^{CO}(cm^{-1} × 10^{-6})	H^{CO}(cm^{-1} × 10^{-9})
010	132.281	−0.495	0.797808	11.4	5.22	−0.14
011	136.135	0.004	0.797813	−6.1	7.09	−0.50
012*	147.651	0.010	0.797827	−6.3	−2.88	1.55
013$_{avg}$	166.696	0.022	0.797844	−10.2	1.07	0.30
014*	193.043	0.051	0.797863	−26.2	1.62	−0.46
015	226.375	0.062	0.797873	−51.7	1.77	−0.11
016	266.311	0.000	0.797857	174.0	1.79	−1.57
017	312.467	0.024	0.797794	50.1	1.76	0.22
018	364.559	−0.001	0.797666	32.8	1.72	0.23
019	422.500	−0.032	0.797470	29.1	1.67	0.35
0110	486.436	−0.042	0.797216	19.6	1.64	0.25
020	140.719	0.000	0.79126	6.6	12.75	−0.55
021	146.053	0.001	0.798187	−8.5	0.14	0.78
022$_{av}$	157.536	−0.002	0.798219	−4.6	1.54	0.30
023	174.922	−0.008	0.798177	−24.1	1.55	0.37

024*	198.182	-0.028	0.798067	-46.2	1.48	-0.31
025avg	227.513	0.058	0.797902	275.2	1.44	-0.20
026	263.264	-0.072	0.797703	184.0	1.44	-1.19
027	305.819	-0.034	0.797491	96.6	1.47	-0.07
028	355.521	-0.009	0.797279	66.7	1.53	0.12
029	412.625	0.030	0.797078	46.2	1.62	0.18
0210	477.295	0.129	0.796891	32.5	1.74	—
030	140.719	0.000	0.798126	6.6	12.75	-0.55
031avg	141.906	-0.002	0.798029	-6.3	3.24	0.12
032*	150.001	-0.008	0.797898	1.5	-2.75	0.00
033	165.337	0.000	0.797781	-13.2	-0.12	0.15
034	188.156	0.026	0.797671	-24.3	0.78	0.08
035	218.607	0.102	0.797572	-49.4	1.25	1.06
036	256.747	0.036	0.797483	-77.4	1.53	-0.30
037	302.544	-0.016	0.797403	87.5	1.70	-0.01
038	355.880	0.015	0.797329	32.8	1.80	0.40
039	416.544	0.018	0.797255	20.8	1.85	0.11
0310	484.237	0.045	0.797172	17.5	1.87	—

[a] Asterisk denotes cases in which the accuracy can be improved by including 4th-order terms in the energy expansion (Henningsen, 1981).

a. *The Q Band.* The Q band constitutes an extremely crowded region of the infrared spectrum with hundreds of lines squeezed into a spectral range of only a few cm^{-1}. This makes the probability of coincidence with pump lines anomalously high, and many of the strong FIR emission lines of CH_3OH are in fact pumped by Q transitions. Figure 36a shows a portion of the spectrum as given in unpublished work by Sattler *et al.*, and Fig. 36b shows the spectrum as predicted by the Q-band table of Appendix C. The intensity of a line is proportional to (Townes and Schawlow, 1955)

$$f \frac{2J + 1}{J(J + 1)} K^2 \exp\left(- \frac{W(n\tau K) + B(n\tau K)J(J + 1)}{kT}\right), \qquad (32)$$

where the nuclear spin factor f is 2 for E states and 4 for unsplit A states. Note in particular the presence in the spectrum of transitions to the upper of the split 025 states. Transitions to the lower components are present with about the same intensity at lower frequencies.

The Q band constitutes an important check on the identification of the R-band transitions. R-line intensities are proportional to

$$f \frac{(J + 1)^2 - K^2}{J + 1} \exp\left(- \frac{W(n\tau K) + B(n\tau K)J(J + 1)}{kT}\right), \qquad (33)$$

and it can therefore be difficult to unambiguously identify the lowest components of a band, where $J \simeq K$. Q lines, on the other hand, are strongest

TABLE VII

(025, J) Level Splitting

J	$\Delta(025,J)$ (cm^{-1})
5	?
6	0.832
7	1.036 618
8	1.240 092
9	1.439 325
10	1.637 417
11	1.848
12	2.039
13	2.226
14	2.455
15	2.660
16	2.847
17	3.108
18	3.340
19	3.577
20	3.823

TABLE VIII

A-STATE SYMMETRY SPLITTING PARAMETERS IN MHz

K	$S^0(K)$ Parameters from Table III	$S^{co}(K)$ Parameters from Table III	$S^{co}(K)$ Data from Sattler et al. (1979)	$S^{co}(K)$ Data from Arimondo et al. (1980)
1	417.3	426.0	432	
2	0.0935	0.1028	0.099	
3	2.435×10^{-5}	2.621×10^{-5}	2.58×10^{-5}	
4	2.214×10^{-9}	2.309×10^{-9}		2.14×10^{-9}

at the bottom of the band, and since $R(J)$ and $Q(J + 1)$ are related through a ground-state transition, it is straightforward to check any assignment for the required correlation.

As a further check on the consistency of the analysis, the calculated Q-band spectrum has led to assignment of several absorption lines which produce FIR emission. The $P_9(31)$ sequence band line (Danielewicz and Weiss, unpublished), pumping at 1033.303 cm^{-1} is close to the $Q(015,8)$ absorption line at 1033.301 cm^{-1}, and this assignment is verified by the observed FIR emission frequencies as indicated in Fig. 37, where the calculated frequencies are given in brackets. Similarly, two FIR lines pumped by a $P_9(36)$ waveguide laser line at 1031.475 cm^{-1} (Inguscio et al., 1981) are found to originate from $Q(0110,17)$ at a calculated frequency of 1031.471 cm^{-1}.

4. Torsionally Excited C–O Stretch States

The Q-absorption band for torsional $n = 1$ transitions, when calculated from the model derived for $n = 0$ states, is expected to have its high-

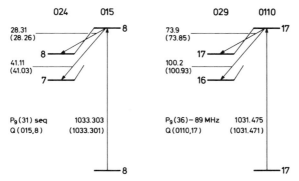

FIG. 37   Verification of Q-band assignment by FIR emission frequencies. Calculated values are given in brackets.

frequency edge at about 1042 cm^{-1} and to extend downward over a somewhat broader range than the $n = 0$ Q band. Close examination of Woods
spectrum (Fig. 24a) reveals some modulation of the $n = 0$ structure
between 1038 and 1045 cm^{-1}, but in view of the fact that about 30% of all
molecules should be thermally excited to $n = 1$, this modulation is surprisingly weak. This seems to indicate that the $n = 1$ states are distributed
over a wider frequency range than predicted by the $n = 0$ model. An analysis similar to that performed for the $n = 0$ states is not yet possible since
too few pump transitions have been found that can unambiguously be associated with $n = 1$. Nevertheless, the optical pumping has provided very
useful information.

The most well-known $n = 1$ pump transition is associated with $P_9(34)$,
which produces an unusually large number of emission lines. Four of
these are associated with torsional $n = 1 \rightarrow 0$ transitions as shown in the
diagram of Fig. 38. This interpretation (Henningsen, 1977) has been subject to some criticism (Forber and Feld, 1979) since the pump frequency
leads to $\nu_{vib}(125) = 1034.374$, which is very far from the 1041.749 cm^{-1}
predicted by the $n = 0$ model. However, with recent frequency measurements (Petersen *et al.*, 1980) and using the $n = 0$ model together with

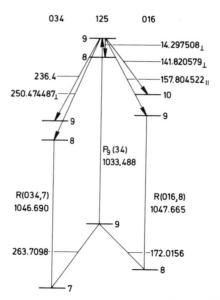

FIG. 38 Combination loops verifying assignments of $P_9(34)$ high-frequency emission
lines.

ground-state calculations, the assignment can be unambiguously proved by combination relations. Three loops can be evaluated. The two which involve the $Q(125,9)$ pump transition add up to -0.0343 cm^{-1} (with $K = 4$ lines) and -0.0351 cm^{-1} (with $K = 6$ lines), while the combined loop, which avoids the pump transition, adds up to 0.0008 cm^{-1}. From this it follows that the residuals of the small loops must be caused by an error of 0.035 cm^{-1} in the calculation of the $(125,9)^0$ energy. This corresponds to the experimental uncertainty of the torsional transition frequencies on which the model is based and, therefore, is to be expected. The residual of the big loop is even smaller than the assumed 0.003 cm^{-1} uncertainty of the R transitions, which are the least accurate elements of the loop. Note also that the four $n = 1 \rightarrow 0$ lines indicated in Fig. 38 are precisely those expected on the basis of line strength considerations (Fig. 29b), that $(n\tau K) = (125)$ gives optimum strength of $n = 1 \rightarrow 0$ transitions (Fig. 27), and that no competing K-changing transitions are possible within the $n = 1$ manifold since (125) states are at a local energy minimum as a function of K.

Although numerous pump lines produce far-infrared emission in a frequency range that is indicative of transitions from $n = 1$, only one other

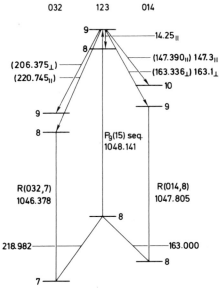

FIG. 39 Combination loop verifying assignments of $P_9(15)$ emission lines.

case has until now been suggested. With the $P_9(15)$ sequence band pump line, three emission lines are observed, which can be interpreted according to the scheme of Fig. 39 (Danielewicz and Weiss, unpublished). Again, the measured $\nu_{vib}(123)$ of 1034.548 cm^{-1} is far from the predicted 1041.933 cm^{-1} of the $n = 0$ model, and again combination relations can be invoked to prove the assignment. In this case the loop adds up to 0.24 cm^{-1}, which is consistent with the 0.5 cm^{-1} uncertainty of the measured emission frequencies. The $\Delta K = -1$ lines were not observed, and the numbers given in brackets in Fig. 39 are predicted from combination relations in connection with Appendix C. Although the redundancy is presumably large enough to prove the assignment correct, a comparison with the accuracy obtained in the $P_9(34)$ case in a convincing way shows the advantage of frequency measurements.

No definite conclusions can be drawn on the basis of the two identifications. The $n = 1$ states of the C–O stretch are close in energy to the OH bending A' mode (see Table II), and a significant interaction is expected. A quick calculation shows that the pump frequencies for the two cases correspond to a C–O stretch energy barrier of about 380 cm^{-1}, which is quite close to the ground-state value of 373.21 cm^{-1}. This would place the $n = 1$ Q band essentially on top of the $n = 0$ band. On the other hand, the analysis of the $n = 0$ Q band seems to account for almost all lines in this range, and the $n = 1$ states thus still represent a problem.

5. Optical Pumping of Unidentified States

a. *Pumping by Parallel Transition.*   When eliminating from the data base everything that can with some confidence be associated with various torsional levels of the C–O stretch, one is still left with a surprisingly large number of lines. A particularly intriguing set is produced by the $P_9(38)$ pump line. Chang *et al.* (1970) observed two pairs of emission lines that are unusual in several respects. No three of them can be grouped together in a triad, and they peak at about 60 mtorr and disappear slightly above 100 mtorr—pressures considerably below what is usually found in CH$_3$OH. The total absence of high-frequency lines speaks against torsionally excited levels, and hot band pumping would seem to require pressures above normal since in that case very few molecules would be thermally excited to the initial pump level. Although the pressure dependence thus suggests that $P_9(38)$ is pumping a quite low-lying level, it is nevertheless impossible to interpret the four emission lines within the framework of C–O stretch states with $n = 0$. This situation prompted a Stark effect investigation, and in the following we shall present in detail the chain of arguments that lead to an assignment, since it illustrates how Stark data can be used (Henningsen, 1980a).

The most distinctive Stark effect was observed for the 51.78 cm^{-1} line, which developed a splitting into two well-resolved components, moving apart linearly in the electric field at a somewhat different rate for the two modes of pump polarization relative to the static field (Fig. 40). By surveying the possible gain profiles, this is found to imply $(J,K) \rightarrow (J - 1, K - 1)$ in the emission, and the measured Stark coefficients for the two polarizations establish the relation $K/J(J + 1) \simeq 0.0544$ by elimination of the dipole moment parameter α (Section I.C.3). The four lines appear at the same pump detuning, and a well-defined transferred Lamb dip can be observed close to the center of the CO_2 line. Therefore they presumably are pumped by the same transition, and from the polarization of the 51.78-cm^{-1} line and Eq. (9), the pump transition is found to be Q type. If we assume an analogy between the two pairs of lines, we must classify them according to the scheme shown in Fig. 41a. J can be determined by assuming a rotational constant of the order of 0.8 cm^{-1} and using combination relations. Combining $51.78–34.23$ and $50.34–35.87$ leads to $J = 11$ and 9, while combining $51.78–35.87$ and $50.34–34.23$ leads to $J = 10$ in both cases. Pumping two different J would imply two perfectly coincident pump transitions, which is extremely unlikely. We therefore conclude that $J = 10$, and the relation between J and K then yields $K = 6$. Thus the final scheme must be as shown in Fig. 41b. The splitting of the $K = 5$ levels is far too large to be caused by asymmetry splitting, and neither the Q-band frequencies nor the pump frequency agree with the C–O stretch.

A fortunate coincidence makes it possible to obtain considerably deeper understanding of this anomaly. Inspection of the lines pumped by $P_9(34)$ reveals a quartet of lines with precisely the same structure and the same splitting as observed for the $P_9(38)$ lines. These lines were early

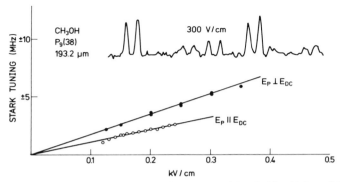

FIG. 40 Stark effect on FIR emission lines pumped by $P_9(38)$. (Adapted from Henningsen, 1980a.)

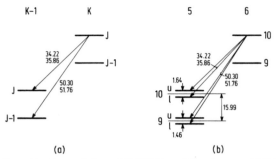

FIG. 41   Interrelation of $P_9(38)$ lines (Henningsen, 1980a).

interpreted as cascade transitions pumped by the strong 141.82-cm^{-1} and 157.80-cm^{-1} transitions (Henningsen, 1977). They were thus, by a completely different reasoning, known to originate from $(J,K) = (10,6)$ and (9,6), respectively, although the original interpretation involved double cascades rather than level splittings. By combining the $P_9(38)$ and $P_9(34)$ data it can be concluded that $P_9(38)$ is pumping a level which we label $(016,10)^x$, the x indicating an as yet unidentified mode. If the split (025) levels are interpreted as hybridized $(025)^x$ and $(025)^{CO}$ states, the strength of the interaction between the two modes can be estimated from the magnitude of the splitting. An interaction of similar strength for (016) can quantitatively account for the shift ΔB of the rotational constant (see Table VI). Some characteristics of the x states are listed in Table IX, where the analogous C–O stretch and ground-state quantities are included for comparison.

The absence of the $(016,10 \rightarrow 9)^x$ emission line from Changs data is understandable since it would be the weakest of the lines in the triad, but it was later identified by heterodyne measurements by Petersen *et al.* (1980). This same work also reports frequency measurements of numerous other lines, and the results are entered in the combined level

TABLE IX

CHARACTERISTICS OF X-STATES

	Ground state	C–O stretch state	x-State
$\nu_{vib}(016)$		1033.777	1030.042
$B(016)$ unperturbed	0.806 550	0.707 857	0.802 832
$B(016)$ perturbed		0.799 597	0.801 092
$D(016)$	1.812×10^{-6}	1.791×10^{-6}	

scheme shown in Fig. 42. The (025) level splittings can now be checked with high precision, and the results, as given in the last column of Table X, are consistent with the experimental uncertainty of the frequency measurements.

The support for the scheme shown in Fig. 42 has so far been gathered from emission lines only. However, the $Q(016,10)^x$ pump frequency can be derived from a simple combination relation as shown in Fig. 43, where $Q(016,10)^{CO}$ is taken from Appendix C. The loop predicts a pump frequency of 1029.442 cm^{-1} in perfect agreement with the $P_9(38)$ line center frequency of 1029.4421 cm^{-1}.

   b. *Pumping by Perpendicular Transitions.* Until now all the pump transitions have been parallel transitions with $\Delta K = 0$, and hence effected by an induced dipole moment directed along the a axis. However, further inspection of data for CH$_3$OH shows that lines pumped by $P_9(10)$ and

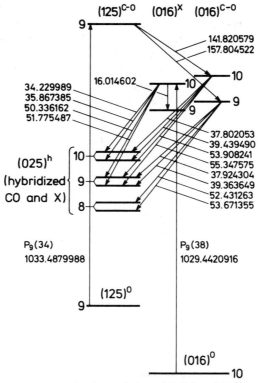

FIG. 42 Level scheme showing interrelation of $P_9(34)$ and $P_9(38)$ emission lines.

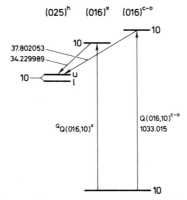

FIG. 43   Determination of $^{Q}Q(016,10)^{x}$ pump transition by combination relation.

$P_9(12)$ have the same pressure dependence and frequency characteristics as the $P_9(38)$ lines, and it proves possible to account for them in exactly the same way by assuming that they are pumped by perpendicular transitions. The complete level scheme, which is shown in Fig. 44, can be verified from several combination loops as shown in Fig. 45. Through the 34.41- and 52.431 263-cm^{-1} emission lines, contact is established to $(016)^{CO}$ states, and from the loop involving $Q(016,9)^{CO}$ the predicted frequency for the $^{R}P(025,9)^{x}$ transition is 1053.92 $\pm$ 0.01 cm^{-1}, in good agreement with

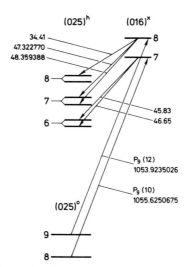

FIG. 44   Level scheme showing x-state pumping by perpendicular transitions.

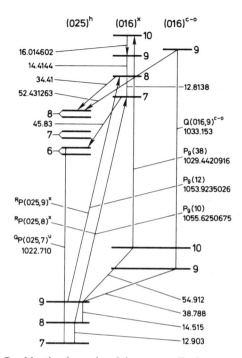

FIG. 45   Combination loops involving perpendicular pump transitions.

the $P_9(12)$ line center frequency of 1053.9235 cm^{-1}. In this case, the uncertainty is dominated by the 34.41-cm^{-1} emission line, which has only been measured interferometrically. A better loop can be constructed by evaluating the $(016,9 \rightarrow 8)^x$ transition from Table IX and closing via $P_9(38)$. This predicts an $^RP(025,9)^x$ frequency of 1053.925 cm^{-1}. The former loop can then be used for predicting a more precise frequency of 34.4124 cm^{-1} for the $(016,8)^x \rightarrow (025,8^u)^h$ transition.

The $P_9(10)$ assignment can be verified by evaluating the average of the two emission frequencies, and by using the $P(025,7)$ transition from Appendix C together with a calculated value for the $(025,8 \rightarrow 7)^0$ transition. This predicts a frequency of 1055.63 cm^{-1} for the $^RP(025,8)^x$ transition, in agreement with the $P_9(10)$ frequency of 1055.625 cm^{-1} to within the 0.01 cm^{-1} accuracy of the emission line frequencies. As a second possibility, the $(016,8 \rightarrow 7)^x$ frequency can be evaluated from Table IX, and a loop can be closed with $P_9(12)$ and the $(025,9 \rightarrow 8)^0$ ground-state transition. This predicts 1055.6247 cm^{-1}, which is 11 MHz below the $P_9(10)$ line center. The measured pump detuning is -20 MHz.

c. *Connection to Microwave Spectroscopy.* From Table VII it follows that the $(025)^h$ level splittings are located in the microwave range, and a further confirmation of the schemes shown in Figs. 42 and 44 is possible by direct observation of absorption lines owing to $(025, J^{l \to u})$ transitions. Such lines are expected to be weak due to the low population of the initial level, and a rather precise a priori knowledge of the frequencies is desirable. For $J = 7$, 8, 9, and 10, the absorption frequencies could be predicted to within 1 MHz on the basis of FIR measurements, and a search was immediately successful (Lees *et al.*, 1981). The results are given in Table X.

d. *Characteristics of the x States.* The most straightforward interpretation of the x states would associate them with highly excited torsional states of the vibrational ground state. However, calculations locate $Q(016 \to 316)^0$ at 698.8 cm^{-1} and $Q(016 \to 416)^0$ at 1436 cm^{-1}, both in manifest disagreement with the observed value of 1030.042 cm^{-1}. Also, the close proximity to the C–O stretch appears to be more than coincidental. However, we shall refrain from speculating about the physical meaning of the x mode and merely summarize what is known from the experiments.

The x levels are nearly degenerate with the C–O stretch levels at $(n\tau K) = (025)$ and are located about 3.7 cm^{-1} below at (016). No pump transitions to the x mode have been identified for K other than 5 or 6, but if the behavior of the C–O stretch effective rotational constant, as displayed in Fig. 34, originates from interaction with x states, then these must be above the C–O stretch states for lower K. It is tempting to associate the x mode with the structure at 1037.5 cm^{-1} observed in matrix isolated CH$_3$OH by both Serallach *et al.* (1974) and Barnes and Hallam (1970), while it requires a rather flexible mind to associate it with the

TABLE X

COMPARISON OF MEASURED MICROWAVE ABSORPTION
FREQUENCIES AND PRECICTED FIR COMBINATION
DIFFERENCES FOR $(025, J^{l \to u})^h$ TRANSITIONS

J	Microwave line (cm^{-1})	FIR combination difference prediction (cm^{-1})	
7	1.036 641	1.036 617	$(016,8)^x \to (025,7)^h$
8	1.240 083	1.240 091	$(016,9)^{cO} \to (025,8)^h$
9	1.439 285	1.439 326	$(016,10)^x \to (025,9)^h$
		1.439 336	$(016,10)^{cO} \to (025,9)^h$
10	1.637 395	1.637 393	$(016,10)^x \to (025,10)^h$
		1.637 440	$(016,10)^{cO} \to (025,10)^h$

1074.5 cm^{-1} structure assigned by both authors as the CH$_3$ in-plane rock (see Table II).

The effective rotational constant $B^x(n\tau K)$ appears to be closer to the ground-state value than to the C–O stretch value. Although no information is yet available about the asymmetry parameter $B-C$ and the interaction constants F_v and G_v, it appears likely that both I_b and I_c are smaller than their C–O stretch values, but larger than the ground-state values.

The transition dipole moment for the x mode is much smaller than for the C–O stretch. This follows from the fact that despite the precise coincidence between $P_9(38)$ and $Q(016,10)^x$, about 10 times as much pump power is needed for achieving FIR laser action than, for example, for $P_9(36)$, which pumps the rather similar transition $Q(018,16)$. Also, the pump transitions to x states appear as very weak lines in the diode laser spectrum, and they would probably have gone unnoticed had they not been known to produce FIR lasing. While no perpendicular transitions have been observed to C–O stretch levels, the x mode apparently can be pumped with about the same efficiency by both parallel and perpendicular transitions. This indicates that the transition dipole moment has sizable components both orthogonal to and parallel to the z axis.

Recently Stark splitting has been resolved for all four $K = 6 \rightarrow 5$ transitions originating from $(016,10)^x$ (Henningsen, 1980b). Although the results are consistent with the (J,K) assignments, they also exhibit a certain asymmetry in the behavior of transitions to the upper and lower component of a given J. Transitions to $J = 10^l$ and 9^u are significantly broader than the corresponding transitions to 10^u and 9^l, and the resulting picture may well turn out to be even more complicated than that presented above.*

C. CONCLUSION

The potential of far-infrared laser emission by optical pumping, viewed as a spectroscopic technique, is associated with its ability to provide high-quality spectroscopic information relating to the levels that are pumped. Its main limitation is associated with the fact that these levels are chosen at random by accidental coincidences between available pump lines and molecular absorption lines. As long as this limitation persists, the amount of data that can be obtained for a given molecule will be limited, and the analysis will require a fair amount of educated guesswork,

* Recent measurements of the permanent dipole moment for the states involved in $P_9(38)$ laser emission suggest that the $(016)^x$ levels should be reassigned as $(025)^x$ and that the hybridizing levels are $(025)^{CO}$ and $(034)^x$ (see Fig. 42). This implies a larger x-state vibrational energy and hence makes the CH$_3$ in-plane rock a more likely candidate (Henningsen, *J. Mol. Spectrosc.*, in press).

the problem being to construct the entire picture of a jigsaw puzzle from which most of the pieces are missing. The potential of the technique can be more fully exploited if it is combined with diode laser spectroscopy, which provides less sophisticated information, but more general coverage. What is really needed is the development of a continuously tunable cw pump laser, which will transform the technique into a routine. Continuously tunable TEA lasers are becoming available, but are not ideal, since the intense pump field disturbs the molecules to a significant extent.

More than 60 different molecules have been pumped by $^{12}C\ ^{16}O_2$ lasers, some of them yielding in excess of 50 emission lines. Considering that a large amount of additional data can be obtained with isotopic lasers and sequence lasers, and by exploiting the increased tunability of waveguide lasers, it is clear that even with state of the art pump lasers, a wealth of information about vibrationally excited molecules is potentially available, ready to be utilized.

Appendix A.   The Data Base

This appendix contains a listing of 156 far-infrared emission lines in CH_3OH reported in the literature before November 1, 1980. Only lines pumped by cw or quasi-cw CO_2 lasers have been included in Table A1. For TEA laser pumped lines, consult Fetterman et al. (w), Izatt et al. (x), Bernard et al. (y), and Gibson (z). Lines reported in Inguscio (g) are waveguide laser pumped. The pressures are given in Pascal and can be converted to mtorr by multiplying by 7.5006. Wave numbers can be converted to frequency, using $c = 2.997\ 924\ 58 \times 10^8$ m/sec for the velocity of light. Pump detunings refer to the operating point of the pump laser, which does not necessarily coincide with the absorption line center. The reader should consult the following references:

(a)	Wagner et al. 1973)	(o)	Weiss (unpublished)
(b)	Dyubko et al. (1974)	(p)	Inguscio et al. (1979c)
(c)	Petersen et al. (1980)	(q)	Domnin et al. (1974)
(d)	Henningsen (1978)	(r)	Radford (1975)
(e)	Petersen et al. (1976)	(s)	Weiss et al. (1977)
(f)	Kon et al. (1975)	(t)	Danielewicz and Weiss (unpublished)
(g)	Inguscio et al. (1981)		
(h)	Landsberg (1980)	(u)	Willenberg (unpublished)
(i)	Tanaka et al. (1975)	(v)	Evenson and Jennings (unpublished)
(j)	Chang et al. (1970)		
(k)	Tanaka et al. (1974)	(w)	Fetterman et al. (1972)
(l)	Hodges et al. (1973)	(x)	Izatt et al. (1975)
(m)	Inguscio et al. (1980)	(y)	Bernard et al. (1981)
(n)	Henningsen (unpublished)	(z)	Gibson (1979)

TABLE A1

CH$_3$OH Measured Lines

Pump line ^{12}C ^{16}O$_2$	Wavelength (μm)	Wave number (cm^{-1})	Relative power	Pressure (Pa)	Pump detuning (MHz)	Polarization	References	Assigned Appendix B No.
$R_9(26)$	151.25369	66.114090	9	23		∥	abc	
	159.67568	62.626943	7	23		∥	bc	
$R_9(22)$	232.78845	42.957458	1	16	+ 15	∥	bdc	
$R_9(18)$	61.613305	162.302607	3	41		⊥	c	
	67.495358	148.158338	0.5	33		⊥	c	
	186.042195	53.751247	10	41	+ 5	∥	ab	
	251.432385	39.772124	4	33	+ 5	∥	bd	
	280.934127	35.595533	2	23	+ 5	∥	bd	
$R_9(16)$	57.1	174.9				⊥	n	
	191.63	52.18				∥	ab	
$R_9(14)$	100.806471	99.199981	9	28		∥	c	3
	194.063191	51.529607	3.6	27		⊥	cd	2
	209.930228	47.634874	3.8	27		∥	bcd	1
$R_9(10)$	96.522394	103.602901	20	80	− 10	∥	abe	6
	164.783246	60.685781	27	73	− 10	⊥	bcf	5
	232.939059	42.929683	20	73	− 10	∥	bc	4
$R_9(8)$	77.405646	129.189543	5	60		∥	c	
	86.239384	115.956302	1	60		∥	bc	
	113.731877	87.926096	1.5	40		⊥	c	
	225.515891	44.342773	0.5	40		∥	bc	
$R_9(2)$	152	65.8				⊥	h	
	176	56.8				⊥	h	
	261	38.3				⊥	h	

(Continued)

TABLE A1 (Continued)

Pump line $^{12}C\ ^{16}O_2$	Wavelength (μm)	Wave number (cm^{-1})	Relative power	Pressure (Pa)	Pump detuning (MHz)	Polarization	References	Assigned Appendix B No.
$P_9(6)$	167.6	59.7				$\parallel$	n	
$P_9(10)$	45.6	219.3	0.2	27			c	
	214.35	46.65	0.1	4	$-\ 20$	$\parallel$	d	7
	218.22	45.83	0.1	4	$-\ 20$	$\parallel$	d	8
	289.70	34.52			$-\ 20$	$\perp$	n	9
$P_9(12)$	206.90	48.33	0.3	4	$-\ 5$	$\parallel$	cd	10
	211.314764	47.322770	1	4	$-\ 5$	$\parallel$	cd	11
	290.62	34.41		5	$-\ 5$	$\perp$	d	12
	164.2	60.90		7	$+\ 85$	$\parallel$	g	15
	261.5	38.24		7	$+\ 85$		g	14
	450.4	22.20		8	$+\ 85$	$\parallel$	g	13
$P_9(14)$	37	270	0.1	5			c	
	117.959536	84.774833	10	23	$-\ 30$	$\parallel$	cd	
	164.507571	60.787476	12	13	$-\ 30$	$\perp$	ci	
	301.994316	33.113206	4	13	$-\ 30$	$\perp$	cd	
	386.339260	25.883986	2	13	$-\ 30$	$\parallel$	cd	
	416.522428	24.008311	5	15	$-\ 30$	$\parallel$	ci	
$P_9(16)$	33	227	2.5	11			c	
	164.600377	60.753202	8	11	$-\ 40$	$\perp$	cj	
	223.5	44.74			$-\ 40$	$\perp$	j	
	369.113688	27.091924	15	20	$-\ 40$	$\parallel$	cj	
$P_9(16)'$	627.34	15.940	50	13	$+\ 30$	$\parallel$	a	16
	570.568643	17.526375			$+\ 30$	$\parallel$	ca	17
	1223.659805	8.172206	0.3	7	$+\ 30$	$\parallel$	ck	18

$P_9(22)$	47.82	209.13				⊥	n	
	57.31	174.5				=	n	
	213.462462	46.846644	0.9	15		=	bc	20
	346.487512	28.861069	1.8	15		=	bc	19
$P_9(24)$	92.543915	108.056807	5	17	0	=	cd	
	133.119557	75.120442	12.5	20	0	=	cd	
	164.697466	60.717388	7.5	13	0	=	bc	
	311.2	32.13				⊥	b	
	602.486981	16.597869	2	13	0	⊥	bc	
	614.285177	16.279084	1.5	13	0	=	cd	
	694.189233	14.405294	2	13	0	=	cd	
$P_9(32)$	37.854207	264.171428	15	33	− 15	=	cl	
	42.159084	237.196807	10	53	− 15	⊥	cl	
	274.3	36.46		33	− 15	=	mn	
$P_9(34)$	39.924226	250.474487	10	17	+ 25		cl	25
	42.30	236.4			+ 25		lo	24
	63.369540	157.804522	33	10	+ 25		cl	23
	70.511629	141.820579	45	13	+ 25	⊥	cj	22
	80.3	124.5			+ 25		l	26
	180.676389	55.347575		8	+ 25		c	32
	185.500395	53.908241	10	9	+ 25	⊥	cj	33
	186.319126	53.671355		7	+ 25	⊥	c	29
	190.725903	52.431263	7	9	+ 25	⊥	cj	28
	237.6	42.08				=	j	27
	253.552974	39.439490	6	9	+ 25	=	cj	34
	254.041488	39.363649			+ 25	=	j	30
	263.683151	37.924304			+ 25	=	j	31
	264.535897	37.802053		8	+ 25	=	cj	35
	292.5	34.19				=	j	36
	699.422584	14.297508	27	15	+ 25	⊥	cj	21

(Continued)

TABLE A1 (Continued)

Pump line $^{12}C\ ^{16}O_2$	Wavelength (μm)	Wave number (cm^{-1})	Relative power	Pressure (Pa)	Pump detuning (MHz)	Polarization	References	Assigned Appendix B No.
$P_9(34)'$	205.6	48.64			+120	⊥	pg	37
	208.3	48.01			>130		vpg	38
$P_9(36)$	118.834092	84.150935	160	17	+ 25	⊥	cj	41
	170.576376	58.624765	58	8	+ 25	=	cj	40
	202.4	49.41			+ 25	=	cj	44
	392.068716	25.505733	35	19	+ 25	⊥	cj	39
	417.8	23.935			+ 25	⊥	j	43
	418.082677	23.918714	0.8	9	+ 25	⊥	c	42
$P_9(36)'$	110.60	90.42		7	− 80	⊥	g	46
	162.12	61.68		11	− 80	=	g	45
$P_9(36)''$	99.28	100.73		7	− 90		g	48
	135.71	73.69		11	− 90	=	g	47
$P_9(36)'''$	118.84	84.15		7	+115	⊥	g	
	170.59	58.62		7	+115	=	g	
$P_9(38)$	193.141592	51.775487	2.2	11	0	⊥	cj	50
	198.664332	50.336162	2.4	11	0	⊥	cj	51
	278.804825	35.867385	2.4	11	0	=	cj	52
	292.141490	34.229989	2	11	0	=	cj	53
	624.430129	16.014602	0.6	7	0	=	c	49
$P_9(40)$	55.370042	180.603079	1	15	− 10	⊥	cd	
	60.173273	166.186738	0.7	9	− 10	=	cd	
	73.306425	136.413691	0.1	20	− 10	=	cd	
	85.600931	116.821159	0.1	20	− 10	=	cd	

98

$R_{10}(48)$	164	61.0					m	
	286	35.0					im	
$R_{10}(46)$	274	36.5					i	
$R_{10}(44)$	121	82.6					i	
	251	39.8					i	
$R_{10}(40)$	97.518535	102.544609	14	37	0	⊥	ci	
	167.586993	59.670502	7	27		=	c	
$R_{10}(38)$	163.033529	61.337076	30	13	+ 25	=	cq	56
	251.113983	39.818455	4	13	+ 25	=	cq	54
	469.023308	21.320902	50	13	+ 25	⊥	cq	55
$R_{10}(36)$	43.1	232.0		10–40	+ 25		m	
	53.88	185.6		10–40	+ 25	⊥	nm	
$R_{10}(34)$	43.47	230.0					d	
	92.664285	107.916443	1	13	− 10	=	cd	
	129.549716	77.190443	5	13	− 10	=	cf	
	242.472689	41.241758	10	17	− 10	=	ci	
	250.781290	39.875383	11	13	− 10	=	cr	
	267.443164	37.391122	7	15	− 10	⊥	cd	
$R_{10}(32)$	99.2	100.8		27	− 45	=	n	
	145.5	68.7		5	− 45	⊥	m	
	242.8	41.19	0.1			=	a	
$R_{10}(16)$	62.965967	158.815951	10	16	0	=	c	
	69.679557	143.514116	6	20	0	⊥	cd	
	77.904888	128.361651	10	25	0	=	ci	
	695.349944	14.381248	1	17		=	ci	
$R_{10}(10)$	191.619608	52.186726	50	23	− 15	=	ca	
	293.821673	34.034249	5	27	− 15	=	cr	
$R_{10}(4)$	179.727913	55.639660	11	20		=	c	
	191.2	52.30				=	a	

(Continued)

TABLE A1 (*Continued*)

Pump line $^{12}C\ ^{16}O_2$	Wavelength (μm)	Wave number (cm⁻¹)	Relative power	Pressure (Pa)	Pump detuning (MHz)	Polarization	References	Assigned Appendix B No.
	211.262893	47.334389	16	35		∥	ci	
	495	20.20	0.5	20		⊥	ci	
$R_{10}(2)$	178	56.2					h	
$P_{10}(16)$	84.91	117.8				∥	n	
	99.85	100.2				∥	n	
	123.64	80.88				⊥	n	
$P_9(13)$ seq	107.7	92.85		27		∥	t	
	147.9	67.61		27		∥	t	
	393.2	25.43				⊥	st	
$P_9(15)$ seq	61.2	163.3		20		⊥	t	59
	67.9	147.3		27		∥	t	58
	702.0	14.25		20		∥	t	57
$P_9(17)$ seq	376.0	26.60		16		∥	t	62
	720.0	13.89		13		∥	t	61
	782.2	12.78		13		⊥	t	60
$P_9(19)$ seq	902	11.09				∥	u	64
	480	20.83				∥	u	63
	85.3	117.2				∥	u	
$P_9(21)$ seq	57.5	173.9		27		∥	t	
	61.5	162.6		20		∥	t	
	80.7	123.9		54		⊥	st	
$P_9(31)$ seq	159.3	62.78		27		⊥	st	
	230.2	43.45		16		∥	t	66
	243.3	41.11		20		∥	t	
	353.3	28.31		20		⊥	t	65
$R_{10}(33)$ seq	126.4	79.1					u	
	250	40.0					u	

Appendix B.   Assigned Lines

This appendix contains assignments for 66 of the lines listed in Appendix A. The pump wave numbers refer to the pump line center, and in some cases the line offsets are given below. Parallel pump transitions to the C–O stretch are identified as $P(n\tau K, J)$, $Q(n\tau K, J)$, and $R(n\tau K, J)$ for $J \to J - 1$, $J \to J$, and $J \to J + 1$, respectively, where n and τ are defined in Section II.A.2.a. Only lines for which all quantum numbers have been identified are included. Pump transitions to the x states, which can be pumped by both parallel and perpendicular transitions, are marked by an additional P, Q, or R superscript, corresponding to $K \to K - 1$, $K \to K$, and $K \to K + 1$, respectively. Emission transitions to split 025 states are provided with superscripts u and l for transitions to the upper and the lower component of a doublet, while $\pm$ superscripts refer to components of asymmetry split A doublets. The reader should consult the following references:

(a) Henningsen (1978)
(b) Danielewicz and Coleman (1977)
(c) Henningsen (1977)
(d) Yano and Kon (unpublished)
(e) Henningsen (1980a)
(f) Henningsen (1981)

(g) Arimondo (1979)
(h) Danielewicz and Weiss (unpublished)
(i) Worchesky (1978)
(j) Inguscio et al. (1981)
(k) Henningsen (unpublished)

Appendix C.   The C–O Stretch Torsional Ground-State Band

This appendix contains calculated R-, Q-, and P-band transitions to torsional ground-state levels of the C–O stretch for the range $0 \le K \le 10$ and $0 \le J \le 35$, based on the parameters of Tables V and VI. For asymmetry split (031), (022), and (034) states, and for the interaction split (025) states, the listed frequencies are average values. The splittings may be computed from Table VIII in conjunction with the expression (24) and from Table VII. For $J \le 25$ the accuracy is generally expected to be better than 0.005 cm^{-1}. An exception is transitions to asymmetry split states for $K = 1$ and 2, since Eq. (24) is an approximation which is not completely adequate at the higher J. For $J > 25$ deviations are expected for all K owing to the higher order terms in the energy expansion. For $(n\tau K) = (012)$, (014), (024), and (032), asymmetry effects are somewhat larger than in general, and in these cases an additional fourth-order term was included in the energy expansion (Henningsen, 1981).

TABLE B1

CH₃OH Assigned Lines

Line number	Pump line	Pump wave number (cm⁻¹)	Symmetry type	Absorbing transition	Emission transition	Emission wave number (cm⁻¹)	Comments	References
1	$R_9(14)$	1074.646490	E_1	$R(039,29)$	$039,30 \to 29$	47.634874	triad	a
2					$039 \to 018,30$	51.529607		a
3					$039,30 \to 018,29$	99.199981		
4	$R_9(10)$	1071.883766	E_1	$R(0210,26)$	$0210,27 \to 26$	42.929683	triad	bcd
5					$0210 \to 039,27$	60.685781		bcd
6					$0210,27 \to 039,26$	103.602901		bcd
7	$P_9(10)$	1055.625066	A	$^RP(025,8)^x$	$016,7 \to 025,6^l$	46.65	x state	e
8					$016,7 \to 025,6^u$	45.83	x state	e
9					$016 \to 025,7^u$	34.52	x state	k
10	$P_9(12)$	1053.923503 +0.0000	A	$^RP(025,9)^x$	$016,8 \to 025,7^l$	48.33	x state	e
11					$016,8 \to 025,7^u$	47.322770	x state	e
12					$016 \to 025,8^u$	34.41	x state	e
13	$P_9(12)$	1053.923503 +0.0030	A	$R(025,13^l)$	$025,14^l \to 13^l$	22.20	triad	f
14					$025,14^l \to 034,14$	38.24		k
15					$025,14^l \to 034,13$	60.90		f
16	$P_9(16)$	1050.441282 +0.0021	A	$R(010,10)$	$010,10 \to 9$	15.940	cascade	c
17					$010,11 \to 10$	17.526375		c
18					$010,11 \to 031^+,10$	8.172206		c
19	$P_9(22)$	1045.021669	E_1	$R(033,17)^{CO \to 2CO}$	$033,18 \to 17$	28.861069	?	g
20					$033,18 \to 012,17$	46.846644	?	g

21	$P_9(34)$	1033.487999 +0.0008	A	$Q(125,9)$	$125,9 \rightarrow 8$	14.297508	$n = 1 \rightarrow 1$	c
22					$125,9 \rightarrow 016,10$	141.820579	$n = 1 \rightarrow 0$	c
23					$125 \rightarrow 016,9$	157.804522	$n = 1 \rightarrow 0$	c
24					$125 \rightarrow 034,9$	236.4	$n = 1 \rightarrow 0$	c
25					$125,9 \rightarrow 034,8$	250.474487	$n = 1 \rightarrow 0$	c
26					$234,8 \rightarrow 125,9$	124.5	refill, $v = 0$	c
27					$134,9 \rightarrow 125,9$	42.09	refill, $v = 0$	c
28					$016,9 \rightarrow 025,8^l$	52.431263	cascade	e
29					$016,9 \rightarrow 025,8^l$	53.671355	cascade	k
30					$016,9 \rightarrow 025,9^l$	39.363649	cascade	e
31					$016,9 \rightarrow 025,9^u$	37.924304	cascade	d
32					$016,10 \rightarrow 025,9^l$	55.347575	cascade	d
33					$016,10 \rightarrow 025,9^u$	53.908241	cascade	d
34					$016,10 \rightarrow 025,10^l$	39.439490	cascade	d
35					$016,10 \rightarrow 025,10^u$	37.802053	cascade	d
36					$034,8 \rightarrow 013,7$	34.19	cascade	c
37	$P_9(34)'$	1033.487999 +0.0040	A	$Q(016,6)$	$016,6 \rightarrow 025,5^l$	48.64		f
38					$016,6 \rightarrow 025,5^u$	48.01		f
39	$P_9(36)$	1031.477430 +0.0008	E_1	$Q(018,16)$	$018,16 \rightarrow 15$	25.505733	triad	bcd
40					$018 \rightarrow 027,16$	58.624765		bcd
41					$018,16 \rightarrow 027,15$	84.150935		bcd
42					$018,15 \rightarrow 14$	23.918714	cascade	bcd
43					$027,15 \rightarrow 14$	23.935	cascade	bcd
44					$027,15 \rightarrow 036,15$	49.41	cascade	cd
45	$P_9(36)'$	1031.477430 −0.0027	A	$Q(0310,18)$	$0310 \rightarrow 019,18$	61.68		jk
46					$0310,18 \rightarrow 019,17$	90.42		jk

(Continued)

103

TABLE B1 (*continued*)

Line number	Pump line	Pump wave number (cm^{-1})	Symmetry type	Absorbing transition	Emission transition	Emission wave number (cm^{-1})	Comments	References
47	$P_9(36)''$	1031.477430	E_2	$Q(0110,17)$	$0110 \to 029,17$	73.69		jk
48		−0.0030			$0110,17 \to 029.16$	100.73		jk
49	$P_9(38)$	1029.442092	A	$^qQ(016,10)^x$	$016,10 \to 9$	16.014602	x state	e
50		+0.0000			$016,10 \to 025,9^i$	51.775487	x state	e
51					$016,10 \to 025,9^u$	50.336162	x state	e
52					$016,10 \to 025,10^l$	35.867385	x state	e
53					$016,10 \to 025,10^u$	34.229989	x state	e
54	$R_{10}(38)$	986.567352	A	$P(034^+,26)$	$034^+,25 \to 24$	39.818455	⎱ triad	bcd
55		+0.0009			$034^+,25 \to 013^-,25$	21.320902	⎰	bcd
56					$034^+,25 \to 013^+,24$	61.337076		bcd
57	$P_9(15)$	1048.141	E_2	$R(123,8)$	$123,9 \to 8$	14.25	$n = 1 \to 1$	h
58					$123,9 \to 014,10$	147.3	$n = 1 \to 0$	h
59					$123,9 \to 014,9$	163.3	$n = 1 \to 0$	h
60	$P_9(17)$	1046.380	E_2	$R(032,7)$	$032,8 \to 7$	12.78	⎱ triad	h
61					$032 \to 011,8$	13.89	⎰	h
62					$032,8 \to 011,7$	26.60		h
63	$P_9(19)$	1044.591	A	$R(031^+,6)$	$031^+,7 \to 6$	11.09	predicted	i
64					$031^+,7 \to 010,6$	20.83	predicted	i
65	$P_9(31)$	1033.303	E_1	$Q(015,8)$	$015 \to 024,8$	28.31		f
66					$015,8 \to 024,7$	41.11		f

TABLE C1

CALCULATED R-, Q-, AND P-BAND TRANSITIONS

K	J	R(01K,J)	R(02K,J)	R(03K,J)
0	0	1035.491	1035.302	1035.302
0	1	1037.069	1036.881	1036.881
0	2	1038.629	1038.442	1038.442
0	3	1040.171	1039.984	1039.984
0	4	1041.694	1041.508	1041.508
0	5	1043.200	1043.013	1043.013
0	6	1044.687	1044.499	1044.499
0	7	1046.155	1045.966	1045.966
0	8	1047.604	1047.413	1047.413
0	9	1049.034	1048.840	1048.840
0	10	1050.445	1050.247	1050.247
0	11	1051.836	1051.634	1051.634
0	12	1053.208	1053.000	1053.000
0	13	1054.560	1054.346	1054.346
0	14	1055.892	1055.671	1055.671
0	15	1057.204	1056.976	1056.976
0	16	1058.496	1058.260	1058.260
0	17	1059.767	1059.523	1059.523
0	18	1061.018	1060.766	1060.766
0	19	1062.248	1061.988	1061.988
0	20	1063.457	1063.190	1063.190
0	21	1064.646	1064.373	1064.373
0	22	1065.813	1065.537	1065.537
0	23	1066.959	1066.682	1066.682
0	24	1068.084	1067.809	1067.809
0	25	1069.187	1068.919	1068.919
0	26	1070.270	1070.013	1070.013
0	27	1071.331	1071.092	1071.092
0	28	1072.371	1072.157	1072.157
0	29	1073.390	1073.209	1073.209
0	30	1074.388	1074.250	1074.250
0	31	1075.365	1075.282	1075.282
0	32	1076.321	1076.307	1076.307
0	33	1077.257	1077.327	1077.327
0	34	1078.173	1078.344	1078.344
0	35	1079.070	1079.362	1079.362
1	1	1037.545	1036.711	1037.082
1	2	1039.104	1038.272	1038.642
1	3	1040.644	1039.816	1040.185
1	4	1042.165	1041.342	1041.709
1	5	1043.668	1042.851	1043.215
1	6	1045.151	1044.342	1044.703
1	7	1046.615	1045.816	1046.172
1	8	1048.059	1047.273	1047.623
1	9	1049.483	1048.713	1049.056
1	10	1050.886	1050.135	1050.470
1	11	1052.270	1051.540	1051.865
1	12	1053.632	1052.928	1053.241
1	13	1054.974	1054.299	1054.598
1	14	1056.294	1055.652	1055.936
1	15	1057.593	1056.987	1057.255
1	16	1058.871	1058.305	1058.554

(*Continued*)

TABLE C1 (*Continued*)

K	J	R(01K,J)	R(02K,J)	R(03K,J)
1	17	1060.127	1059.605	1059.833
1	18	1061.361	1060.887	1061.093
1	19	1062.574	1062.150	1062.332
1	20	1063.764	1063.393	1063.551
1	21	1064.934	1064.617	1064.750
1	22	1066.081	1065.820	1065.928
1	23	1067.208	1067.002	1067.084
1	24	1068.313	1068.161	1068.219
1	25	1069.398	1069.297	1069.332
1	26	1070.463	1070.408	1070.423
1	27	1071.508	1071.492	1071.491
1	28	1072.535	1072.548	1072.536
1	29	1073.543	1073.575	1073.557
1	30	1074.535	1074.568	1074.555
1	31	1075.511	1075.528	1075.527
1	32	1076.473	1076.450	1076.474
1	33	1077.421	1077.332	1077.395
1	34	1078.358	1078.170	1078.290
1	35	1079.286	1078.961 ·	1079.156
2	2	1039.036	1038.172	1038.842
2	3	1040.579	1039.716	1040.384
2	4	1042.105	1041.242	1041.909
2	5	1043.613	1042.751	1043.416
2	6	1045.103	1044.243	1044.906
2	7	1046.576	1045.716	1046.378
2	8	1048.031	1047.173	1047.834
2	9	1049.468	1048.611	1049.272
2	10	1050.887	1050.032	1050.693
2	11	1052.287	1051.435	1052.097
2	12	1053.667	1052.820	1053.485
2	13	1055.028	1054.187	1054.856
2	14	1056.368	1055.537	1056.211
2	15	1057.688	1056.868	1057.550
2	16	1058.986	1058.180	1058.872
2	17	1060.262	1059.475	1060.177
2	18	1061.516	1060.751	1061.466
2	19	1062.747	1062.008	1062.738
2	20	1063.955	1063.245	1063.992
2	21	1065.141	1064.464	1065.228
2	22	1066.305	1065.662	1066.444
2	23	1067.449	1066.840	1067.638
2	24	1068.574	1067.998	1068.809
2	25	1069.684	1069.134	1069.954
2	26	1070.780	1070.249	1071.070
2	27	1071.870	1071.341	1072.153
2	28	1072.957	1072.410	1073.198
2	29	1074.051	1073.454	1074.199
2	30	1075.160	1074.474	1075.150
2	31	1076.295	1075.467	1076.043
2	32	1077.471	1076.434	1076.867
2	33	1078.702	1077.372	1077.613
2	34	1080.009	1078.280	1078.268

(*Continued*)

TABLE C1 *(Continued)*

K	J	R(01K,J)	R(02K,J)	R(03K,J)
2	35	1081.414	1079.156	1078.816
3	3	1040.489	1039.716	1040.558
3	4	1042.012	1041.240	1042.081
3	5	1043.518	1042.747	1043.585
3	6	1045.005	1044.235	1045.072
3	7	1046.474	1045.706	1046.541
3	8	1047.925	1047.158	1047.991
3	9	1049.357	1048.592	1049.424
3	10	1050.772	1050.008	1050.838
3	11	1052.168	1051.406	1052.234
3	12	1053.545	1052.786	1053.612
3	13	1054.904	1054.147	1054.971
3	14	1056.245	1055.489	1056.313
3	15	1057.567	1056.812	1057.635
3	16	1058.870	1058.117	1058.940
3	17	1060.154	1059.402	1060.226
3	18	1061.418	1060.668	1061.493
3	19	1062.663	1061.914	1062.741
3	20	1063.888	1063.139	1063.970
3	21	1065.093	1064.344	1065.180
3	22	1066.278	1065.528	1066.371
3	23	1067.441	1066.690	1067.542
3	24	1068.584	1067.830	1068.694
3	25	1069.704	1068.947	1069.825
3	26	1070.801	1070.041	1070.935
3	27	1071.875	1071.109	1072.025
3	28	1072.925	1072.152	1073.094
3	29	1073.949	1073.169	1074.141
3	30	1074.948	1074.157	1075.165
3	31	1075.920	1075.117	1076.167
3	32	1076.863	1076.046	1077.145
3	33	1077.777	1076.942	1078.100
3	34	1078.660	1077.805	1079.029
3	35	1079.511	1078.632	1079.933
4	4	1041.893	1041.338	1042.238
4	5	1043.398	1042.840	1043.741
4	6	1044.885	1044.325	1045.225
4	7	1046.354	1045.790	1046.690
4	8	1047.805	1047.237	1048.137
4	9	1049.236	1048.666	1049.566
4	10	1050.650	1050.076	1050.976
4	11	1052.044	1051.468	1052.367
4	12	1053.420	1052.842	1053.740
4	13	1054.778	1054.197	1055.093
4	14	1056.116	1055.535	1056.428
4	15	1057.436	1056.854	1057.745
4	16	1058.738	1058.156	1059.042
4	17	1060.022	1059.441	1060.320
4	18	1061.290	1060.708	1061.580
4	19	1062.541	1061.957	1062.820
4	20	1063.777	1063.189	1064.041
4	21	1065.001	1064.403	1065.243

(Continued)

TABLE C1 (*Continued*)

K	J	R(01K,J)	R(02K,J)	R(03K,J)
4	22	1066.214	1065.600	1066.425
4	23	1067.420	1066.779	1067.588
4	24	1068.621	1067.939	1068.731
4	25	1069.823	1069.081	1069.854
4	26	1071.032	1070.202	1070.956
4	27	1072.252	1071.303	1072.039
4	28	1073.494	1072.381	1073.101
4	29	1074.765	1073.435	1074.142
4	30	1076.077	1074.462	1075.162
4	31	1077.442	1075.460	1076.161
4	32	1078.876	1076.426	1077.138
4	33	1080.396	1077.355	1078.093
4	34	1082.021	1078.244	1079.025
4	35	1083.773	1079.086	1079.935
5	5	1043.257	1043.263	1043.899
5	6	1044.739	1044.792	1045.379
5	7	1046.203	1046.308	1046.840
5	8	1047.648	1047.813	1048.281
5	9	1049.074	1049.306	1049.703
5	10	1050.481	1050.786	1051.105
5	11	1051.869	1052.254	1052.488
5	12	1053.239	1053.710	1053.850
5	13	1054.589	1055.154	1055.193
5	14	1055.920	1056.585	1056.514
5	15	1057.233	1058.005	1057.814
5	16	1058.527	1059.412	1059.093
5	17	1059.801	1060.808	1060.349
5	18	1061.057	1062.192	1061.582
5	19	1062.294	1063.564	1062.791
5	20	1063.513	1064.924	1063.974
5	21	1064.713	1066.273	1065.132
5	22	1065.894	1067.612	1066.262
5	23	1067.058	1068.939	1067.362
5	24	1068.202	1070.256	1068.432
5	25	1069.329	1071.562	1069.468
5	26	1070.439	1072.859	1070.469
5	27	1071.530	1074.147	1071.432
5	28	1072.605	1075.426	1072.355
5	29	1073.662	1076.696	1073.234
5	30	1074.703	1077.959	1074.066
5	31	1075.727	1079.215	1074.846
5	32	1076.736	1080.465	1075.571
5	33	1077.729	1081.709	1076.237
5	34	1078.707	1082.949	1076.837
5	35	1079.671	1084.186	1077.368
6	6	1044.677	1044.857	1045.346
6	7	1046.178	1046.358	1046.802
6	8	1047.665	1047.845	1048.239
6	9	1049.138	1049.319	1049.656
6	10	1050.598	1050.778	1051.053
6	11	1052.043	1052.224	1052.430
6	12	1053.476	1053.656	1053.788

(*Continued*)

TABLE C1 (*Continued*)

K	J	R(01K,J)	R(02K,J)	R(03K,J)
6	13	1054.896	1055.075	1055.126
6	14	1056.303	1056.482	1056.444
6	15	1057.699	1057.876	1057.743
6	16	1059.083	1059.258	1059.023
6	17	1060.458	1060.630	1060.284
6	18	1061.823	1061.991	1061.525
6	19	1063.181	1063.342	1062.748
6	20	1064.532	1064.686	1063.952
6	21	1065.878	1066.022	1065.138
6	22	1067.222	1067.353	1066.306
6	23	1068.566	1068.680	1067.456
6	24	1069.911	1070.006	1068.590
6	25	1071.262	1071.332	1069.707
6	26	1072.621	1072.660	1070.808
6	27	1073.992	1073.995	1071.893
6	28	1075.379	1075.338	1072.965
6	29	1076.787	1076.694	1074.023
6	30	1078.221	1078.065	1075.068
6	31	1079.685	1079.458	1076.101
6	32	1081.188	1080.875	1077.124
6	33	1082.735	1082.322	1078.138
6	34	1084.333	1083.805	1079.144
6	35	1085.991	1085.330	1080.143
7	7	1046.047	1046.585	1046.873
7	8	1047.511	1048.054	1048.339
7	9	1048.959	1049.508	1049.788
7	10	1050.391	1050.945	1051.221
7	11	1051.805	1052.366	1052.638
7	12	1053.202	1053.771	1054.038
7	13	1054.582	1055.160	1055.422
7	14	1055.946	1056.533	1056.788
7	15	1057.292	1057.890	1058.138
7	16	1058.620	1059.230	1059.472
7	17	1059.931	1060.554	1060.788
7	18	1061.224	1061.863	1062.088
7	19	1062.499	1063.155	1063.372
7	20	1063.756	1064.430	1064.638
7	21	1064.994	1065.690	1065.887
7	22	1066.214	1066.934	1067.120
7	23	1067.414	1068.162	1068.336
7	24	1068.594	1069.374	1069.535
7	25	1069.755	1070.570	1070.717
7	26	1070.895	1071.751	1071.882
7	27	1072.014	1072.916	1073.030
7	28	1073.111	1074.065	1074.162
7	29	1074.186	1075.200	1075.276
7	30	1075.238	1076.319	1076.374
7	31	1076.265	1077.424	1077.455
7	32	1077.268	1078.514	1078.518
7	33	1078.245	1079.590	1079.565
7	34	1079.195	1080.652	1080.596
7	35	1080.117	1081.700	1081.609

(*Continued*)

TABLE C1 (*Continued*)

K	J	R(01K,J)	R(02K,J)	R(03K,J)
8	8	1047.502	1048.284	1048.285
8	9	1048.946	1049.730	1049.724
8	10	1050.373	1051.159	1051.145
8	11	1051.784	1052.571	1052.548
8	12	1053.177	1053.966	1053.934
8	13	1054.552	1055.344	1055.301
8	14	1055.910	1056.705	1056.651
8	15	1057.251	1058.049	1057.981
8	16	1058.573	1059.375	1059.294
8	17	1059.878	1060.684	1060.587
8	18	1061.165	1061.975	1061.861
8	19	1062.433	1063.249	1063.115
8	20	1063.682	1064.505	1064.350
8	21	1064.913	1065.742	1065.564
8	22	1066.124	1066.962	1066.756
8	23	1067.316	1068.163	1067.928
8	24	1068.488	1069.345	1069.076
8	25	1069.639	1070.509	1070.202
8	26	1070.769	1071.653	1071.304
8	27	1071.877	1072.778	1072.380
8	28	1072.963	1073.882	1073.431
8	29	1074.026	1074.967	1074.454
8	30	1075.065	1076.031	1075.448
8	31	1076.079	1077.073	1076.413
8	32	1077.068	1078.094	1077.346
8	33	1078.029	1079.093	1078.245
8	34	1078.963	1080.069	1079.109
8	35	1079.868	1081.021	1079.936
9	9	1049.048	1049.945	1049.645
9	10	1050.474	1051.368	1051.065
9	11	1051.882	1052.774	1052.467
9	12	1053.273	1054.162	1053.851
9	13	1054.646	1055.533	1055.217
9	14	1056.002	1056.885	1056.566
9	15	1057.340	1058.220	1057.896
9	16	1058.659	1059.537	1059.208
9	17	1059.960	1060.836	1060.502
9	18	1061.242	1062.117	1061.777
9	19	1062.506	1063.379	1063.034
9	20	1063.750	1064.622	1064.272
9	21	1064.974	1065.847	1065.492
9	22	1066.178	1067.052	1066.692
9	23	1067.361	1068.238	1067.874
9	24	1068.523	1069.404	1069.035
9	25	1069.663	1070.550	1070.178
9	26	1070.779	1071.675	1071.300
9	27	1071.872	1072.779	1072.402
9	28	1072.940	1073.862	1073.484
9	29	1073.983	1074.922	1074.545
9	30	1074.998	1075.960	1075.584
9	31	1075.985	1076.974	1076.602
9	32	1076.943	1077.964	1077.598

(*Continued*)

TABLE C1 (*Continued*)

K	J	R(01K,J)	R(02K,J)	R(03K,J)
9	33	1077.869	1078.929	1078.571
9	34	1078.763	1079.868	1079.521
9	35	1079.622	1080.780	1080.447
10	10	1050.680	1051.608	1051.003
10	11	1052.084	1053.010	1052.406
10	12	1053.471	1054.394	1053.791
10	13	1054.841	1055.760	1055.158
10	14	1056.192	1057.109	1056.507
10	15	1057.525	1058.439	1057.839
10	16	1058.840	1059.752	1059.153
10	17	1060.137	1061.046	1060.448
10	18	1061.415	1062.323	1061.726
10	19	1062.674	1063.582	1062.986
10	20	1063.914	1064.822	1064.228
10	21	1065.135	1066.045	1065.452
10	22	1066.335	1067.250	1066.658
10	23	1067.516	1068.436	1067.845
10	24	1068.676	1069.604	1069.015
10	25	1069.814	1070.754	1070.166
10	26	1070.931	1071.886	1071.299
10	27	1072.025	1073.000	1072.414
10	28	1073.097	1074.095	1073.510
10	29	1074.144	1075.172	1074.588
10	30	1075.167	1076.230	1075.648
10	31	1076.164	1077.271	1076.690
10	32	1077.134	1078.293	1077.713
10	33	1078.076	1079.296	1078.717
10	34	1078.989	1080.281	1079.703
10	35	1079.872	1081.247	1080.671

(*Continued*)

TABLE C1 (*Continued*)

K	J	Q(01K,J)	Q(02K,J)	Q(03K,J)
1	1	1034.354	1033.518	1033.890
1	2	1034.318	1033.483	1033.855
1	3	1034.264	1033.431	1033.802
1	4	1034.191	1033.361	1033.731
1	5	1034.101	1033.274	1033.642
1	6	1033.992	1033.169	1033.536
1	7	1033.865	1033.047	1033.411
1	8	1033.720	1032.908	1033.269
1	9	1033.556	1032.752	1033.109
1	10	1033.373	1032.578	1032.932
1	11	1033.172	1032.388	1032.736
1	12	1032.952	1032.180	1032.522
1	13	1032.713	1031.956	1032.291
1	14	1032.455	1031.714	1032.041
1	15	1032.178	1031.455	1031.774
1	16	1031.882	1031.179	1031.488
1	17	1031.567	1030.886	1031.183
1	18	1031.234	1030.574	1030.861
1	19	1030.881	1030.245	1030.520
1	20	1030.509	1029.897	1030.160
1	21	1030.119	1029.530	1029.781
1	22	1029.710	1029.144	1029.383
1	23	1029.284	1028.738	1028.967
1	24	1028.839	1028.311	1028.530
1	25	1028.378	1027.861	1028.074
1	26	1027.899	1027.388	1027.598
1	27	1027.405	1026.891	1027.102
1	28	1026.895	1026.368	1026.585
1	29	1026.371	1025.817	1026.047
1	30	1025.834	1025.237	1025.488
1	31	1025.284	1024.625	1024.906
1	32	1024.723	1023.979	1024.303
1	33	1024.153	1023.296	1023.676
1	34	1023.575	1022.573	1023.026
1	35	1022.991	1021.808	1022.352
2	2	1034.248	1033.383	1034.054
2	3	1034.195	1033.331	1034.001
2	4	1034.124	1033.261	1033.929
2	5	1034.035	1033.174	1033.840
2	6	1033.929	1033.070	1033.732
2	7	1033.804	1032.949	1033.607
2	8	1033.661	1032.810	1033.464
2	9	1033.500	1032.654	1033.304
2	10	1033.319	1032.480	1033.126
2	11	1033.119	1032.290	1032.931
2	12	1032.897	1032.082	1032.718
2	13	1032.654	1031.856	1032.490
2	14	1032.388	1031.613	1032.245
2	15	1032.096	1031.353	1031.984
2	16	1031.777	1031.075	1031.709
2	17	1031.428	1030.780	1031.420
2	18	1031.046	1030.467	1031.119

(*Continued*)

TABLE C1 (*Continued*)

K	J	Q(01K,J)	Q(02K,J)	Q(03K,J)
2	19	1030.626	1030.136	1030.806
2	20	1030.166	1029.786	1030.483
2	21	1029.658	1029.419	1030.152
2	22	1029.098	1029.033	1029.816
2	23	1028.478	1028.628	1029.476
2	24	1027.791	1028.203	1029.136
2	25	1027.028	1027.759	1028.798
2	26	1026.178	1027.295	1028.467
2	27	1025.231	1026.810	1028.145
2	28	·1024.173	1026.303	1027.839
2	29	1022.991	1025.774	1027.553
2	30	1021.669	1025.223	1027.292
2	31	1020.189	1024.647	1027.063
2	32	1018.533	1024.047	1026.873
2	33	1016.679	1023.420	1026.730
2	34	1014.604	1022.767	1026.641
2	35	1012.284	1022.085	1026.616
3	3	1034.107	1033.332	1034.176
3	4	1034.035	1033.261	1034.104
3	5	1033.946	1033.173	1034.014
3	6	1033.838	1033.066	1033.905
3	7	1033.712	1032.942	1033.778
3	8	1033.568	1032.800	1033.633
3	9	1033.407	1032.640	1033.470
3	10	1033.227	1032.463	1033.289
3	11	1033.030	1032.267	1033.090
3	12	1032.814	1032.054	1032.872
3	13	1032.580	1031.823	1032.636
3	14	1032.329	1031.573	1032.382
3	15	1032.059	1031.306	1032.110
3	16	1031.770	1031.021	1031.819
3	17	1031.463	1030.717	1031.509
3	18	1031.138	1030.394	1031.181
3	19	1030.794	1030.053	1030.834
3	20	1030.430	1029.692	1030.468
3	21	1030.048	1029.312	1030.083
3	22	1029.646	1028.913	1029.679
3	23	1029.224	1028.493	1029.256
3	24	1028.781	1028.052	1028.812
3	25	1028.318	1027.590	1028.349
3	26	1027.833	1027.106	1027.866
3	27	1027.327	1026.599	1027.362
3	28	1026.798	1026.069	1026.837
3	29	1026.245	1025.514	1026.290
3	30	1025.668	1024.933	1025.722
3	31	1025.066	1024.326	1025.132
3	32	1024.438	1023.691	1024.518
3	33	1023.782	1023.027	1023.881
3	34	1023.098	1022.331	1023.220
3	35	1022.383	1021.603	1022.534
4	4	1033.917	1033.363	1034.264
4	5	1033.827	1033.271	1034.172

(*Continued*)

TABLE C1 (*Continued*)

K	J	Q(01K,J)	Q(02K,J)	Q(03K,J)
4	6	1033.720	1033.161	1034.062
4	7	1033.594	1033.032	1033.933
4	8	1033.450	1032.886	1033.786
4	9	1033.288	1032.721	1033.621
4	10	1033.108	1032.538	1033.437
4	11	1032.909	1032.337	1033.234
4	12	1032.691	1032.118	1033.014
4	13	1032.454	1031.882	1032.774
4	14	1032.197	1031.630	1032.517
4	15	1031.919	1031.360	1032.240
4	16	1031.619	1031.075	1031.946
4	17	1031.297	1030.774	1031.632
4	18	1030.952	1030.459	1031.300
4	19	1030.581	1030.130	1030.949
4	20	1030.183	1029.789	1030.580
4	21	1029.756	1029.436	1030.192
4	22	1029.297	1029.074	1029.784
4	23	1028.805	1028.704	1029.358
4	24	1028.275	1028.328	1028.913
4	25	1027.705	1027.948	1028.448
4	26	1027.090	1027.567	1027.964
4	27	1026.426	1027.188	1027.460
4	28	1025.708	1026.814	1026.936
4	29	1024.930	1026.448	1026.392
4	30	1024.086	1026.094	1025.828
4	31	1023.170	1025.757	1025.244
4	32	1022.173	1025.441	1024.638
4	33	1021.088	1025.153	1024.012
4	34	1019.904	1024.896	1023.364
4	35	1018.613	1024.679	1022.694
5	5	1033.690	1033.656	1034.335
5	6	1033.579	1033.584	1034.222
5	7	1033.449	1033.501	1034.089
5	8	1033.301	1033.405	1033.938
5	9	1033.134	1033.298	1033.767
5	10	1032.949	1033.179	1033.577
5	11	1032.745	1033.048	1033.368
5	12	1032.523	1032.905	1033.140
5	13	1032.282	1032.750	1032.891
5	14	1032.023	1032.584	1032.623
5	15	1031.746	1032.406	1032.335
5	16	1031.450	1032.217	1032.026
5	17	1031.136	1032.016	1031.695
5	18	1030.805	1031.803	1031.343
5	19	1030.455	1031.580	1030.968
5	20	1030.087	1031.345	1030.569
5	21	1029.701	1031.100	1030.146
5	22	1029.298	1030.844	1029.698
5	23	1028.877	1030.578	1029.222
5	24	1028.439	1030.301	1028.718
5	25	1027.983	1030.015	1028.183
5	26	1027.511	1029.720	1027.616

(*Continued*)

TABLE C1 (*Continued*)

K	J	Q(01K,J)	Q(02K,J)	Q(03K,J)
5	27	1027.022	1029.416	1027.015
5	28	1026.517	1029.103	1026.376
5	29	1025.995	1028.783	1025.698
5	30	1025.458	1028.455	1024.977
5	31	1024.906	1028.121	1024.210
5	32	1024.338	1027.781	1023.392
5	33	1023.756	1027.436	1022.520
5	34	1023.160	1027.087	1021.589
5	35	1022.551	1026.734	1020.595
6	6	1033.485	1033.665	1034.194
6	7	1033.388	1033.568	1034.058
6	8	1033.277	1033.457	1033.902
6	9	1033.153	1033.333	1033.727
6	10	1033.015	1033.195	1033.533
6	11	1032.863	1033.043	1033.319
6	12	1032.699	1032.878	1033.086
6	13	1032.522	1032.700	1032.834
6	14	1032.332	1032.510	1032.562
6	15	1032.131	1032.307	1032.272
6	16	1031.919	1032.092	1031.962
6	17	1031.696	1031.866	1031.634
6	18	1031.464	1031.630	1031.288
6	19	1031.224	1031.383	1030.923
6	20	1030.977	1031.129	1030.539
6	21	1030.724	1030.866	1030.138
6	22	1030.467	1030.598	1029.720
6	23	1030.209	1030.324	1029.284
6	24	1029.951	1030.048	1028.832
6	25	1029.697	1029.771	1028.364
6	26	1029.449	1029.495	1027.880
6	27	1029.210	1029.223	1027.381
6	28	1028.984	1028.958	1026.868
6	29	1028.776	1028.702	1026.342
6	30	1028.590	1028.460	1025.803
6	31	1028.430	1028.235	1025.252
6	32	1028.304	1028.031	1024.691
6	33	1028.216	1027.853	1024.120
6	34	1028.174	1027.707	1023.542
6	35	1028.185	1027.597	1022.957
7	7	1033.278	1033.812	1034.104
7	8	1033.147	1033.686	1033.975
7	9	1033.001	1033.544	1033.830
7	10	1032.838	1033.387	1033.668
7	11	1032.658	1033.213	1033.491
7	12	1032.462	1033.024	1033.298
7	13	1032.250	1032.819	1033.088
7	14	1032.022	1032.599	1032.863
7	15	1031.776	1032.363	1032.621
7	16	1031.514	1032.111	1032.364
7	17	1031.236	1031.843	1032.090
7	18	1030.940	1031.560	1031.800
7	19	1030.628	1031.262	1031.495

(*Continued*)

TABLE C1 (*Continued*)

K	J	Q(01K,J)	Q(02K,J)	Q(03K,J)
7	20	1030.298	1030.948	1031.173
7	21	1029.951	1030.618	1030.835
7	22	1029.586	1030.274	1030.482
7	23	1029.204	1029.914	1030.112
7	24	1028.803	1029.539	1029.727
7	25	1028.383	1029.148	1029.325
7	26	1027.945	1028.743	1028.908
7	27	1027.487	1028.324	1028.475
7	28	1027.009	1027.889	1028.026
7	29	1026.511	1027.441	1027.562
7	30	1025.991	1026.978	1027.082
7	31	1025.450	1026.501	1026.586
7	32	1024.886	1026.011	1026.074
7	33	1024.298	1025.507	1025.547
7	34	1023.686	1024.990	1025.005
7	35	1023.049	1024.460	1024.447
8	8	1033.143	1033.926	1033.932
8	9	1032.993	1033.778	1033.778
8	10	1032.827	1033.613	1033.606
8	11	1032.644	1033.431	1033.418
8	12	1032.444	1033.234	1033.212
8	13	1032.228	1033.019	1032.988
8	14	1031.995	1032.788	1032.747
8	15	1031.744	1032.540	1032.489
8	16	1031.477	1032.276	1032.212
8	17	1031.193	1031.995	1031.918
8	18	1030.891	1031.697	1031.605
8	19	1030.572	1031.382	1031.275
8	20	1030.236	1031.050	1030.925
8	21	1029.882	1030.702	1030.556
8	22	1029.509	1030.335	1030.167
8	23	1029.118	1029.952	1029.759
8	24	1028.709	1029.551	1029.330
8	25	1028.280	1029.132	1028.879
8	26	1027.832	1028.695	1028.407
8	27	1027.364	1028.240	1027.911
8	28	1026.875	1027.766	1027.392
8	29	1026.365	1027.273	1026.848
8	30	1025.834	1026.761	1026.278
8	31	1025.279	1026.230	1025.680
8	32	1024.701	1025.678	1025.054
8	33	1024.099	1025.106	1024.397
8	34	1023.472	1024.513	1023.709
8	35	1022.818	1023.898	1022.987
9	9	1033.100	1034.001	1033.703
9	10	1032.932	1033.831	1033.531
9	11	1032.748	1033.644	1033.341
9	12	1032.547	1033.441	1033.134
9	13	1032.329	1033.220	1032.909
9	14	1032.093	1032.982	1032.667
9	15	1031.841	1032.727	1032.408
9	16	1031.571	1032.454	1032.132

(*Continued*)

TABLE C1 (*Continued*)

K	J	Q(01K,J)	Q(02K,J)	Q(03K,J)
9	17	1031.284	1032.165	1031.838
9	18	1030.979	1031.857	1031.526
9	19	1030.656	1031.533	1031.197
9	20	1030.314	1031.190	1030.850
9	21	1029.955	1030.830	1030.486
9	22	1029.576	1030.451	1030.103
9	23	1029.178	1030.054	1029.703
9	24	1028.760	1029.639	1029.285
9	25	1028.321	1029.205	1028.848
9	26	1027.862	1028.751	1028.393
9	27	1027.380	1028.278	1027.919
9	28	1026.876	1027.785	1027.426
9	29	1026.348	1027.272	1026.913
9	30	1025.796	1026.737	1026.382
9	31	1025.218	1026.181	1025.830
9	32	1024.613	1025.602	1025.258
9	33	1023.979	1025.001	1024.665
9	34	1023.316	1024.376	1024.051
9	35	1022.621	1023.726	1023.416
10	10	1033.146	1034.079	1033.471
10	11	1032.958	1033.889	1033.282
10	12	1032.754	1033.682	1033.076
10	13	1032.532	1033.457	1032.853
10	14	1032.293	1033.216	1032.612
10	15	1032.036	1032.957	1032.354
10	16	1031.763	1032.681	1032.079
10	17	1031.471	1032.387	1031.787
10	18	1031.162	1032.077	1031.478
10	19	1030.835	1031.749	1031.152
10	20	1030.490	1031.404	1030.808
10	21	1030.126	1031.042	1030.448
10	22	1029.744	1030.662	1030.070
10	23	1029.343	1030.266	1029.676
10	24	1028.922	1029.852	1029.264
10	25	1028.482	1029.421	1028.835
10	26	1028.022	1028.973	1028.389
10	27	1027.541	1028.508	1027.926
10	28	1027.038	1028.026	1027.446
10	29	1026.513	1027.526	1026.949
10	30	1025.966	1027.010	1026.435
10	31	1025.395	1026.477	1025.904
10	32	1024.800	1025.926	1025.356
10	33	1024.179	1025.358	1024.791
10	34	1023.532	1024.774	1024.209
10	35	1022.857	1024.172	1023.610

(*Continued*)

J. O. HENNINGSEN

TABLE C1 (*Continued*)

K	J	P(01K,J)	P(02K,J)	P(03K,J)
0	1	1032.281	1032.092	1032.092
0	2	1030.650	1030.462	1030.462
0	3	1029.002	1028.814	1028.814
0	4	1027.336	1027.150	1027.150
0	5	1025.653	1025.470	1025.470
0	6	1023.952	1023.775	1023.775
0	7	1022.235	1022.064	1022.064
0	8	1020.500	1020.337	1020.337
0	9	1018.749	1018.596	1018.596
0	10	1016.980	1016.841	1016.841
0	11	1015.195	1015.072	1015.072
0	12	1013.393	1013.289	1013.289
0	13	1011.575	1011.493	1011.493
0	14	1009.739	1009.684	1009.684
0	15	1007.887	1007.863	1007.863
0	16	1006.019	1006.030	1006.030
0	17	1004.134	1004.186	1004.186
0	18	1002.233	1002.332	1002.332
0	19	1000.316	1000.468	1000.468
0	20	998.382	998.595	998.595
0	21	996.432	996.713	996.713
0	22	994.466	994.824	994.824
0	23	992.484	992.929	992.929
0	24	990.487	991.029	991.029
0	25	988.474	989.124	989.124
0	26	986.445	987.217	987.217
0	27	984.401	985.308	985.308
0	28	982.343	983.400	983.400
0	29	980.269	981.493	981.493
0	30	978.181	979.590	979.590
0	31	976.079	977.693	977.693
0	32	973.963	975.804	975.804
0	33	971.834	973.925	973.925
0	34	969.691	972.058	972.058
0	35	967.537	970.208	970.208
1	2	1031.127	1030.291	1030.663
1	3	1029.478	1028.642	1029.014
1	4	1027.811	1026.976	1027.348
1	5	1026.127	1025.293	1025.664
1	6	1024.425	1023.592	1023.963
1	7	1022.706	1021.874	1022.244
1	8	1020.970	1020.139	1020.508
1	9	1019.217	1018.387	1018.755
1	10	1017.446	1016.617	1016.985
1	11	1015.659	1014.831	1015.198
1	12	1013.854	1013.028	1013.394
1	13	1012.033	1011.208	1011.573
1	14	1010.194	1009.371	1009.734
1	15	1008.339	1007.518	1007.879
1	16	1006.467	1005.647	1006.007
1	17	1004.579	1003.759	1004.117
1	18	1002.674	1001.855	1002.211

(*Continued*)

TABLE C1 (*Continued*)

K	J	P(01K,J)	P(02K,J)	P(03K,J)
1	19	1000.753	999.932	1000.288
1	20	998.816	997.993	998.347
1	21	996.864	996.035	996.390
1	22	994.896	994.058	994.415
1	23	992.913	992.062	992.423
1	24	990.916	990.047	990.413
1	25	988.904	988.010	988.386
1	26	986.879	985.953	986.340
1	27	984.842	983.872	984.277
1	28	982.792	981.768	982.196
1	29	980.732	979.637	980.096
1	30	978.662	977.480	977.977
1	31	976.582	975.294	975.839
1	32	974.496	973.076	973.682
1	33	972.403	970.825	971.505
1	34	970.306	968.537	969.307
1	35	968.207	966.211	967.088
2	3	1029.407	1028.542	1029.213
2	4	1027.740	1026.876	1027.546
2	5	1026.055	1025.194	1025.860
2	6	1024.352	1023.494	1024.156
2	7	1022.630	1021.777	1022.434
2	8	1020.890	1020.042	1020.693
2	9	1019.130	1018.291	1018.935
2	10	1017.351	1016.523	1017.158
2	11	1015.551	1014.738	1015.364
2	12	1013.729	1012.937	1013.552
2	13	1011.884	1011.118	1011.723
2	14	1010.014	1009.282	1009.878
2	15	1008.115	1007.430	1008.018
2	16	1006.185	1005.561	1006.144
2	17	1004.219	1003.675	1004.258
2	18	1002.212	1001.772	1002.361
2	19	1000.156	999.852	1000.458
2	20	998.045	997.915	998.550
2	21	995.868	995.960	996.643
2	22	993.615	993.988	994.740
2	23	991.271	991.998	992.848
2	24	988.820	989.990	990.974
2	25	986.245	987.964	989.125
2	26	983.522	985.919	987.310
2	27	980.628	983.855	985.542
2	28	977.534	981.772	983.832
2	29	974.207	979.668	982.194
2	30	970.608	977.542	980.646
2	31	966.698	975.396	979.205
2	32	962.426	973.226	977.894
2	33	957.741	971.033	976.736
2	34	952.580	968.816	975.757
2	35	946.878	966.572	974.988
3	4	1027.654	1026.878	1027.723
3	5	1025.969	1025.194	1026.037

(*Continued*)

TABLE C1 (*Continued*)

K	J	P(01K,J)	P(02K,J)	P(03K,J)
3	6	1024.266	1023.492	1024.333
3	7	1022.545	1021.773	1022.611
3	8	1020.807	1020.036	1020.871
3	9	1019.051	1018.282	1019.112
3	10	1017.277	1016.510	1017.336
3	11	1015.485	1014.721	1015.541
3	12	1013.676	1012.915	1013.728
3	13	1011.849	1011.091	1011.897
3	14	1010.005	1009.249	1010.047
3	15	1008.142	1007.391	1008.179
3	16	1006.262	1005.514	1006.293
3	17	1004.364	1003.620	1004.388
3	18	1002.448	1001.709	1002.464
3	19	1000.513	999.779	1000.522
3	20	998.561	997.832	998.561
3	21	996.590	995.866	996.581
3	22	994.600	993.881	994.582
3	23	992.592	991.877	992.564
3	24	990.564	989.854	990.526
3	25	988.516	987.812	988.468
3	26	986.448	985.748	986.390
3	27	984.359	983.664	984.292
3	28	982.250	981.558	982.173
3	29	980.118	979.430	980.033
3	30	977.964	977.278	977.872
3	31	975.786	975.102	975.689
3	32	973.584	972.901	973.483
3	33	971.357	970.672	971.254
3	34	969.103	968.416	969.002
3	35	966.821	966.130	966.725
4	5	1025.851	1025.297	1026.198
4	6	1024.148	1023.592	1024.493
4	7	1022.428	1021.869	1022.770
4	8	1020.690	1020.128	1021.029
4	9	1018.934	1018.369	1019.269
4	10	1017.160	1016.592	1017.491
4	11	1015.368	1014.798	1015.695
4	12	1013.556	1012.987	1013.881
4	13	1011.725	1011.159	1012.048
4	14	1009.873	1009.315	1010.197
4	15	1007.999	1007.455	1008.328
4	16	1006.102	1005.581	1006.441
4	17	1004.178	1003.692	1004.536
4	18	1002.226	1001.792	1002.612
4	19	1000.242	999.881	1000.670
4	20	998.222	997.962	998.709
4	21	996.161	996.036	996.730
4	22	994.052	994.107	994.733
4	23	991.888	992.178	992.717
4	24	989.660	990.253	990.683
4	25	987.359	988.337	988.630
4	26	984.971	986.435	986.558

(*Continued*)

TABLE C1 (*Continued*)

K	J	P(01K,J)	P(02K,J)	P(03K,J)
4	27	982.484	984.553	984.467
4	28	979.882	982.699	982.357
4	29	977.145	980.881	980.227
4	30	974.252	979.107	978.078
4	31	971.180	977.389	975.910
4	32	967.901	975.738	973.721
4	33	964.385	974.168	971.512
4	34	960.597	972.693	969.283
4	35	956.497	971.331	967.033
5	6	1024.012	1023.978	1024.658
5	7	1022.288	1022.294	1022.932
5	8	1020.546	1020.598	1021.187
5	9	1018.787	1018.890	1019.424
5	10	1017.008	1017.171	1017.641
5	11	1015.212	1015.441	1015.840
5	12	1013.398	1013.699	1014.020
5	13	1011.566	1011.946	1012.181
5	14	1009.716	1010.181	1010.322
5	15	1007.848	1008.405	1008.444
5	16	1005.963	1006.618	1006.546
5	17	1004.060	1004.820	1004.628
5	18	1002.140	1003.011	1002.689
5	19	1000.202	1001.191	1000.729
5	20	998.247	999.361	998.746
5	21	996.275	997.521	996.741
5	22	994.286	995.670	994.712
5	23	992.280	993.810	992.658
5	24	990.258	991.940	990.578
5	25	988.219	990.061	988.469
5	26	986.165	988.173	986.331
5	27	984.094	986.277	984.162
5	28	982.008	984.373	981.959
5	29	979.907	982.461	979.720
5	30	977.791	980.542	977.441
5	31	975.661	978.618	975.121
5	32	973.517	976.688	972.755
5	33	971.359	974.753	970.341
5	34	969.188	972.814	967.873
5	35	967.004	970.872	965.347
6	7	1022.196	1022.376	1022.906
6	8	1020.487	1020.667	1021.158
6	9	1018.764	1018.945	1019.391
6	10	1017.029	1017.209	1017.605
6	11	1015.280	1015.459	1015.799
6	12	1013.519	1013.697	1013.975
6	13	1011.744	1011.922	1012.132
6	14	1009.958	1010.134	1010.270
6	15	1008.160	1008.334	1008.390
6	16	1006.351	1006.523	1006.491
6	17	1004.532	1004.700	1004.574
6	18	1002.703	1002.866	1002.638
6	19	1000.865	1001.022	1000.685

(*Continued*)

TABLE C1 (*Continued*)

K	J	P(01K,J)	P(02K,J)	P(03K,J)
6	20	999.020	999.170	998.714
6	21	997.169	997.309	996.726
6	22	995.313	995.442	994.721
6	23	993.454	993.569	992.699
6	24	991.595	991.693	990.660
6	25	989.737	989.814	988.606
6	26	987.884	987.935	986.537
6	27	986.037	986.058	984.453
6	28	984.202	984.186	982.356
6	29	982.381	982.322	980.245
6	30	980.579	980.468	978.122
6	31	978.799	978.629	975.987
6	32	977.049	976.808	973.842
6	33	975.332	975.009	971.687
6	34	973.655	973.238	969.525
6	35	972.026	971.499	967.356
7	8	1020.378	1020.914	1021.206
7	9	1018.637	1019.177	1019.466
7	10	1016.879	1017.424	1017.710
7	11	1015.105	1015.655	1015.938
7	12	1013.316	1013.871	1014.151
7	13	1011.511	1012.072	1012.348
7	14	1009.689	1010.258	1010.530
7	15	1007.852	1008.429	1008.696
7	16	1005.999	1006.584	1006.847
7	17	1004.130	1004.724	1004.982
7	18	1002.245	1002.849	1003.102
7	19	1000.344	1000.960	1001.207
7	20	998.427	999.055	999.296
7	21	996.493	997.136	997.371
7	22	994.543	995.202	995.430
7	23	992.576	993.253	993.474
7	24	990.593	991.290	991.503
7	25	988.592	989.313	989.517
7	26	986.573	987.322	987.517
7	27	984.537	985.316	985.501
7	28	982.482	983.297	983.471
7	29	980.409	981.265	981.427
7	30	978.316	979.219	979.367
7	31	976.204	977.159	977.294
7	32	974.070	975.087	975.206
7	33	971.916	973.003	973.103
7	34	969.740	970.906	970.987
7	35	967.540	968.797	968.856
8	9	1018.634	1019.419	1019.425
8	10	1016.874	1017.660	1017.661
8	11	1015.097	1015.885	1015.879
8	12	1013.305	1014.094	1014.081
8	13	1011.496	1012.286	1012.266
8	14	1009.670	1010.463	1010.434
8	15	1007.829	1008.623	1008.585
8	16	1005.971	1006.768	1006.719

(*Continued*)

TABLE C1 (*Continued*)

K	J	P(01K,J)	P(02K,J)	P(03K,J)
8	17	1004.097	1004.896	1004.836
8	18	1002.206	1003.008	1002.936
8	19	1000.299	1001.104	1001.019
8	20	998.376	999.184	999.084
8	21	996.435	997.248	997.131
8	22	994.478	995.295	995.160
8	23	992.503	993.326	993.170
8	24	990.511	991.340	991.161
8	25	988.501	989.337	989.133
8	26	986.474	987.318	987.084
8	27	984.427	985.281	985.014
8	28	982.362	983.228	982.923
8	29	980.278	981.157	980.809
8	30	978.173	979.068	978.671
8	31	976.048	976.960	976.509
8	32	973.902	974.835	974.321
8	33	971.733	972.690	972.105
8	34	969.542	970.527	969.861
8	35	967.326	968.343	967.586
9	10	1016.984	1017.887	1017.589
9	11	1015.207	1016.107	1015.807
9	12	1013.413	1014.311	1014.007
9	13	1011.602	1012.498	1012.192
9	14	1009.775	1010.669	1010.359
9	15	1007.932	1008.823	1008.510
9	16	1006.072	1006.961	1006.644
9	17	1004.196	1005.082	1004.761
9	18	1002.302	1003.186	1002.862
9	19	1000.392	1001.273	1000.946
9	20	998.464	999.344	999.013
9	21	996.519	997.398	997.064
9	22	994.556	995.434	995.098
9	23	992.576	993.454	993.114
9	24	990.576	991.456	991.114
9	25	988.558	989.440	989.097
9	26	986.521	987.406	987.063
9	27	984.463	985.355	985.011
9	28	982.384	983.284	982.942
9	29	980.284	981.195	980.855
9	30	978.162	979.086	978.750
9	31	976.016	976.958	976.627
9	32	973.845	974.809	974.486
9	33	971.649	972.639	972.325
9	34	969.426	970.448	970.146
9	35	967.174	968.234	967.946
10	11	1015.424	1016.359	1015.751
10	12	1013.628	1014.561	1013.953
10	13	1011.814	1012.745	1012.138
10	14	1009.984	1010.913	1010.307
10	15	1008.137	1009.064	1008.459
10	16	1006.274	1007.198	1006.595
10	17	1004.394	1005.316	1004.714

(*Continued*)

TABLE C1 (*Continued*)

K	J	P(01K,J)	P(02K,J)	P(03K,J)
10	18	1002.496	1003.417	1002.817
10	19	1000.582	1001.502	1000.904
10	20	998.651	999.571	998.974
10	21	996.702	997.623	997.028
10	22	994.736	995.659	995.067
10	23	992.752	993.678	993.088
10	24	990.749	991.682	991.094
10	25	988.729	989.669	989.084
10	26	986.690	987.640	987.058
10	27	984.631	985.595	985.016
10	28	982.553	983.534	982.958
10	29	980.455	981.457	980.885
10	30	978.336	979.365	978.796
10	31	976.195	977.256	976.691
10	32	974.032	975.132	974.570
10	33	971.845	972.992	972.434
10	34	969.635	970.836	970.282
10	35	967.399	968.665	968.115

ACKNOWLEDGMENTS

This work has drawn on results and stimuli from a large number of colleagues around the world, and I wish to acknowledge all those who have permitted the use of their material. In particular, I wish to thank J. P. Sattler, T. L. Worchesky, and W. A. Riessler for sharing with me their published and unpublished diode laser spectra, and Y. Y. Kwan for providing the computer program, which is the cornerstone of the numerical work. I am indebted to K. M. Evenson, F. R. Petersen, and D. A. Jennings for their tutoring in the art of frequency measurements, to E. J. Danielewicz and C. O. Weiss for communicating unpublished results as well as for exchange of ideas, to R. M. Lees for sharing his expertise on the theory of the CH_3OH molecule, to J. Heppner for valuable comments, and to M. Feld, M. Inguscio, and F. Strumia for lively discussions which—although consensus was not always reached—invariably deepened my insight. Last but not least, I wish to thank J. Beyer Nielsen for his valuable programming assistance and Grethe Bjørndal for her skilled and rapid typing of the manuscript.

REFERENCES

Abrams, R. L. (1974). *Appl. Phys. Lett.* **25**, 304–306.
Arimondo, E. (1979). *IEEE J. Quantum Electron.* **QE-15**, 1081–1083.
Arimondo, E., and Inguscio, M. (1979). *J. Mol. Spectrosc.* **75**, 81–86.
Arimondo, E., Inguscio, M., Moretti, A., Pellegrino, M., and Strumia, F. (1980). *Opt. Lett.* **5**, 496–498.
Barnes, A. J., and Hallam, H. E. (1970). *Trans. Faraday Soc.* **66**, 1920–1931.
Bedwell, D. J., Duxbury, G., Herman, H., and Orengo, C. A. (1978). *Infrared Phys.* **18**, 453–460.

Berger, W., Siemsen, K., and Reid, J. (1977). *Rev. Sci. Instrum.* **48**, 1031–1033.
Bernard, P., Izatt, J. R., and Mathieu, P. (1981). *Int. J. Infrared and Millimeter Waves* **2**, 65–70.
Bionducci, G., Inguscio, M., Moretti, A., and Strumia, F. (1979). *Infrared Phys.* **19**, 297–308.
Biron, D. G., Temkin, R. J., Lax, B., and Danly, B. G. (1979). *Int. Conf. Infrared Millimeter Waves Their Appl., 4th, Miami Beach, Florida* Digest pp. 213–214.
Bleaney, T. G., Knight, D. J. E., and Murray Lloyd, E. K. (1978). *Opt. Commun.* **25**, 176–178.
Borden, A., and Barker, E. F. (1938). *J. Chem. Phys.* **6**, 553–563.
Burkhard, D. G., and Dennison, D. M. (1959). *J. Mol. Specrosc.* **3**, 299–334.
Busse, G., and Renk, K. F. (1978). *Infrared Phys.* **18**, 517–519.
Busse, G., Basel, E., and Pfaller, A. (1977). *Appl. Phys.* **12**, 387–389.
Carter, G. M., and Marcus, S. (1979). *Appl. Phys. Lett.* **35**, 129–130.
Chang, T. Y. (1977). *IEEE J. Quantum Electron.* **QE-13**, 937–942.
Chang, T. Y., and Bridges, T. J. (1970). *Opt. Commun.* **1**, 423–426.
Chang, T. Y., Bridges, T. J., and Burkhardt, E. G. (1970). *Appl. Phys. Lett.* **17**, 249–251.
Danielewicz, E. J., and Coleman, P. D. (1977). *IEEE J. Quantum Electron.* **QE-13**, 485–490.
Danielewicz, E. J., Plant, T. K., and De Temple, T. A. (1975). *Opt. Commun.* **13**, 366–369.
Degnan, J. J. (1976). *Appl. Phys.* **11**, 1–33.
Domnin, Y. S., Tatarenkov, U. M., and Shumyatskii, P. S. (1974). *Sov. Phys.-Quant. Electron.* **4**, 401–402.
Drozdowicz, Z., Temkin, R. J., and Lax, B. (1979). *IEEE J. Quantum Electron.* **QE-15**, 170–178.
Dyubko, S. F., Svich, V. A., Fesenko, L. D. (1974). *Sov. Phys.-Tech. Phys.* **18**, 1121.
Evenson, K. M. *et al.* (1977). *IEEE J. Quantum Electron.* **QE-13**, 442–444.
Fetterman, H. R., Schlossberg, R., and Waldman, J. (1972). *Opt. Commun.* **6**, 156–159.
Fetterman, H. R., Clifton, B. J., and Tannenwald, P. E. (1974). *Appl. Phys. Lett.* **24**, 70–72.
Feld, M. S. (1979). *Int. Conf. Infrared Millimeter Waves Their Appl., 4th, Miami Beach, Florida* Digest Suppl., pp. 36–37.
Forber, R., and Feld M. S. (1979). *Int. Conf. Infrared Millimeter Waves Their Appl., 4th, Miami Beach, Florida* Digest Suppl. pp. 34–35.
Freed, C., and Javan, A. (1970). *Appl. Phys. Lett.* **17**, 53–56.
Freed, C., Ross, A. H., and O'Donnell, R. G. (1974). *J. Mol. Spectrosc.* **49**, 439–453.
Gamble, E. B., and Danielewicz, E. J. (1981). *IEEE J. Quantum Electron.*
Gastaud, C., Sentz, A., Redon, M., and Fourrier, M. (1980). *IEEE J. Quantum Electron.* **QE-16**, 1285–1287.
Gibson, R. B. (1979). Thesis, MIT, Cambridge Massachusetts.
Gullberg, K., Hartmann, B., and Kleman, B. (1973). *Phys. Scripta* **8**, 177–182.
Hadni, A., Thomas, R., Mangin, J., and Bagard, M. (1978). *Infrared Phys.* **18**, 663–668.
Henningsen, J. O. (1977). *IEEE J. Quantum Electron.* **QE-13**, 435–441.
Henningsen, J. O. (1978). *IEEE J. Quantum Electron.* **QE-14**, 958–962.
Henningsen, J. O. (1980a). *J. Mol. Spectrosc.* **83**, 70–93.
Henningsen, J. O. (1980b). *Int. Conf. Infrared Millimeter Waves, 5th Würzburg* Digest, pp. 223–224.
Henningsen, J. O. (1981). *J. Mol. Spectrosc.* **85**, 282–300.
Henningsen, J. O., and Jensen, H. G. (1975). *IEEE J. Quantum Electron.* **QE-11**, 248–252.
Henningsen, J. O., and Petersen, J. C. (1978). *Infrared Phys.* **18**, 475–479.
Henningsen, J. O., Petersen, J. C., Petersen, F. R., Jennings, D. A., and Evenson, K. M. (1979). *J. Mol. Spectrosc.* **77**, 298–309.

Heppner, J., and Weiss, C. O. (1978). *Appl. Phys. Lett.* **33**, 590–592.
Heppner, J., Weiss, C. O., Hübner, U., and Schinn, G. (1980). *IEEE J. Quantum Electron.* **QE-16**, 392–402.
Hodges, D. T., and Hartwick, T. S. (1973). *Appl. Phys. Lett.* **23**, 252–253.
Hodges, D. T., Reel, R. D., and Barker, D. H. (1973). *IEEE J. Quantum Electron.* **QE-9**, 1159–1160.
Hodges, D. T., Foote, F. B., and Reel, R. D. (1976). *Appl. Phys. Lett.* **29**, 662–664.
Hodges, D. T., Foote, F. B., and Reel, R. D. (1977). *IEEE J. Quantum Electron.* **QE-13**, 491–494.
Hughes, R. H., Good, W. E., and Coles, D. K. (1951). *Phys. Rev.* **84**, 418–425.
Inguscio, M., Moretti, A., and Strumia, F. (1979a). *Opt. Commun.* **30**, 355–360.
Inguscio, M., Minguzzi, P., Moretti, A., Strumia, F., and Tonelli, M. (1979b). *Appl. Phys.* **18**, 261–270.
Inguscio, M., Moretti, A., and Strumia, F. (1979c). *Int. Conf. Infrared Millimeter Waves Their Appl., 4th, Miami Beach, Florida* Digest, pp. 205–206.
Inguscio, M., Moretti, A., and Strumia, F. (1980). *Opt. Commun.* **32**, 87–90.
Inguscio, M., Ioli, N., Moretti, A., Moruzzi, G., and Strumia, F. (1981). *Opt. Commun.* **37**, 211–216.
Ioli, N., Moruzzi, G., and Strumia, F. (1980). *Lett. Nuovo Cimento* **28**, 257–264.
Ivash, E. V., and Dennison, D. M. (1953). *J. Chem. Phys.* **21**, 1804–1816.
Izatt, J. R., and Mathieu, P., (1978). *Infrared Phys.* **18**, 509–516.
Izatt, J. R., Bean, B. L., and Caudle, G. F. (1975). *Opt. Commun.* **14**, 385–387.
Jacobs, R. R., Prosnitz, D., Bischel, W. K., and Rhodes, C. K. (1976). *Appl. Phys. Lett.* **29**, 710–712.
Jennings, D. A., Petersen, F. R., and Evenson, K. M. (1979). *Opt. Lett.* **4**, 129–130.
Keilmann, F., Sheffield, R. L., Leite, J. R. R., Feld, M. S., and Javan, A. (1975). *Appl. Phys. Lett.* **26**, 19–22.
Kimmitt, M. F. (1970). "Far-Infrared Techniques." Pion, London.
Knight, D. J. E. (1981). National Physical Lab. Rep. No. Qu 45, Teddington, Middlesex, England.
Koepf, G. A., and McAvoy, N. (1977). *IEEE J. Quantum Electron.* **QE-13**, 418–420.
Kogelnik, H., and Li, T. (1966). *Proc. IEEE* **54**, 1312–1329.
Kon, S., Yano, T., Hagiwara, E., and Hirose, H. (1975). *J. Appl. Phys.* **14**, 1861–1862.
Koo, K. P., and Claspy, P. C. (1979). *Appl. Opt.* **18**, 1314–1321.
Kwan, Y. Y., and Dennison, D. M. (1972). *J. Mol. Spectrosc.* **43**, 291–319.
Landsberg, B. M. (1980). *IEEE J. Quantum Electron.* **QE-16**, 704–706.
Lees, R. M. (1972). *J. Chem. Phys.* **57**, 2249–2252.
Lees, R. M., and Baker, J. G. (1968). *J. Chem. Phys.* **48**, 5299–5318.
Lees, R. M., Young, C., Van der Linde, J., and Oliver, B. A. (1979). *J. Mol. Spectrosc.* **75**, 161–167.
Lees, R. M., Walton, M. A., and Henningsen, J. O. (1981). *J. Mol. Spectrosc.* **88**, 90–94.
Lengfellner, H., and Renk, K. F. (1977). *IEEE J. Quantum Electron.* **QE-13**, 421–424.
Lund, M. W., Cogan, J. N., and Davis, J. A. (1979). *Rev. Sci. Instrum.* **50**, 791–792.
Malk, E. G., Niesen, J. W., Parsons, D. F., and Coleman, P. D. (1978). *IEEE J. Quantum Electron.* **QE-14**, 544–550.
Marcatili, E. A. J., and Schmeltzer, R. A. (1964). *Bell Syst. Tech. J.* **43**, 1783.
McDonald, D. G., Risley, A. S., Cupp, J. D., Evenson, K. M., and Ashley, J. R. (1972). *Appl. Phys. Lett.* **20**, 296–299.
McDowell, R. S., Patterson, C. W., Jones, C. R., Buchwald, M. I., and Telle, J. M. (1979). *Opt. Lett.* **4**, 274–276.

Panock, R. L., and Temkin, R. J. (1977). *IEEE J. Quantum Electron.* **QE-13**, 425–434.

Petersen, F. R., McDonald, G. D., Cupp, J. D., and Danielson, B. L. (1974). "Laser Spectroscopy" (R. G. Brewer and A. Mooradian, eds.), pp. 555–569. Plenum Press, New York.

Petersen, F. R., Evenson, K. M., Jennings, D. A., Wells, J. S., Goto, K., and Jiménez, J. J. (1975). *IEEE J. Quantum Electron.* **QE-11**, 838–843.

Petersen, F. R., Evenson, K. M., Jennings, D. A., Wells, J. S., Goto, K., and Jiménez, J. J. (1976). *IEEE J. Quantum Electron.* **QE-12**, 86–87.

Petersen, F. R., Evenson, K. M., Jennings, D. A., and Scalabrin, A. (1980). *IEEE J. Quantum Electron.* **QE-16**, 319–323.

Petersen, J. C., McCombie, J., and Duxbury, G. (1980). *Int. Conf. Infrared Millimeter Waves, 5th, Würzburg* Digest, pp. 140–141.

Petuchowsky, S. J., Rosenberger, A. T., and De Temple, T. A. (1977). *IEEE J. Quantum Electron.* **QE-13**, 476–481.

Prosnitz, D., Jacobs, R. R., Bischel, W. K., and Rhodes, C. K. (1978). *Appl. Phys. Lett.* **32**, 221–222.

Putley, E. H. (1966). *Proc. Inst. Electron. Electron. Eng.* **54**, 1096.

Radford, H. E. (1975). *IEEE J. Quantum Electron.* **QE-11**, 213–214.

Radford, H. E., Petersen, F. R., Jennings, D. A., and Mucha, J. A. (1977). *IEEE J. Quantum Electron.* **QE-13**, 92–94.

Redon, M., Gastaud, C., and Fourrier, M. (1979). *IEEE J. Quantum Electron.* **QE-15**, 412–414.

Reid, J., and Siemsen, K. (1976). *J. Appl. Phys.* **29**, 250–251.

Sattler, J. P., Worchesky, T. L., and Riessler, W. A. (1978). *Infrared Phys.* **18**, 521–528.

Sattler, J. P., Worchesky, T. L., and Riessler, W. A. (1979). *Infrared Phys.* **19**, 217–224.

Serrallach, A., Meyer, R., and Günthard, Hs.H. (1974). *J. Mol. Spectrosc.* **52**, 94–129.

Siemsen, K., and Whitford, B. G. (1977). *Opt. Commun.* **22**, 11–16.

Steffen, H., and Kneubühl, F. K. (1968). *IEEE J. Quantum Electron.* **QE-4**, 992–1008.

Stern, V., Belorgeot, C., Kachmarsky, J., and Möller, K. D. (1977). *J. Mol. Spectrosc.* **67**, 244–264.

Tanaka, A., Tanimoto, A., Murata, N., Yamanaka, M., and Yoshinaga, K. (1974). *Jpn. J. Appl. Phys.* **13**, 1491–1492.

Tanaka, A., Yamanaka, M., and Yoshinaga, H. (1975). *IEEE J. Quantum Electron.* **QE-11**, 853–854.

Tobin, M. S. (1979). *Int. Conf. Infrared Millimeter Waves Their Applications, 4th, Miami Beach, Florida* Digest, pp. 169–170.

Tobin, M. S., and Jensen, R. E. (1976). *Appl. Opt.* **15**, 2023–2024.

Tobin, M. S., and Koepf, G. A. (1980). *Int. Conf. Infrared Millimeter Waves, 5th, Würzburg* Digest, pp. 306–307.

Townes, C. H., and Schawlow, A. L. (1955). "Microwave Spectroscopy," p. 96. McGraw-Hill, New York.

Wagner, R. J., Zelano, A. J., and Ngai, L. H. (1973). *Opt. Commun.* **8**, 46–47.

Weiss, C. O. (1976). *IEEE J. Quantum Electron.* **QE-12**, 580–584.

Weiss, C. O. (1977). *Appl. Phys.* **13**, 383–385.

Weiss, C. O., and Heppner, J. (1979). *J. Phys. E: Sci. Instrum.* **12**, 67–68.

Weiss, C. O., and Kramer, G. (1976). *Proc. Int. Conf. At. Masses Fundamental Constants, 5th* pp. 383–389. Plenum Press, New York.

Weiss, C. O., Grinda, M., and Siemsen, K. (1977). *IEEE J. Quantum Electron.* **QE-13**, 892–895.

Willenberg, G. D., Weiss, C. O., and Jones, H. (1980). *Appl. Phys. Lett.* **37**, 133–135.

Woods, D. R. (1970). Thesis, Univ. of Michigan, Ann Arbor, Michigan.

Worchesky, T. L. (1978). *Opt. Lett.* **3**, 232–234.

Yamanaka, M. (1976a). *Rev. Laser Eng.* **3**, 253–294.

Yamanaka, M. (1976b). *J. Opt. Soc. Am.* **67**, 952–958.

Yamanaka, M., Homma, Y. Tanaka, A., Takada, M., Tanimoto, A., and Yoshinaga, H. (1974). *Jpn. J. Appl. Phys.* **13**, 843–850.

Yamanaka, M., Tsuda, H., and Mitani, S. (1975). *Opt. Commun.* **15**, 426–428.

Young, C., Lees, R. M., Van der Linde, J., and Oliver, B. A. (1979). *J. Appl. Phys.* **50**, 3808–3810.

CHAPTER 3

Stark Spectroscopy and Frequency Tuning in Optically Pumped Far-Infrared Molecular Lasers

F. Strumia and M. Inguscio

Istituto di Fisica dell'Università
Pisa, Italy

	I.	INTRODUCTION	130
	II.	STARK EFFECT IN POLAR MOLECULES	133
		A. *Symmetric and Asymmetric Tops*	133
		B. *Stark Effect in CH_3OH*	134
		C. *Stark Effect in Ammonia*	136
		D. *Stark Effect on Rotovibrational Transitions*	137
	III.	LASER GAIN CURVES IN THE PRESENCE OF AN ELECTRIC FIELD	139
		A. *Stark Frequency Shifts*	139
		B. *Relative Intensities*	141
		C. *Symmetric Top Stark Laser Line Shapes*	142
		D. *Methanol Stark Laser Expected Line Shapes*	148
	IV.	GENERAL PERFORMANCES OF OPTICALLY PUMPED FIR LASERS	155
		A. *High Efficiency and Stability FIR Systems*	155
		B. *Conventional and Stark Waveguides*	159
		C. *Electrical Breakdown in the Active Medium*	166
		D. *FIR Power and Frequency Measurements*	169
		E. *Saturation Effects in the FIR Output*	171
	V.	EXPERIMENTAL RESULTS—DC ELECTRIC FIELDS	176
		A. *Off-Resonant Pumping of Molecules*	176
		B. *Evidence for Stark-Induced Frequency Shifts in FIR Emission*	177
		C. *Large Frequency Tuning by Stark Effect on FIR Lines of CH_3F and CH_3OH*	
		D. *Small Frequency Tuning*	178
		E. *IR–FIR Transferred Lamb Dip Spectroscopy: Stark Effect in the Pump Transition*	187
		F. *Electric Field Effects on FIR Output Level*	188
		G. *Optoacoustic Stark Spectroscopy Investigation of Pump Saturation*	192
			196

H. *High-Resolution Spectroscopy in Stark-FIR*
    Lasers: New Techniques   201
VI. EXPERIMENTAL RESULTS—AC ELECTRIC FIELDS   204
  A. *Effects on the Output Power*   205
  B. *High-Speed FM Modulation*   206
VII. *Conclusion*   208
  References   210

I. Introduction

Rotational transitions in molecules are associated with the permanent dipole moment and generally occur in the far-infrared (FIR) region, from about 30 μm to 2 mm, and in the microwave region. Microwave molecular spectra were widely investigated since the development of klystron and magnetron oscillators and the waveguide techniques. The experimental studies were generally limited to the lower molecular levels because of the lack of suitable sources in the FIR region. Before 1970, the FIR portion of the electromagnetic spectrum was only very sparsely covered by coherent monochromatic sources. Glow discharges in gases had been used to obtain FIR emission from polyatomic molecules. They had been successful with few molecular species and a rather poor list of cw laser FIR lines was available (Coleman, 1973; Chantry and Duxbury, 1974; Kneubühl and Sturzenegger, 1980). The situation changed dramatically since Chang and Bridges (1970) discovered the FIR laser action in optically pumped CH_3F molecules. A rich spectrum of laser lines throughout the 30 μm to 2 mm region is now available. Output levels of many milliwatts are easily obtained in cw operation. Power fluctuations can be limited to a few percent and frequency stability can be better than 1 part to 10^9 for carefully designed laser systems. As a consequence optically pumped FIR molecular lasers have many of the requirements to be a powerful tool for spectroscopic investigations.

The fundamental method of spectroscopy is to observe molecular response as a function of the frequency of the applied radiation. Unfortunately the frequency tunability of optically pumped molecular lasers is limited within the width of the lines, which is some 10^{-6} of their center wavelength. As a consequence up to now spectroscopy measurements have been carried out bringing molecular transitions into resonance with the fixed laser frequency by applying an external field to the investigated molecules (Uehara *et al.*, 1968; Evenson *et al.*, 1968, 1977). The usefulness of the optically pumped FIR coherent sources was expected to be significantly increased if a simple technique for frequency tuning and modulation was available. In the last years much effort has been devoted to frequency tuning FIR laser sources by applying a Stark field to the active medium. The method was suggested by Weitz and Flynn (1971); independently Inguscio *et al.* (1975, 1976) explicitly calculated the frequency tunabilities of the levels involved in the laser action and predicted the possibility of a wide frequency tuning of the FIR emission. The effectiveness of the method was demonstrated in 1977 (Inguscio *et al.*, 1977; Stein *et al.*, 1977). More recently, an improvement of 2 orders of magnitude in the frequency tunability of optically pumped FIR lasers has been reported (Strumia, 1978; Inguscio *et al.*, 1978a, 1979a, b). The method of Stark tuning the frequency of optically pumped FIR molecular lasers is quite simple and can be understood by reference to the general operating principles of these coherent sources, recently reviewed by Chang (1977) and Hodges (1978). In optically pumped FIR lasers, the radiation from a powerful IR laser is focused into a second laser resonator where it is absorbed by the molecular gas.

A partial energy-level diagram for a polar molecule showing possible FIR laser transitions and the corresponding pump transition is shown in Fig. 1.

Absorption of 10-μm CO_2 laser radiation leads to inversion between rotational levels J and $J - 1$ in an excited vibrational state. Collisions occurring at a rate of $P \cdot \tau^{-1} \approx 10^8$ torr/sec (Birnbaum, 1967; Schmidt *et al.*, 1973) tend to thermalize the population among rotational levels. For simplicity, it is assumed that the same relaxation rate τ_R^{-1} applies to both $\Delta K = 0$ processes and (normally slower) $\Delta K \neq 0$ processes (Oka, 1973). Rotational relaxation out of the lower level is therefore sufficiently rapid to prevent destruction of the inversion.

An obstacle to efficient cw operation of an optically pumped FIR laser

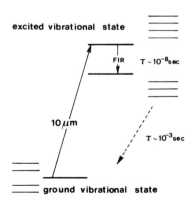

FIG. 1 Schematic energy level diagram of a polar molecule. The absorption of the 10-μm pump radiation leads to inversion between rotational levels in an excited vibrational state.

is the low rate of vibrational deactivation relative to rotational relaxation $[\tau_v \approx 10^{-3}$ sec torr, (Weitz *et al.*, 1972)]. This suggested the use of a waveguide resonator of a small cross section, the collisions with the walls being relied on to remove population from the lower laser level. In this case the deactivation time is determined by the molecular diffusion time across the tube radius ($\tau_{diff} \approx 10^{-3}$ sec torr cm). The waveguide configuration was first introduced by Hodges and Hartwick (1973) and was proved to be more efficient than conventional open resonator structures. Efficient laser action can be observed also in a rectangular metal–dielectric hybrid waveguide (Tobin and Jensen, 1976a). This structure also allows the application of an electric field (dc or ac) on the lasing medium. The molecules active in the FIR are polar and hence their levels can be shifted by the Stark effect. Operation of optically pumped molecular lasers in the presence of several kV/cm electric fields has been reported in the literature. Low electric fields can enhance the IR absorption of the gas and increase the efficiency of the FIR laser. High electric fields allow a wide frequency tuning of many strong FIR laser lines. ac electric fields are effective in producing high-speed amplitude and frequency modulation of the laser output. Optically pumped FIR molecular lasers are by themselves a very useful spectroscopic tool for studying excited vibrational states of polar molecules. The application of the electric fields extends the technique to the Stark spectroscopy of the excited states. The structures of the lines can be resolved and direct evidence for nonrandomizing $\Delta M = \pm 1$ collisions in high-vibrational levels is obtained.

The observed Stark line shapes critically depend on the quantum numbers and on the selection rules involved in the laser cycle. As a consequence, the Stark technique can provide complementary and sometime exclusive information on the molecular transitions' assignments. Saturation effects are also described. A Lamb dip, transferred from the IR pump transition, is observed in the FIR emission. The resolution and sensitivity of this new spectroscopic technique allow accurate measurements on the molecular vibrational transitions.

As a final introductory remark, it is important to stress that optically pumped FIR sources can be considered MASER oscillators. In fact the emitted lines are in general homogeneously broadened and the linewidths of the resonator modes are larger than the gain curve of the active medium. As a consequence, the frequency of the emitted radiation is determined by the molecular transitions and is only slightly perturbed by the cavity tuning via pulling effects.

This makes it possible to use FIR sources in high-resolution spectroscopic investigations without the need of active frequency stabilization systems.

II. Stark Effect in Polar Molecules

A. SYMMETRIC AND ASYMMETRIC TOPS

A general discussion of the Stark effect in molecular spectra can be found in several standard textbooks (Townes and Schwalow, 1955; Wollrab, 1967; Gordy and Cook, 1970). In this section we summarize the points which are relevant to the operation of the FIR lasers.

The polar molecules possess a permanent electric dipole moment of the order of unity when expressed in Debye (1 D = 503.44 MHz/kV cm^{-1}). For rotational and rotovibrational lines observed in the far-infrared and infrared regions, the Stark energies, at least for the electric field strengths that are usually applicable, are small compared to the level spacings and can be accurately evaluated by perturbation treatment.

For a symmetric top molecule, the Stark energies, up to second order in the applied field, are

$$\Delta W = -\mu E \frac{MK}{J(J+1)} + \frac{\mu^2 E^2}{2Bh} \left(\frac{(J^2 - K^2)(J^2 - M^2)}{J^3(2J-1)(2J+1)} \right.$$
$$\left. - \frac{[(J+1)^2 - K^2][(J+1)^2 - M^2]}{(J+1)^3(2J+1)(2J+3)} \right), \tag{1}$$

where μ is the electric dipole moment of the level, directed along the symmetry axis of the molecule, J is the total angular momentum, K and M its components, respectively, directed along the symmetry axis of the molecule and the external field E, and B is the rotational constant of the level. The second-order energy is generally much smaller than the first-order one, so that the second term on right-hand side of Eq. (1) can be neglected for the electric field strengths usually applicable. It is worth noting that a linear molecule can be treated as a special case of symmetric top with $K = 0$. As a consequence, linear molecules exhibit only a small quadratic Stark effect. The presence of a nuclear quadrupole coupling can introduce new structures in the spectrum, but the order of magnitude of the Stark energy is not affected.

The case of asymmetric top molecules is more complex since K is no longer a good quantum number. In general the levels are not degenerate and the Stark effect is of the second-order type proportional to E^2.

The general expression for the Stark energy is

$$\Delta W = (A + BM^2)E^2, \tag{2}$$

where A and B are functions of the molecular constants. A first-order Stark effect can be observed also in asymmetric top molecules when (i) a degeneration between $+K$ and $-K$ is restored by further interactions (e.g., internal rotation), (ii) in slightly asymmetric rotors, the K splitting is small enough to be neglected, and (iii) some accidental degeneracy occurs between levels with $\Delta J = 1$.

Methyl alcohol, which plays an important role in conventional and Stark-tuned optically pumped FIR lasers, is a typical example of a slightly asymmetric molecule exhibiting both linear and quadratic Stark effect.

B. STARK EFFECT IN CH₃OH

Methyl alcohol is a slightly asymmetric-top molecule; it is also one of the lightest molecules with hindered internal rotation. The hydrogen attached to oxygen (Fig. 2) has three possible positions of equal energy with respect to the CH_3 group. The O–H group may rotate around the symmetry axis of the CH_3 group and is subject to a threefold degenerate hindering potential with a barrier height of about 400 cm⁻¹. As a consequence the energies of the rotational levels of CH_3OH are largely modified by the hindered rotation–torsional motion and new quantum numbers must be introduced in order to fully identify the states.

Following the notations by Ivash and Dennison, (1953), Lees and Baker, (1968), and Kwan and Dennison (1972), we shall label the rotation–torsional levels by $(n, \tau, K, J)^v$, where n is the torsional quantum number, $\tau = 1, 2, 3$ identifies the tunneling state of the internal rotation, K and J are the usual momentum quantum numbers, and v denotes the vibrational state. The states belong to A, E_1, E_2 symmetries for $\tau + K = 3N + 1, 3N, 3N + 2$, respectively, N being integral. The E states are still

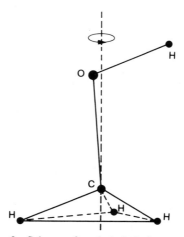

FIG. 2   Scheme of methyl alcohol structure.

doubly degenerate on K, while the A states display a K splitting increasing with J, but dramatically decreasing with increasing K. The lowest order Stark effect may therefore be either linear or quadratic. The linear effect is shown by E states and by the A states for which the K splitting cannot be resolved. The quadratic Stark effect occurs for A states when the K splitting is larger than the Stark energy. The actual evaluation of the Stark shift for CH_3OH levels is complicated by the presence of a transverse component of the permanent electric dipole moment of the molecule. As usual, we shall denote the electric dipole component along the CH_3 symmetry axis by μ_a, where μ_b denotes the dipole component orthogonal to this axis. The μ_a contribution to the Hamiltonian is diagonal in all the quantum numbers n, τ, K, J, and the energy shift of the M sublevels is given by

$$\Delta W = -\mu_a E \frac{kM}{J(J+1)}. \tag{3}$$

This term would be the only one appearing for a symmetric top. Off-diagonal terms are introduced by μ_b. It can be shown (Burkhard and Dennison, 1951; Ivash and Dennison, 1953) that also these off-diagonal terms lead to a first-order Stark shift:

$$\Delta W_b = \mu_b \frac{EM}{J(J+1)} f(K, \tau, n, J, C, D_{ab}, B), \tag{4}$$

where f is a function of the quantum numbers defining the state and of the quantities C, B, and D_{ab}, which are related to the moments of inertia and to the cross moment of inertia of the molecule (Kwan and Dennison, 1972). An explicit expression for the function f can be found in Henningsen (1980a). For methyl alcohol, ΔW_b is a small fraction (a few percent) of the contribution through Eq. (3) of μ_a. By summing the two linear terms, we have for the Stark energies (Henningsen, 1980a):

$$W(n, \tau, K, J) = -\frac{EMK}{J(J+1)} \mu(n, \tau, K, J), \tag{5}$$

where $\mu(n, \tau, K, J) = A + BJ(J+1) \approx \mu_a$. Thus neglecting μ_b corresponds to an error of the same order as the uncertainty on the known values of the dipole moments (see Table I). Therefore Eq. (5) is a satisfactory approximation for the linear Stark shift, especially for the excited levels, where the accuracy of the dipole values is lower.

An A state with K doubling is split, at zero electric field, into two sublevels with energy U_1 and U_2, respectively. The energy difference

TABLE I

PERMANENT ELECTRIC DIPOLE MOMENT EXPERIMENTAL VALUES FOR SOME OF THE
MOLECULES USED IN OPTICALLY PUMPED STARK FIR LASERS

Vibrational state	CH_3F μ (Debye)	References
ν_3 band		
0	1.858 40 (2)	Marshall and Muenter (1980)
1	1.9077 (10)	Brewer (1973)
1	1.905 36 (56)	Freund et al. (1974)

	NH_3	
ν_2 band		
0	1.47	Shimizu (1969)
1	1.25	Shimizu (1969)

	CH_3OH		
CO stretch	μ_a (Debye)	μ_b (Debye)	
0	0.893	1.435	Burkhard and Dennison (1951)
0	0.885	1.44	Ivash and Dennison (1951)
1	0.90		Lees (1972)
1	0.912		Bedwell et al. (1978)

$U_1 - U_2$ is given by

$$U_1 - U_2 = S(K)(J + K)!/(J - K)!, \tag{6}$$

where the constants $S(K)$ decrease very rapidly with increasing K (Ivash and Dennison, 1953; Arimondo et al., 1980). Diagonalization of the Hamiltonian in the presence of an electric field leads to the energy eigenvalues,

$$W = \frac{U_1 + U_2}{2} \pm \left[\left(\frac{U_1 - U_2}{2} \right)^2 + \left(\mu_a \frac{EKM}{J(J + 1)} \right)^2 \right]^{1/2}. \tag{7}$$

C. STARK EFFECT IN AMMONIA

Equation (7) gives also the Stark energies of the M sublevels in the case of a symmetric top molecule with an inversion spectrum like NH_3. In this case the difference $U_1 - U_2$ is the inversion splitting of the states. In the ammonia ground state the inversion energy is about 0.8 cm^{-1}, of the same order of the quantity $\mu EMK/[J(J + 1)]$ and as a consequence large Stark effects can be easily observed. On the contrary, in the ν_2 excited state responsible for the FIR laser emissions, the energy inversion splitting is about 36 cm^{-1} and the Stark effect is negligible. In the case of PH_3, which also is a FIR lasing molecule (Malk *et al.*, 1978), the inversion frequency is very small (a few MHz) and the Stark effect is linear as in a usual symmetric top.

D. STARK EFFECT ON ROTOVIBRATIONAL TRANSITIONS

The frequency shift for any transition as a consequence of the Stark effect can thus be evaluated from the energy shift of its upper and lower level, according to the quantum numbers and symmetries of the involved levels. It is worth noting that the upper and lower levels of a transition can exhibit different μ values, μ_e and μ_o, respectively; this difference can be important when the vibrational quantum numbers of the two levels are different.

Different behaviors are observed for the frequency shift according to the energy shifts of the levels. When both levels display a linear shift, the resulting frequency shift is linear. When both levels display K doubling and the Stark energy does not exceed the splitting energy of either level, the resulting frequency shift is quadratic. When both levels display K doubling but the Stark energy can exceed the splitting energies, the dependence of the frequency shift on the electric field becomes more complex, however, it can still be easily computed by use of Eq. (7). An interesting case occurs when one of the levels displays a linear Stark shift, while the energy of the other is given by Eq. (7).

The selection rules for the pump vibrational transitions and the FIR lasing transitions are as follows:

(i) *Symmetric top*
   Rotational lines:

$$\Delta J = \pm 1, \Delta K = 0, \Delta M = 0, \pm 1.$$

   Vibrational lines:

$$\Delta J = 0, \pm 1; \Delta M = 0, \pm 1;$$
$$\Delta K = 0 \text{ for parallel bands;}$$
$$\Delta K = \pm 1 \text{ for perpendicular bands.}$$

(ii) *Ammonialike molecules*
   Rotational lines:

$$\Delta K = 0, a \leftrightarrow s, \Delta M = 0, \pm 1;$$
$$\Delta J = 0 \ (K \neq 0) \quad \text{inversion transitions;}$$
$$\Delta J = \pm 1 \quad \text{rotation–inversion transitions.}$$

   Vibrational lines:

$$\Delta J = 0, \pm 1; \Delta K = 0; a \leftrightarrow s; \Delta M = 0, \pm 1 \quad \text{for parallel bands.}$$

In presence of a Stark field, a large mixing occurs in the ground state inversion levels and the $a \leftrightarrow s$ selection rule can be violated.

(iii) *Methyl alcohollike molecules*
Vibrational lines:

$$\Delta J = 0, \pm 1; \ \Delta K = 0; \ \Delta M = 0, \pm 1; \ \Delta n = 0; \ \Delta \tau = 0$$

for parallel bands

Rotational lines:

$$\Delta J = 0, \pm 1; \ \Delta M = 0, \pm 1; \ A \leftrightarrow A; \ E_1 \leftrightarrow E_1; \ E_2 \leftrightarrow E_2;$$
$$\Delta n = \text{any and } \Delta K = \pm 1 \text{ or } \Delta n = 0 \text{ and } \Delta K = 0, \pm 1.$$

The selection rule $\Delta K = \pm 1$ is owing to the presence of a component of the electric dipole moment orthogonal to the symmetry axis (see Table I).

The number of possible FIR lasing transitions for each pump coincidence increases from case (i) to case (iii), as shown in Fig. 3.

The liberality of the selection rules makes methyl alcohol the most interesting molecule both for conventional and Stark operation.

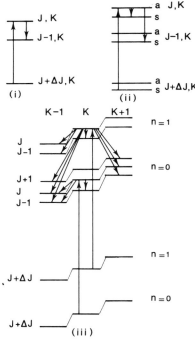

FIG. 3 Schematic illustration of the possible IR and FIR transitions: (i) symmetric top, (ii) ammonialike, (iii) methyl alcohollike molecules.

III. Laser Gain Curves in the Presence of an Electric Field

A. STARK FREQUENCY SHIFTS

Although the frequency shifts of the various line M components are completely determined by the expression given in the Section II, the analysis of the Stark laser operation also requires knowledge of the line intensities. In particular the frequency dependence of a small signal gain of a FIR lasing transition in presence of a Stark field is a function of many parameters like pump and FIR lasing selection rules: the relative M intensity, the intensity and polarization of the pump radiation, the FIR laser cavity losses, the active medium density and temperature, etc. However, we will show that general behavior of each lasing line can be grouped in a few typical cases.

The result is relevant to the problem of the assignment of the unknown transitions. From the previous discussion it follows that for most cases the line frequency shift can be computed using Eq. (5).

If the levels involved in the transition belong to different vibrational states the μ values are different by typically a few percent (3%–10%) and

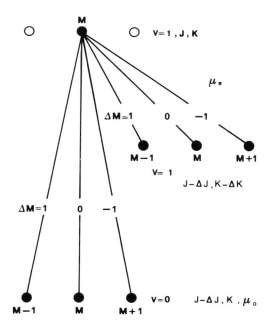

FIG. 4 Labeling scheme adopted for the M sublevels in the ground and in the excited states of lasing molecules.

this has to be taken into account. For levels belonging to the same vibrational state the variations in μ are much smaller for symmetric top molecules, and of the order of 1%–2% in the case of CH_3OH, because of the contribution of the μ_b component. Hence in CH_3OH we can neglect, to first order in μ, the variation of μ whose contribution will be taken into account as a small correction to the computed gain profile.

We consider also only parallel bands ($\Delta K = 0$) for the pump transition since this is the case for all experimental results. We are dealing with a three-level system, and an unambiguous labeling of the levels is necessary. From now on we will follow the scheme of Fig. 4, where all the quantum numbers are referred to the upper level.

For the pump line Stark frequency shift we then have:

(i) $J - 1 \rightarrow J$ (^aR branch):

$$\Delta\nu = E \frac{K}{J(J^2 - 1)} \{[\mu_0(J + 1) - \mu_e(J - 1)]M - \mu_0(J + 1)\Delta M\}; \quad (8)$$

(ii) $J \rightarrow J$ (^aQ branch):

$$\Delta\nu = E \frac{K}{J(J + 1)} [(\mu_0 - \mu_e)M - \mu_0\Delta M]; \quad (9)$$

and
(iii) $J + 1 \rightarrow J$ (^aP branch):

$$\Delta\nu = E \frac{K}{J(J + 1)(J + 2)} \{[\mu_0 J - \mu_e(J + 2)]M - \mu_0 J\Delta M\}. \quad (10)$$

For the FIR lasing lines we have seven possible cases

(i) $\Delta K = 0; J \rightarrow J - 1$:

$$\Delta\nu = \mu E \frac{K}{J(J^2 - 1)} [2M - (J + 1)\Delta M]; \quad (11)$$

(ii) $K \rightarrow K - 1; \Delta J = 0$:

$$\Delta\nu = \frac{-\mu E}{J(J + 1)} [M + (K - 1)\Delta M]; \quad (12)$$

(iii) $K \rightarrow K + 1; \Delta J = 0$:

$$\Delta\nu = \frac{\mu E}{J(J + 1)} [M - (K + 1)\Delta M]; \quad (13)$$

(iv) $K \rightarrow K - 1; J \rightarrow J + 1$:

$$\Delta\nu = \frac{-\mu E}{(J + 1)(J + 2)} \left(\frac{J + 2K}{J} M + (K - 1)\Delta M\right); \quad (14)$$

(v) $K \rightarrow K + 1; J \rightarrow J - 1$:

$$\Delta\nu = \frac{\mu E}{J(J - 1)} \left(\frac{J + K + 1}{J + 1} M - (K + 1)\Delta M \right); \qquad (15)$$

(vi) $K \rightarrow K - 1; J \rightarrow J - 1$:

$$\Delta\nu = \frac{\mu E}{J(J - 1)} \left(\frac{2K - J - 1}{J + 1} M - (K - 1)\Delta M \right); \qquad (16)$$

(vii) $K \rightarrow K + 1; J \rightarrow J + 1$:

$$\Delta\nu = \frac{\mu E}{(J + 1)(J + 2)} \left(\frac{J - 2K}{J} M - (K + 1)\Delta M \right). \qquad (17)$$

For symmetric-top molecules we have only the case of Eq. (11), while the most common cases for CH_3OH are that of Eqs. (11), (12), and (16).

Of particular interest are the last two cases since for $K \approx J/2$ the first term in the bracket becomes negligible and all the M components collapse in only two close packets corresponding to $\Delta M = \pm 1$. In this case, the FIR lasing line is seen to split into only two components when a Stark field is applied to the active medium.

This is the case of many of the strongest laser lines of CH_3OH, as will be discussed in the Section II.B.

B. RELATIVE INTENSITIES

As a second step we have to consider the relative strength of the M components of a given transition. This depends only on the quantum numbers J and M and a general equation can be given for each selection rule as shown in Table II.

Should be population of the upper state be the same for all the M sublevels, the small signal gain of each M transition would be proportional to the quantities given in Table II and only the line for which the gain is above the threshold would lase. This is the case when a rapid collisional mixing between the M sublevels keeps the populations equalized. The experimental results are in disagreement with this assumption and clearly

TABLE II

INTENSITY FACTORS FOR VARIOUS ΔJ, ΔM MOLECULAR
ROTATIONAL AND VIBRATIONAL TRANSITIONS

	$\Delta M = 0$	$\Delta M = \mp 1$
$J \rightarrow J$	M^2	$(J \pm M)(J \mp M + 1)$
$J \rightarrow J - 1$	$J^2 - M^2$	$(J \pm M)(J \pm M - 1)$
$J \rightarrow J + 1$	$(J + 1)^2 - M^2$	$(J \mp M + 1)(J \mp M + 2)$

demonstrate that J changing collisions have a larger cross section as shown in Section V.C. As a consequence, each M sublevel participates with the optically pumped three-level system without any sensible mixing with the other M sublevels with the exception of that allowed by the electric dipole radiative transition selection rules $\Delta M = 0, \pm 1$. The excitation rate in the upper level is then proportional to the equations of Table II and to the pump saturation, which is also dependent upon the relative intensities as given in Table II. In conclusion, the FIR laser gain of each M Stark component is also a function of the polarization, relative to the Stark field, and of the intensity of the pump radiation. In the case of FIR lasers optically pumped with a cw CO_2 or N_2O laser, the pump power intensity inside the FIR cavity is small (a few W/cm^2) and the saturation degree is also small with the exception of the strongest M components. As a first approximation, we can disregard the saturation and compute the gain of the FIR laser lines by multiplying the relative intensity of the pump and FIR lines as given by Table II and taking into account the correct selection rules regarding the polarization of the IR and FIR electromagnetic fields with respect to the Stark field. The pump saturation and other effects can be considered at a better approximation by writing the correct rate equation for each M sublevel. However, this second step is unnecessary for understanding the general behavior of the Stark effect in FIR lasers and in the following examples will be disregarded. As will be discussed in Section IV.B, Stark operation of FIR lasers has been achieved by using parallel plates waveguide or a hybrid waveguide. In both cases the polarization of the electric field of FIR radiation is forced to be orthogonal to that of the static electric field. As a consequence, only the selection rules $\Delta M = \pm 1$ are allowed for the FIR lasing transition, while the pump radiation can be either parallel or orthogonal to the Stark field. Therefore only two orthogonal cases are observed experimentally: (i) pump, $\Delta M = 0$; FIR, $\Delta M = \mp 1$; and (ii) pump, $\Delta M = \mp 1$; FIR, $M = \mp 1$. The absorption of the pump radiation in the FIR cavity is larger in case (i) (see Section III.B), and FIR laser action is generally easier in case (ii).

C. Symmetric Top Stark Laser Line Shapes

As an example of the behavior of a symmetric top molecule, we will consider the case of the popular 496-μm CH_3F FIR laser line optically pumped by the 9–P(20) line of the CO_2 laser. The pump transition is assigned as $\nu_3 \leftarrow 0$, $Q(12,2)$. The Stark effect is given by Eq. (9). Substituting the quantum numbers' values and the dipole values reported in Table I we have

$$\Delta\nu(Q_{12,2}) = (-0.302M - 11.996\Delta M) \quad kHz/V\ cm^{-1}. \tag{18}$$

The Stark tunability is shown in Fig. 5. For each ΔM value, only the Stark components corresponding to the extremal M values are shown, the other lines being equally spaced between them. The *rotational transition* is assigned as ν_3, $(J, K):(12, 2 \rightarrow 11, 2)$.

The first-order Stark tunability can be computed using Eq. (11). It is

$$\Delta\nu_{\text{FIR}} = (2.236M - 14.534\Delta M) \quad \text{kHz/V cm}^{-1}. \tag{19}$$

The tunability is schematically shown in Fig. 6. For each ΔM value, only the Stark components corresponding to the extremal M values are shown. For $\Delta M = 0$ one obtains 23 lines. For $\Delta M = \mp 1$ one obtains 46 lines, but the lines with $\Delta M = -1$, $-12 < M < -1$ have the same frequency shifts as the lines with $\Delta M = +1$, $+1 < M < 12$, and only 34 lines with $\Delta M = \mp 1$ are resolved.

From Eqs. (18) and (19) it follows that the frequency shift in the vibrational lines is the smaller one, by about 1 order of magnitude, as a consequence of the term in $(\mu_0 - \mu_e)$. This result is important because, considering also that the Doppler broadening is larger for the vibrational transition, shows that efficient CO_2 pumping is possible even in presence of a strong electric field.

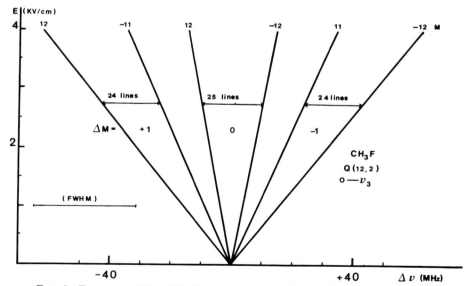

FIG. 5 Frequency shifts of the Stark components of the ν_3 fundamental line of CH_3F. Only the lines corresponding to the extremal M values are shown. The others are equally spaced between them.

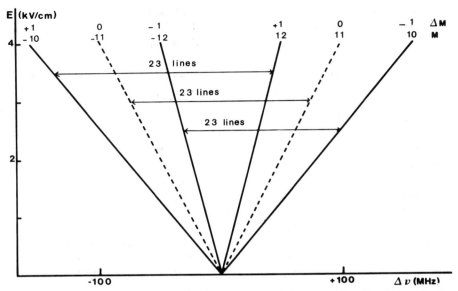

FIG. 6   Frequency shifts of the Stark components of the $\nu_3 = 1$, $J = 12-11$, $K = 2$ rotational line of CH_3F. Only the lines corresponding to the extremal M values are shown.

The small signal gain of the FIR laser for each M sublevel in absence of saturation can be computed by using the equations of Table II.

If we assume a fast collisional randomization between the M sublevels of the excited state, the intensities of the different components are proportional to

$$I(M \rightarrow M \mp 1) \propto (J \pm M)(J \pm M - 1) \qquad (20)$$

and are independent of the polarization direction of the CO_2 pump radiation. If no randomization occurs, we must take into account the polarization of the pump radiation. For simplicity we consider only the two orthogonal cases of the pump radiation linearly polarized parallel or orthogonal to the Stark E field. The gain is then proportional to

$$I(M \rightarrow M \mp 1) \propto M^2(J \pm M)(J \pm M - 1) \qquad (21)$$

for the case $E\,(CO_2) \parallel E$ and

$$I(M \rightarrow M \mp 1) \propto [J(J + 1) - M^2](J \pm M)(J \pm M - 1) \qquad (22)$$

for the case $E\,(CO_2) \perp E$.

In the first case the Stark effect in the pump transition is small [Eq. (18)], much smaller than the Doppler linewidth, so that we can neglect the frequency dependence of the absorption coefficients.

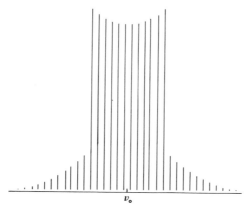

FIG. 7 Relative intensity of the small signal gain of the Stark components assuming a complete randomization between the M states of the $J, K = 12, 2$ level. $J = 12–11$; $\Delta M = \mp 1$.

On the contrary care must be taken in the second case. Then sum over $\Delta M = -1$ and $\Delta M = +1$ pump selection rules must be used for moderate Stark fields, but the IR absorbing line is tuned out from its unperturbed frequency as the Stark field is increased (Fig. 5) so that for high field only one is effective in the laser action.

In Figs. 7, 8, and 9 we show the small signal gain corresponding to Eqs.

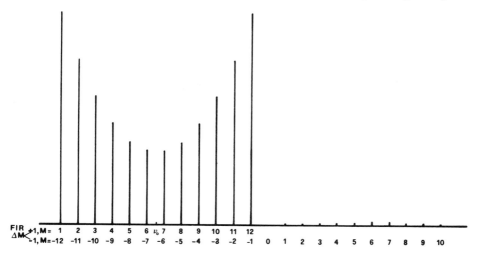

FIG. 8 Relative intensity of the small signal gain of the Stark components assuming no randomization between the M states and a CO_2 pump absorption with the selection rule $\Delta M = 0$. The structure is symmetric around ν_0 but the low-intensity components with $M < 0$ are not reported for sake of simplicity.

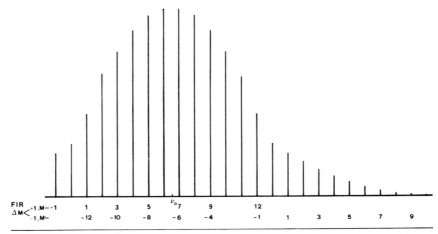

FIG. 9 Relative intensity of the small signal gain of the Stark components assuming no randomization between the M states and a CO_2 pump absorption with the selection rule $\Delta M = \mp 1$. The components with $M < 0$ are not shown.

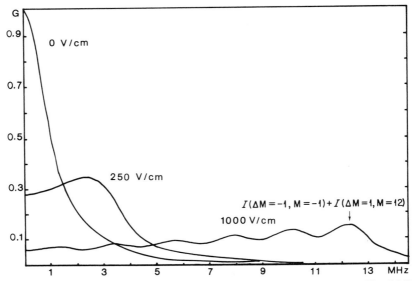

FIG. 10 Computed gain profile of the CH_3F laser line at 496 μm, (J, K): $(12,2) \rightarrow$ $(12, 2) \rightarrow (11, 2)$, in presence of electric field; $\Delta M = 0$ pump transition and $\Delta M = \pm 1$ FIR transition selection rules. A homogeneous linewidth of 2 MHz (FWHM) has been assumed. The curves are symmetric around ν_0, and only one-half is shown. The gain is normalized to 1 for zero-field curve.

(20), (21), and (22), respectively. In the last case we have summed over both possible pump selection rules $\Delta M = \mp 1$. As we will see in Section V.C, the experimental results are in agreement with the hypothesis of no collisional mixing between M sublevels. This is a very important result since it is the key for explaining many of the effects associated with the Stark operation of FIR lasers. The small signal gain of the FIR laser can also be computed by taking into account the FIR line width which is homogeneous as a consequence of pressure broadening (~ 40 MHz/torr for CH_3F). In Fig. 10 we show the result for pump selection rule $\Delta M = 0$ and for three values of the Stark field. A comment is necessary in the case of Fig. 10: for relatively small electric fields, it is possible to observe experimentally a splitting of the FIR line in a doublet, depending on the threshold level of the system. This misleading result can be corrected by recording the Stark profile at higher fields.

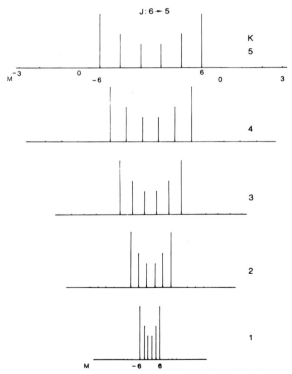

FIG. 11 Computed Stark patterns assuming a ^aQ branch, $\Delta M = 0$ pump transition and $J \rightarrow J - 1, \Delta K = 0$ FIR transition selection rules. The patterns are computed for $J = 6$ and different K values.

D. Methanol Stark Laser Expected Line Shapes

The selection rule $\Delta K = \mp 1$ is also allowed in this case, and the Stark gain pattern is more varied than with a symmetric top. Obviously K does not enter in the relative intensities of the M components but only in their frequencies, nevertheless, it causes some remarkable effects. On the other hand, the Stark pattern is independent of the quantum numbers n and τ. For simplicity we will consider only the case where all the involved levels have $n = 0$. Three FIR laser lines are then normally observed for each absorbing transition with the selection rules as depicted in Fig. 4. We have considered as an example the case of an upper level with $J = 6$.

In the case of a Q branch pumping line ($\Delta J = 0$), the gain Stark pattern is shown in Figs. 11, 12, and 13 when the pump rule is $\Delta M = 0$ and the FIR emission is $J \to J - 1$, $\Delta K = 0$ (Fig. 11); $\Delta J = 0$, $K \to K - 1$ (Fig. 12); and $J \to J - 1$, $K \to K - 1$ (Fig. 13), respectively.

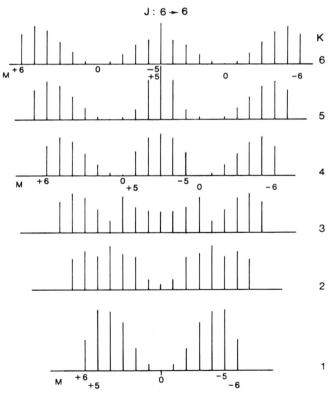

$J : 6 \to 6$

FIG. 12   As in Fig. 11, but with $\Delta J = 0$, $K \to K - 1$ FIR transition selection rules.

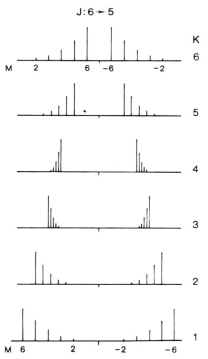

FIG. 13   As in Fig. 11, but with $J \to J - 1$, $K \to K - 1$ FIR transition selection rules.

In Figs. 14, 15, and 16, we have considered the same cases but when the pump polarization is orthogonal to the Stark field. In particular we have considered for the pump only the $\Delta M = +1$ case (see Fig. 4). The other case (pump width $\Delta M = -1$) is simply symmetric and, for a small electric field when the pump is the sum of both cases, the correct Stark intensity pattern can be obtained by summing to the patterns in Figs. 14, 15, and 16 their symmetricals. The most interesting result is that of Figs. 13 and 16 since for $K \sim J/2$ the M components are grouped together in two groups and the frequency separation between the components of each group is much smaller than the separation between the groups. As a consequence, the FIR laser emission in presence of a Stark field splits in only two components whose frequency corresponds to that of the strongest M transition of each group. This situation is very favorable for obtaining an easy Stark frequency tuning of the FIR laser radiation. The lines with the $\Delta J = \Delta K = \mp 1$ selection rules are also the most intense in the CH_3OH spectrum and are subject to the largest enhancement effect when the pump

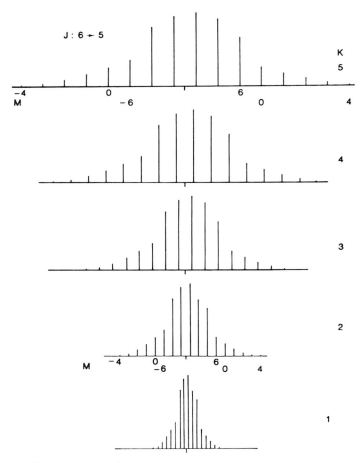

FIG. 14   As in Fig. 11, but with $\Delta M = +1$ pump selection rule.

selection rule is $\Delta M = \mp 1$ (see Section V.F). When $\Delta J = \Delta K = -1$, J odd and $\Delta J = \Delta K = +1$, J even, the degeneracy is complete to the first order as shown in Fig. 17 for the case $J = 5$ and $\Delta M = +1$ pump selection rules. From Figs. 13, 16, and 17, we can also observe that the value of the doublet splitting of the FIR radiation depends significantly on the K value when the pump is $\Delta M = 0$, while it is practically independent of it when the pump is $\Delta M = \mp 1$. As a consequence the ratio of the splittings in the two cases is a significant help for the correct assignment of the K value of

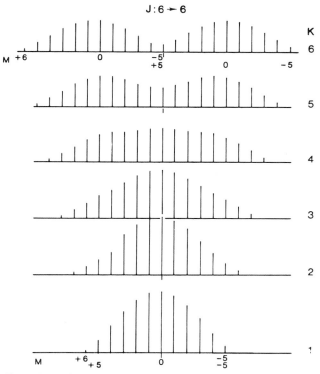

FIG. 15   As in Fig. 12, but with $\Delta M = +1$ pump selection rule.

the transition. When the pump line belongs to a P or R branch, the relative intensity of the M components is different. However the intensity patterns for the FIR transitions are similar to that shown for the Q branch pump. The most relevant difference is that the patterns for the $\Delta M = 0$ case of P and R branch lines is similar to that of the $\Delta M = \mp 1$ case of the Q branch lines and the contrary. In Figs. 18, 19, and 20, we show a few cases for a better evidence. Any particular case can be easily computed as a product of the two pertinent terms in Table II.

The quadratic Stark effect causes only very small frequency shifts. Large effects can be observed only when the two terms under the radical in Eq. (7) become of the same order of magnitude for practical values of the electric field. This is, for example, the case of the ground vibrational state of NH_3 and of some of the A states of CH_3OH. When the second

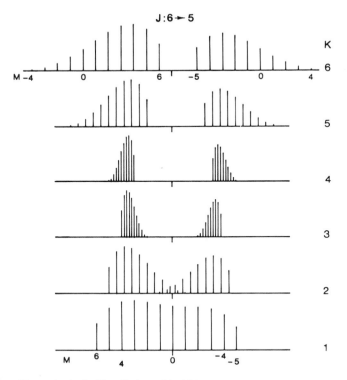

FIG. 16   As in Fig. 13, but with $\Delta M = +1$ pump selection rule.

term under radical sign becomes much larger than the first, the Stark effect is asymptotically linear.

In conclusion, the behavior of the FIR laser lines when a Stark field is applied on the active medium can be classified in a few general cases:

(a)   No Stark effect at all, either in frequency or intensity. This happens when $K = 0$ on all the three levels involved in the laser cycle or when there is a large splitting in all the levels (>1 cm^{-1}).

(b)   The lines split into two Stark components. Beside the obvious case $J: 1 \rightarrow 0$; $K: 1 \rightarrow 0$, which normally corresponds to transitions in the microwave region, this is the case when $\Delta J = \Delta K = \mp 1$ and $K \sim J/2$, the second condition being more and more relaxed as the J value increases. When the pump selection rule is $\Delta M = \mp 1$, a large enhancement effect is always observed (Section V.F).

(c)   The Stark pattern is resolved in more than two lines. This behavior

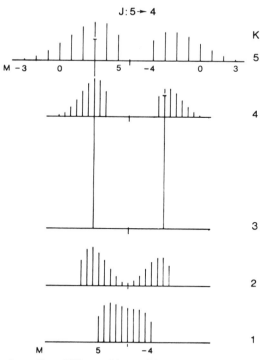

FIG.17 Stark patterns for a FIR transitions with $J: 5 \rightarrow 4$, $K \rightarrow K - 1$, and $\Delta M = +1$ selection rules.

is independent of the values of K, ΔK, and ΔJ and is displayed whenever the J values involved in the laser transition are relatively small to preserve enough gain when the most intense M components are completely resolved. The number of the observed laser lines is smaller than the expected one since the gain of the weaker components is normally below the threshold level.

(d) No frequency shift or splitting of the laser line is observed, and the emitted power decreases rather rapidly with the applied electric field. This behavior is observed for high J values and whenever one of the following conditions is present: (i) any K, $\Delta J = 1$, $\Delta K = 0$; (ii) any K, $\Delta J = 0$, $\Delta K = 1$; or (iii) K far from $J/2$, $\Delta K = \mp 1$.

(e) A quadratic Stark effect is observed whenever zero field level splittings are present, provided that the matrix elements of the Stark–

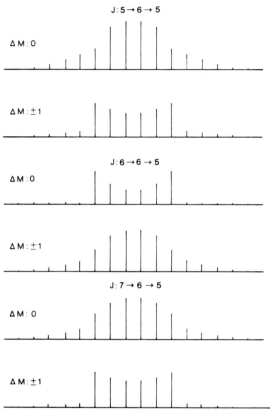

FIG. 18 Comparison between Stark patterns for ^qP, ^qQ, and ^qR pump transitions and J:
$6 \rightarrow 5$; K: $3 \rightarrow 3$ for the FIR transition.

Hamiltonian between the doublet state are nonzero and the energy splitting is moderate (<1 cm^{-1}).

Typical examples are some A states of CH$_3$OH (K doubling). It is also worthwhile to recall again that for certain cases a doublet splitting can be observed for low Stark fields also in the cases (c) and (d). However, in this case the splitting can be observed with only one pump polarization and by increasing the field we obtain very soon cases (c) or (d).

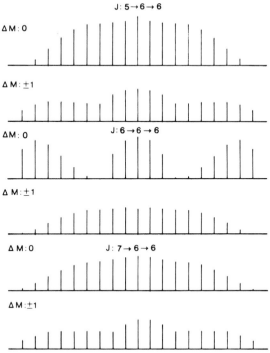

FIG. 19   As in Fig. 18, but J: $6 \to 6$ and K: $4 \to 3$ for the FIR transition.

IV.   General Performances of Optically Pumped FIR Lasers

A.   HIGH EFFICIENCY AND STABILITY FIR SYSTEMS

Most of the improvements in the performances of cw FIR lasers can be ascribed to increases in cavity efficiency at the pump and far-infrared frequencies. The most commonly used laser resonators have been reviewed recently by Chang (1977), Yamanaka (1977) and Hodges (1978). The conventional Fabry–Pérot open resonator structure was introduced in the original work by Chang and Bridges (1970). Hodges and Hartwick (1973) operated for the first time an optically pumped FIR laser in waveguide resonator structures. Their relative compactness, favorable dimensions for wall deexcitation collisions, and allowable variety of output coupling techniques are among the advantages of the waveguide structures. The input coupling hole for the infrared pump radiation is a common

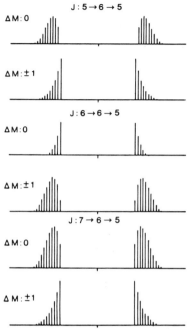

FIG. 20   As in Fig. 18, but J: $6 \rightarrow 5$ and K: $4 \rightarrow 3$ for the FIR transition.

experimental feature. In addition to the hole configuration, hybrid output coupling techniques were introduced by Danielewicz et al. (1975).

It is worth noting that good performance in terms of power has been reported by Fesenko and Dyubko (1976), using the simple hole output coupling configuration in a Fabry–Perot resonator. The hole output coupling has the advantage of low losses in the submillimeter region and allows operation over a broad spectral range. On the other hand, the output from the hole can be highly diverging and it could be difficult to optimize the laser output by changing the diameter of the coupling hole because the laser mode can distort or switch to a higher order transverse mode (Danielewicz et al., 1975). These difficulties were partially overcome offcentering the output hole and increasing its diameter (Stein et al., 1977). The output coupling at a given wavelength can be optimized by changing the hole diameter and offcentering it, as discussed by Inguscio (1979). Performances in terms of power comparable with those obtained by hybrid output configurations have been reported (Inguscio et al., 1979b) for a hollow cylindric waveguide configuration using holes for the

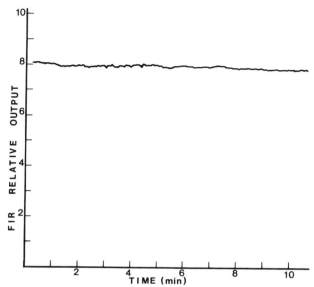

FIG. 21 Typical output power stability for the 1-m long dielectric waveguide laser. The pump line is the $9-P(36)$ (9 W) and the CH_3OH FIR output at 119 μm is about 10 mW at optimum pressure. The time constant of the detector is 0.3 sec. Both CO_2 and FIR laser are free running (Inguscio et al., 1979b).

output coupling. It is difficult to compare data available in the literature and then select an optimum laser geometry for a particular system. A competent analysis has been performed by Hodges et al. (1977).

High efficiency and stability are necessary features for a FIR laser system to be used in Stark or conventional spectroscopy. Low thresholds are required since (see Sections III.B,C) the laser gain curve is significantly broadened and lowered by the application of an electric field. Satisfactory designs of the FIR cavities are reported to have been obtained using invar frames and thermally compensated mirror mountings. A typical laser stability performance is shown in Fig. 21. Amplitude fluctuations were less than 4%. It is worth noting that both the CO_2 and FIR lasers were operated free running. The frequency instability of the CO_2 laser affects especially the FIR laser long-term stability. Power drift can be dramatically reduced by frequency stabilizing the pump CO_2 laser, as shown in Fig. 22. For that measurement the FIR cavity was set to maximum output power and then allowed to run freely in normal laboratory environmental conditions. Poor sensitivity of the pyroelectric detector mainly

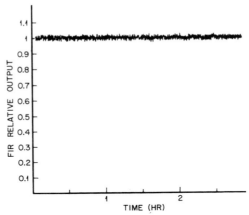

FIG. 22 Typical output power long term stability for a 1.5-m long dielectric waveguide CH$_3$F laser at 496 μm. The detector time constant is 0.3 sec. An active feedback is used to control pump frequency. Pump power is 10 W and the FIR output level is 1.5 mW (Bionducci *et al.*, 1979).

caused the fast signal noise. As for spectral purity, short-term frequency fluctuations can be limited to less than 10^{-9} in free-running laser systems.

A typical result for CH$_3$F laser at 496 μm is shown in Fig. 23. Frequency instabilities down to 2×10^{-12} have been demonstrated in apparatus especially designed for metrologic purposes (Plainchamp, 1979).

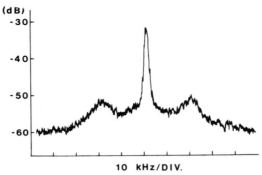

10 kHz/DIV.

FIG. 23 Beat note obtained by mixing a CH$_3$F laser with the eighth harmonic of a 75.5-GHz klystron in a point contact diode (beat width, 1 kHz). The full width of the signal at -3 dB is about 2 KHz. The klystron is phase locked to a 5-MHz low noise crystal oscillator and the small noise sidebands come from phase noise in the microwave source. Total scanning time is 10 sec (Inguscio *et al.*, 1979b).

The frequency stability performances of optically pumped FIR lasers have been recently reviewed by Jimenez (1979).

B. CONVENTIONAL AND STARK WAVEGUIDES

The field configurations and the propagation constants of the modes of hollow circular dielectric or metal waveguides were first analyzed by Marcatili and Schmeltzer (1964). Low-loss waveguide modes require small Fresnel numbers and a large ratio of diameter to wavelength. In a waveguide laser the radiation does not obey the laws of free-space propagation. For instance, phase shifts and losses depend on the diameter of the tube and on the complex refraction index of the walls, while in the conventional two-mirror resonator these quantities essentially depend on diffraction. A review on waveguide lasers was published by Degnan (1976) and Abrams (1979). More recently, the waveguides for submillimeter-wave lasers were discussed by Kneubühl (1977), Yamanaka (1977), and by Kneubühl and Affolter (1979).

For Stark operation of optically pumped FIR lasers, the three solutions schematically shown in Fig. 24 have been adopted: (i) hollow dielectric waveguide with external Stark plates, (ii) two parallel plates inside a Fabry–Perot cavity, and (iii) rectangular hybrid metal–dielectric

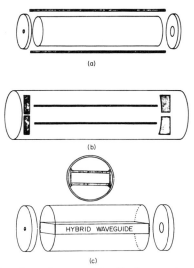

FIG. 24   Schemes of configurations used for Stark operation of optically pumped FIR lasers.

waveguide. In hollow dielectric waveguides the EH_{11} mode has the E field linearly polarized with no preferential direction as a consequence of the cylindrical symmetry. Consequently, the molecules can emit in the FIR cavity with a linear polarization which is fixed according to the Chang rule (Chang, 1974). This situation is not affected if an electric field is added from outside by placing the dielectric waveguide between two parallel metal plates. A Stark effect was actually observed with this configuration by Inguscio et al. (1977) and Benedetti et al. (1977). Unfortunately, even with a low electric field of a few hundred V/cm, a charge accumulation was observed on the inside wall of the waveguide with rough compensation of the applied electric field (Inguscio et al., 1977). The configuration with two parallel plates placed inside the Fabry–Perot cavity was introduced by Fetterman et al. (1973). This resonator design is rather simple but the cavity must be terminated by a spherical mirror of proper radius in order to prevent lateral escape of the pumping radiation. Moreover, the plates must be properly designed in order to obtain high electric field operation in presence of the suitable gas pressure. (It is worth noting that the optimum pressure in the FIR lasers is about 1 order of magnitude larger than that used in CO_2 or N_2O laser Stark spectroscopy experiments.)

Polarization of the submillimeter output is always perpendicular to the Stark field, irrespective of the CO_2 pump polarization. This is caused by the different attenuation suffered by the two polarizations, parallel and orthogonal, in propagating through the waveguide (Fetterman et al., 1973). The hybrid rectangular waveguide was introduced by Adam and Kneubühl (1975), and first used in optically pumped FIR lasers by Tobin and Jensen (1976a). The mode configurations can be calculated referring to a rectangular cylinder of infinite length limited by a dielectric on two sides and by an ideal conductor on the other two sides (see Fig. 25). For hybrid waveguides it can be demonstrated that it is impossible to separate the field components into E- or H-type modes. However, solutions can be represented by linear combinations of the two types of modes. Solutions with vanishing electric field components normal to the dielectric interface possess only longitudinal components: they are called longitudinal section electric (LSE) modes ($E_x = 0$). Solutions containing no magnetic field component normal to the interface are called longitudinal section magnetic (LSM) modes ($H_x = 0$).

The complete solution of the wave equation for E and H components can be found in Adam and Kneubühl (1975). Waveguide losses can be computed by adding separately the contribution of the wide metallic walls and of the narrow dielectric walls. The low-loss modes will have E field polarization parallel to the metal walls: they will be TE as seen by the

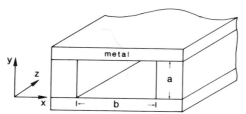

FIG. 25   Idealized metal–dielectric rectangular waveguide structure.

metal walls, while the dielectric side walls will experience this polarization as a TM mode. An expression for the attenuation of the lowest loss mode is given by Garmire *et al.* (1976)

$$\alpha_{10}(E_\parallel) = \frac{\lambda^2}{2a^3} \left(\frac{\omega\varepsilon_0}{2\sigma}\right)^{1/2} + \frac{\lambda^2}{2b^3} \operatorname{Re} \frac{\nu^2}{(\nu^2 - 1)^{1/2}}, \tag{23}$$

where a and b are defined in Fig. 25, σ is the conductivity of the metal sides, ε_0 is the electric susceptibility of the internal medium, ν is the complex refractive index of the dielectric. Assuming an aluminum insulator hybrid waveguide, we have

$$\left(\frac{\omega\varepsilon_0}{2\sigma}\right)^{1/2} \simeq (1.53 \times 10^{-5})\lambda^{-1/2}, \tag{24}$$

where λ is given in m, for the metal, and $\operatorname{Re} \nu^2/(\nu^2 - 1)^{1/2} \simeq 2$–2.3 for the conventional low losses insulators in the submillimeter region. We have thus

$$\alpha_{10}(\mathrm{m}^{-1}) \simeq (7.6 \times 10^{-6}) \frac{\lambda^{3/2}}{a^3} + \frac{\lambda^2}{b^3}. \tag{25}$$

The larger contribution to the losses comes from the dielectric, and $\alpha_{10}(E_\parallel)$ depends more critically on the width of the metal plates than on their spacing. In Fig. 26 α_{10} is plotted versus the wavelength, for fixed dimensions of the waveguide. It is worth noting that at shorter wavelengths, such as 10 μm, the microwave expression for the metal resistivity [Eq. (24)] is no more a good approximation. The complex refractive index of the metal must be introduced (Garmire, 1976; Kneubühl and Affolter, 1979).

In agreement with theory the FIR laser electric field has always been observed to be polarized parallel to the metal waveguide sides, irrespective of the CO_2 pump polarization. In presence of an electric field, FIR polarization corresponds to the selection rule $\Delta M = \pm 1$. In cw optically

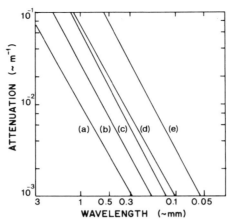

FIG. 26   Computed values of α_{10} ($E_\parallel$) as a function of λ for fixed dimensions of a aluminium–Teflon hybrid rectangular waveguide: (a) $b = 5 \times a = 0.5$ cm, (b) 3.4×0.5 cm, (c) 2.54×1.27 cm, (d) 2.5×0.3 cm, (e) 1.5×0.3 cm.

pumped lasers, the absorption coefficient of the active medium is low and multipass excitation is required for efficient operation. As a consequence the FIR cavity must also be a good resonator for the infrared pumping radiation. This condition has of course to be met in Stark waveguide structures too. At 10 μm a hybrid waveguide resonator can resonate with polarization either parallel or orthogonal to the metal sides, even though in the second case losses are higher.

Typical multipass transmission at 10 μm from a Stark hybrid waveguide resonator with $b = 35$ mm is shown in Fig. 27. For a 150 cm long waveguide cavity the ratio between the two transmissions resulted $I_\parallel/I_\perp \sim 2.5$ for $a = 5$ mm. When a was increased to 10 mm a ratio of $I_\parallel/I_\perp \sim 1.5$ could be obtained for a 1-m long waveguide (Bionducci et al., 1979). When the cavity length was translated, CO_2 resonances were observed. Typical results are shown in Fig. 27. The different modes are better resolved for CO_2 polarization parallel to the metal walls in agreement with the better quality factor of the cavity.

Both the two plates and the hybrid configurations (Fig. 24b,c) provide a sufficiently uniform distribution of the electric field. With the two plates configuration Fetterman et al. (1973), reported FIR laser action in the presence of an electric field of 21 kV/cm. With the hybrid metal dielectric configuration, laser operation in the presence of an electric field higher than 70 kV/cm was recently reported by Redon et al. (1979a,b,c). Cavity finesse in the presence of a strong electric field was investigated for the

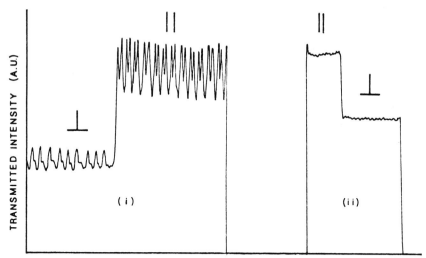

FIG. 27 Multipass transmission at 10 μm from hybrid waveguide for radiation polarization orthogonal (⊥) and parallel (‖) to the metal walls. The plate separation was (i) 0.5 cm or (ii) 1 cm. In (i) the Stark resonator cavity length was scanned and resonances of the IR radiation were observed. The CO_2 line was the 9.5 μm $P(20)$ (Bionducci et al., 1979).

metal–dielectric hybrid laser configuration by Inguscio et al. (1979b) and Bionducci et al. (1979).

In Fig. 28 the cavity length of a Stark CH_3F FIR laser at 496 μm was scanned in the presence of an electric field of 600 V/cm. Cavity length

FIG. 28 Interferogram obtained by scanning the cavity length in a CH_3F Stark laser (free spectral range 98 MHz). In presence of an electric field of 600 V/cm, lasing was obtained for single M components. Pump polarization was orthogonal to the static field and gas pressure was 27 mtorr (Inguscio et al., 1979b).

was 154 cm, corresponding to a free spectral range of 98 MHz. Spectral resolution of about one hundredth of the free spectral range was typical. Single-mode operation was easily obtained in hybrid waveguide configuration. This is an important feature for clear quantitative spectroscopic investigations. Electric field homogeneity was tested at 6.5 kV/cm by recording an interferogram of the 119-μm line emitted by a CH_3OH laser. The result is shown in Fig. 29. As can be seen, there is no broadening or distortion of the laser line. The efficiency of the hybrid waveguide FIR

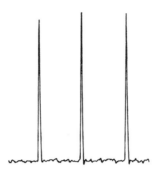

FIG. 29　Typical interferograms of the CH_3OH 119-μm line obtained by scanning the cavity length in presence of an electric field of 6.5 kV/cm. Resonator-free spectral range was 98 MHz (Bionducci *et al.*, 1979).

TABLE III

OUTPUT LEVELS AVAILABLE AT DIFFERENT WAVELENGTHS FOR
CONVENTIONAL AND STARK FIR LASERS[a]

| Line (μm) | Molecule | Dielectric waveguide 40 mm I.D. | | |
		Length (m)	FIR power (mW)	Pump power (W)
119	CH_3OH	1	20	16
70.5	CH_3OH	1	10	17
496	CH_3F	1	7	15
119	CH_3OH	1.5	18	9
Line (μm)	Molecule	Hybrid waveguide − l = 1 mt		
		Spacing (mm)	FIR power (mW)	Pump power (W)
119	CH_3OH	5	1	5
119	CH_3OH	10	2	5
70.5	CH_3OH	10	1	5
496	CH_3F	10	0.8	6

[a] Input and output powers were not corrected for vacuum windows' transmission. FIR power was measured using the device described by Evenson *et al.* (1977).

lasers is comparable to that of conventional hollow dielectric cylindrical waveguide lasers. Typical output levels obtained with the same pump apparatus, but with conventional or Stark waveguide in the FIR resonator, are listed in Table III.

The pump threshold of the FIR laser depends also on the particular line observed. In general threshold for Stark operation is not much higher than for conventional operation. For instance emission at 119 μm was obtained from a CH_3OH laser (Pyrex tube with 40-mm internal diameter and 150-cm length) with a pump power of about 400 mW. The threshold was increased to about 1 W for operation with the metal–dielectric hybrid waveguide (150-cm length, 10-mm metal side separation, pump polarization parallel to the metal plates). Note that the application of a small electric field can increase in some cases the output power, as discussed in Section V.F.

The output power stability for the Stark waveguide laser is similar to that for conventional operation. A typical result for the CH_3F 496-μm line is shown in Fig. 30 without and with a voltage applied to the Stark plates. When a low electric field was applied orthogonal to the CO_2 pump polarization, FIR output was increased and the good power stability was not affected.

Finally, the solution for the electrodes adopted by Stein *et al.*, (1977) should be mentioned. They simply glued two strips of metal foil on the inside of a hollow cylindrical waveguide. The solution is very quick and allowed the authors to observe frequency shift and modulations induced by

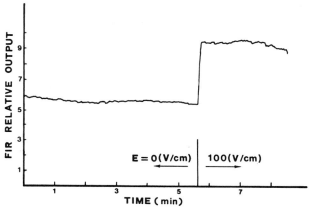

FIG. 30 Typical output power stability for the 154-cm long rectangular hybrid waveguide laser at 496 μm. Both the CO_2 and FIR lasers are free running. When a low electric field is applied orthogonal to the CO_2 pump polarization, FIR output is increased and the good power stability is not affected. (Cosmovici *et al.*, 1979).

Stark effect. Unfortunately, the metal strips change the mode polarization and increase cavity losses with a reduction of output power. Furthermore the electric field is not homogeneous and high field operation is not possible.

C. ELECTRICAL BREAKDOWN IN THE ACTIVE MEDIUM

Electric breakdown of the gas is an effect that can limit the Stark operation of optically pumped FIR lasers. In all gases there are free charges created from the interaction with the environment. These charges can be collected by applying an electric voltage between two electrodes and a very small current flows in the gas (dark current). The electric field in the gas is proportional to the applied voltage. When the strength of the field is increased above a critical value (breakdown field), an avalanche multiplication of the charged particles leads to a very large increase of the current density (discharge). Any further increase of the voltage from the power supply increases the discharge current and not the electric field in the gas.

Paschen (1889) discovered that in the case of a uniform field, the breakdown voltage is a function of the product of pressure (p) and gap length (d) only. The analytic expression for the Paschen law is (Von Engel, 1965)

$$V_B = Bpd/[C + \ln(pd)], \tag{26}$$

where the constants B and C depend on the gas and electrodes. From Eq. (26) it follows that the breakdown voltage has a minimum for a critical value of (pd). At low pressure, below the $(pd)_{min}$, the slope of the Paschen curve is negative and the spark discharge takes place on the longest of the possible paths, the longer path requiring the lower breakdown voltage. This effect should be carefully considered for a good laser main frame design.

In Fig. 31, interferograms from a CH_3OH Stark laser at 119 μm are shown. The waveguide was hybrid aluminum–Pyrex with 5-mm electrodes spacing. As a consequence of the Stark effect the 119-μm line splits

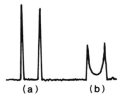

(a)   (b)

FIG. 31   Electrical discharge effects on the interferogram of the laser output. When breakdown occurred (b) doublet frequency separation was not increased by a further rise in the applied electric voltage. Finesse of the Fabry–Perot cavity was also deteriorated. CH_3OH pressure was 120 mtorr and the discharge occurred between 1000 and 1200 V/cm.

in a doublet (see Section V.C) and the splitting is proportional to the applied electric field. When breakdown occurs, the doublet frequency separation is not increased by a further rise in the power supply voltage (right side registrations). Finesse is also deteriorated. This dependence of breakdown on pressure in the same FIR cavity is also shown in Fig. 32, where the measured frequency splittings are reported versus the electric field for four different operation pressures. Above a given value, which increases by decreasing the pressure, the observed splitting remained constant showing that the plateau of the $V = V(i)$ curve was reached. These results demonstrate that in the FIR laser cavity the value (pd) was smaller than the value of $(pd)_{min}$. This is typical for the gas pressure and the electrode separation generally used in the Stark FIR lasers. As a consequence the discharge toward the laser main frame (larger path) is more likely to occur than between the Stark plates. Once one has remedied to the possibility of discharges over the longer external path, the maximum electric field that one can actually apply to the gas between the Stark plates is determined by the plate separation and by the Paschen curve of the gas. Measured Paschen curves for some of the organic molecules used in the

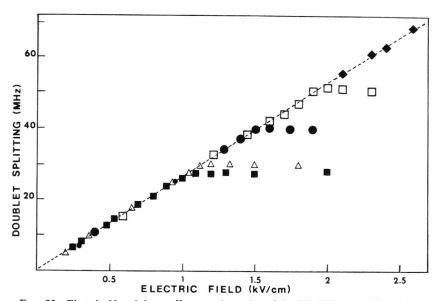

FIG. 32 Electrical breakdown effects on the tuning of the CH_3OH 119-μm line. Pressure (mtorr): ■, 120; △, 90; ●, 75; □, 60. The electric field actually applied on the gas could be increased by lowering the pressure. The Stark plates were separated of 0.5 cm. The points in the upper part of the figure were obtained with the end mirrors kept very close to the waveguide (Bionducci *et al.*, 1979).

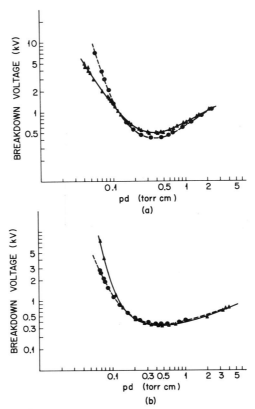

FIG. 33   Paschen curve for (a) CH₃OH and (b) CH₃F (Bionducci *et al.*, 1979).

lasers are given by Bionducci *et al.* (1979). Dependence on the product (*pd*) was verified for *d* from 0.5 to 1.5 cm. Aluminum and stainless steel electrodes showed, respectively, the highest and the lowest breakdown voltages (Fig. 33).

For copper and brass electrodes intermediate results were observed. The result for CH₃OH is also reported in Fig. 34. The curve is compared with the breakdown voltages (Curve a) actually measured with Stark laser of unoptimized design (the same which the results of Fig. 32 refer to). The measurements of Curve c were taken with the end mirrors of the FIR cavity closer to the hybrid waveguide. The decrease in the possible external paths length causes an increase in the breakdown voltage. By carefully insulating the waveguide from the mainframe it is possible to obtain

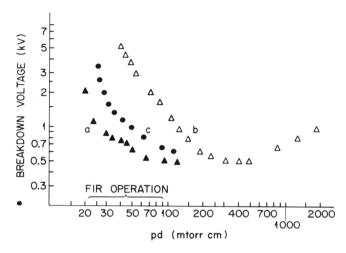

FIG. 34 Breakdown voltage as a function of *pd* (gas pressure × electrodes spacing) for CH$_3$OH; *d* = 0.5 cm curve *a* refers to the Stark laser cavity used for the measurements reported in Fig. 31. Curve *b* is the measured Paschen curve of the gas. Curve *c* refers to an optimized Stark laser cavity (Bionducci *et al.*, 1979).

Stark laser breakdown curves very close to the Paschen curve of the active gas. Very low operation pressures allow electric field intensities up to several tens of kV/cm, as reported for NH$_3$ by Fettermann *et al.* (1973) and Redon *et al.* (1979a).

D. FIR POWER AND FREQUENCY MEASUREMENTS

As discussed in the previous sections, the amplitude level and stability of the output of cw Stark FIR lasers are similar to that of conventional systems. As a consequence usual FIR detectors such as pyroelectrics, Golay cells, bolometers, etc. can be used in the Stark systems. Power and energy absolute measurements in the far infrared are not easy. The use of detectors calibrated under different conditions makes it difficult to compare the results obtained in different experimental apparatus. A satisfactory broad-band power meter was described by Evenson *et al.* (1977). Precise frequency measurements of optically pumped FIR molecular lasers were made possible since the development of nonlinear devices capable of harmonic generation and of sum—or difference—frequency generation from two or more incident radiations over the spectral region of interest. Metal–insulator–metal (MIM) diodes, Schottky barrier diodes, and Josephson junctions (JJ) are generally used. A recent summary of the performance characteristics of the frequency detectors and mixer devices was reported by Jimenez (1979). The frequency of the submillimeter lasers is generally self-stable and resettable to within a fraction of a megahertz, and is essentially determined by the active molecular transition. Hence frequency measurements can provide very useful information on the spectroscopy of polar molecules in highly excited rotovibrational states. As an example, we discuss the case of the popular laser emission at 496 μm from optically pumped CH_3F. In the original work by Chang and Bridges (1970), the frequency was measured by beating the FIR output with harmonics of a tunable microwave signal in a point contact diode. The reported accuracy (3 MHz) allowed a first precise determination of the rotational constants in the ν_3 state. Since then heterodyne measurements accurate to 0.2/0.5 MHz were performed in different experimental apparatus. The results are listed in Table IV.

Kramer and Weiss (1976) used a Fabry–Perot open resonator configuration for the FIR cavity and a W–Ni point contact diode to mix the laser radiation with a phase-locked millimeter-wave Klystron. Radford *et al.* (1977) and Bava *et al.* (1977) used a Schottky barrier diode as a mixer and a waveguide configuration for the FIR resonator. The measurements are in good agreement within the experimental errors. The precision (a few parts in 10^7) is comparable to that of the microwave experiments. High-rotational quantum numbers are involved in FIR transitions and an appreciable contribution to the energy separation comes from high-order rotovibration interactions in the molecule. Arimondo and Inguscio (1979) included the precise FIR data in a least-squares fit of all the microwave, laser Stark and far-infrared laser data related to the $\nu_3 \to 0$ band. Accurate values for the CH_3F $\nu_3 \to 0$ band rotovibrational constants were given. Absolute or relative precise FIR frequency measurements make it possible to investigate even small effects. In the second case a resolution of few KHz can be easily obtained in free-running operation (Fig. 23). For instance, this was the case of the first experimental evidence of a fre-

TABLE IV

FREQUENCY MEASUREMENTS OF THE ν_3, (J,K): $(12,2) \to (11,2)$ TRANSITION OF CH_3F USING DIFFERENT FIR EXPERIMENTAL APPARATUS

f_{CH_3F}(MHz)	Cavity	Reference
604293(3)	Fabry–Pérot open resonator	Chang and Bridges (1970)
604297.3(5)	Fabry–Pérot open resonator	Kramer and Weiss (1976)
604297.1(5)	waveguide	Radford *et al.* (1977)
604297.5(2)	waveguide	Bava *et al.* (1977)

quency shift induced by Stark effect (Stein *et al.*, 1977; Inguscio *et al.*, 1977).

E. SATURATION EFFECTS IN THE FIR OUTPUT

A large majority of the theoretical investigations on optically pumped molecular lasers have been performed using rate equation models (Tucker, 1974; Henningsen and Jensen, 1975; De Temple and Danielewicz, 1976; Hodges *et al.*, 1976; Temkin and Cohn, 1976; Temkin, 1977). More recently, Koepf and Smith (1978) have extended the theory to account for the pump radiation traveling in both directions in the laser tube. Quantum-mechanical predictions for the gain of an optically pumped cw FIR laser were presented by Seligson *et al.* (1977) for cases in which one or both pump and FIR transitions were pressure or Doppler broadened. They also verified some of the quantum-mechanical predictions in CH_3OH.

From the quantum-mechanical point of view optically pumped molecular lasers can be considered as a coupled three-level system interacting with applied radiation fields. As schematically shown in Fig. 35, the theoretical model considers a three-level system composed of an IR pump transition, $2 \rightarrow 0$, center frequency ω_2 coupled to a FIR transition $0 \rightarrow 1$, center frequency ω_1, via the common level 0. The IR pump is always inhomogeneously Doppler broadened, while the lasing FIR transition can be either Doppler or homogeneously pressure broadened, depending on wavelength and pressure. The active medium in the resonator interacts simultaneously with forward and backward propagating pumping laser beams. When a detuning Δ_2 is introduced between the pump frequency

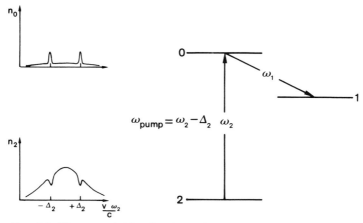

FIG. 35   Three-coupled levels scheme for optically pumped FIR lasers.

and the absorbing transition line center, in the case of fully Doppler-broadened system, the FIR gain curve was predicted (Feldman and Feld, 1972; De Temple and Danielewicz, 1976) to split into the sum of two Lorentzian curves centered at frequencies

$$\Omega(\pm) = \omega_1(1 \pm \Delta_2/\omega_2). \tag{27}$$

An asymmetry in the forward $(+)$ backward $(-)$ gain was also predicted (Feldman and Feld, 1972). For the short wavelength CH_3OH lines the effect was actually observed (Seligson et al., 1977; Heppner et al., 1977; Lourtioz et al., 1979) by scanning the FIR frequency in the presence of fixed pump detunings. A more general effect was expected when the pump detuning Δ_2 was scanned at fixed FIR cavity length. When $\Delta_2 = 0$ a Lamb dip is generated in the Doppler-broadened IR molecular transition. Since the overall FIR laser intensity depends on the number of excited molecules (Feldman and Feld, 1975), a dip was expected also in the FIR laser power emission. This effect was also predicted for the 496-μm line of CH_3F by Koepf and Smith (1978), using a rate equation model.

The first experimental evidence of a saturation effect in the FIR laser output was reported by Fetterman et al. (1976). They used a dc electric field to bring a NH_3 vibrational absorption into coincidence with a CO_2 laser line, hence obtaining population inversion and FIR emission. They demonstrated the presence of saturation effects by slowly sweeping the vibrational absorption through the CO_2 laser frequency (essentially

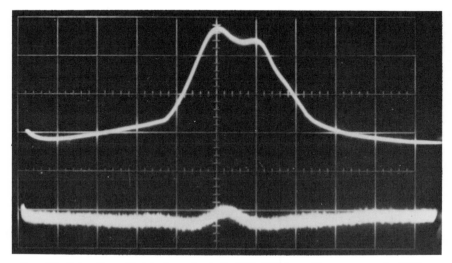

FIG. 36   Saturation effect in the output of a NH_3 FIR laser (Fetterman et al., 1976).

changing Δ_2). Under these conditions, Fig. 36 shows the simultaneous output of the far-infrared laser (upper trace) and the reflected CO_2 power from the cavity (lower trace). The effect was successively reported for transitions that do not require an electric field to be pumped by Duxbury and Herman (1978), and Inguscio *et al.* (1979c), and Lund *et al.* (1979). In those cases, the detuning Δ_2 was changed by piezoelectric (PZT) scanning the CO_2 laser cavity length. An experimental investigation of the spectroscopic application of the IR–FIR transferred Lamb dip in optically pumped molecular lasers can be found in Inguscio *et al.* (1979d, e). A typical result for the CH_3OH laser at 70.5 μm is shown in Fig. 37b.

The Lamb dip could also be detected directly in the IR absorbing transitions using pump intensities and gas pressures typical for FIR laser operation. For instance, Fig. 37a shows the optoacoustic signal recorded in CH_3OH by PZT frequency scanning the $9-P$ (34) line of the CO_2 laser. Nearly the full Doppler profile of the absorbing transition could be recorded. In that case counterpropagating laser beams were present in the

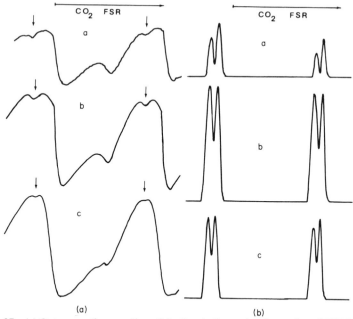

FIG. 37   (a) Optoacoustic recording of the Lamb dip on the absorption of CH_3OH by frequency tuning the $9-P(34)$ CO_2 line: a, 24 mtorr; b, 65 mtorr; c, 300 mtorr. (b) Transferred Lamb dip simultaneously observed in the FIR laser emission at 70.5 μm: a, 80 mtorr; b, 100 mtorr; c, 130 mtorr (Inguscio *et al.*, 1979d).

absorption cell and the Lamb dip could be observed at lower pressures. At pressures higher than 300 mtorr the Lamb dip was not observed because of the too large collisional broadening and of the reduced saturation. In Fig. 37b is shown the corresponding FIR output at 70.5 μm as a function of the CO_2 9–P (34) line tuning. The Lamb dip transferred from the absorbing line was observed. The dip contrast was much larger than in the optoacoustic signal as a consequence of the nonlinear detection through the FIR oscillator, by increasing the laser threshold it was possible to obtain a contrast of nearly 100%. The dip detected in the FIR emission had nearly the same width as in the optoacoustic signal. The homogeneous broadening depends on the collisions and on the pump in-

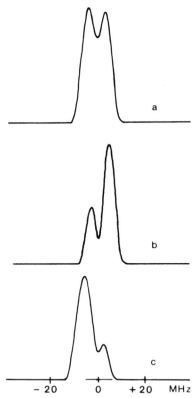

FIG. 38   TLD on the 70.5 μm line. (a) FIR cavity tuned nearly exactly at the center of the gain curve. Opposite sign FIR cavity detunings were introduced in (b) and (c) (Inguscio *et al.*, 1979d).

tensity I. It is given by

$$\Delta\nu = \Delta\nu_{\text{pres}}(1 + I/I_s)^{1/2}, \tag{28}$$

where I_s is the saturation intensity.

The measured values were consistent with the computed ones according to data given by Seligson *et al.* (1977) and Weiss (1976) and with the measurements by Forber *et al.* (1980). The linewidth did not change significantly with pressure because lowering the latter reduced collision broadening but increased saturation broadening.

The IR–FIR transferred Lamb dip (TLD) has been observed for FIR emissions ranging from 37.5 to 1222 μm. The frequency of the dip center depends only on pump transition offset with respect to the CO_2 line. In agreement with the theoretical predictions the FIR cavity tuning affects only the symmetry of the recorded signal as shown in Fig. 38. A simultaneous recording of the FIR output, with the TLD, and of the CO_2 pump mode allowed the determination of pump offset. A typical recording is shown in Fig. 39.

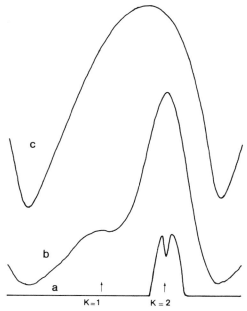

FIG. 39   Recording of the FIR output from CH_3F (pressure-18 mtorr) at 496 μm (a) together with the mode of the 9–$P(20)$ pump radiation from a waveguide CO_2 laser (c) (free-spectral range = 240 MHz) and the optoacoustic absorption signal (b).

As a consequence TLD is an accurate and sensitive method for the precise determination of the structure of the absorption line responsible for FIR laser emission. Its importance, compared to other methods, is based on the following points:

(i)   The signal is recorded by measuring FIR laser output and as a consequence the recorded TLD is the one associated to the pump transition. In case of many closely spaced absorbing lines only one TLD is observed, while with conventional saturation spectroscopy many Lamb dips are recorded without a priori correlation to the FIR laser line.

(ii)   A TLD can be recorded for each FIR laser line and it is possible to control, within the homogeneous linewidth, if the FIR laser lines share or not a common pump transition. This point is important for assignments.

(iii)   A good signal can be obtained directly from conventional FIR laser apparatus, without any further equipment. The resolution is very high, even higher than homogeneous linewidth (Inguscio et $al.$, 1980c; also see Section V.H). The absolute frequency accuracy is limited by the calibration of the CO_2 frequency scale, which in conventional CO_2 lasers is of a few MHz as a consequence of the pressure broadening and shift of the lines. It is worth noting that high absolute accuracy can be obtained by a simultaneous heterodyning of the CO_2 pump radiation with that of an actively stabilized reference CO_2 laser.

V.   Experimental Results — dc Electric Fields

A.   Off-Resonant Pumping of Molecules

Operation of optically pumped FIR lasers is usually carried on in gases with vibrational absorptions closely resonant with the pump laser frequency. In the case of high-power TEA lasers, it is possible to obtain off-resonant pumping even for large detunings.

In a cw system the limitations caused by a need for close-pump resonance can be overcome by either increasing the frequency tunability of the pumping laser or Stark tuning the molecular absorption.

A larger frequency tunability of the pump can be obtained by increasing the pressure of the gas mixture used in the CO_2 laser, thus broadening the gain profile of the active medium. cw high-pressure operation of CO_2 lasers has been made possible with the introduction of the hollow waveguide configuration. The low power available (less than 1 W) in single-line, single-mode operation has limited the possible applications of these high-pressure lasers to conventional IR spectroscopy. Recently, however, output powers as high as 3 and 8 W were reported from waveguide CO_2 cw lasers in single mode, single-line operation, respectively, by Evans *et al.* (1980) and Ioli *et al.* (1980). Ioli *et al.* (1980) successfully applied a waveguide CO_2 cw high-pressure laser to optical pumping of FIR lasers. Optical pumping of FIR lasers via previously inaccessible absorption lines yielded new strong laser emission from CH_3OH (Ioli *et al.*, 1980; Inguscio *et al.*, 1980b). Recently Tobin and Koepf (1980) reported FIR action from CDF_3 optically pumped with a rf excited waveguide laser. The possibility of Stark tuning the molecular absorption was for the first time demonstrated by Fetterman *et al.* (1973). Using the configuration of two parallel plate inside a Fabry–Perot resonator described in Section IV.B, they obtained cw laser action from some NH_3 levels that are also close to, but not in resonance with, pump laser lines. Inguscio *et al.* (1979b) reported off-resonant optical pumping of CH_3OH by Stark tuning the molecular absorption around the CO_2 9–P (34) laser line. In that case a large tunability of the FIR emissions was also observed and a spectroscopic investigation could be performed (Inguscio *et al.*, 1979g).

More recently one of the FIR emissions (at 205 μm) has been detected by pumping with a TEA CO_2 laser (Bluyssen *et al.*, 1980; Izatt *et al.*, 1979) and with a cw CO_2 waveguide laser (Ioli *et al.*, 1980). Redon *et al.* (1979a) used the Stark tuning to obtain new FIR emissions from NH_3 in hybrid Stark waveguide configuration. Redon *et al.* (1979b) demonstrated that in ammonia the use of "forbidden" transitions, becoming partially allowed by the mixing of the wave functions owing to the electric field, is a practical means of obtaining new FIR lasing lines that cannot be obtained by other pumping schemes. Pump detunings up to 13 GHz could be overcome by Stark fields up to 70 kV/cm.

B. EVIDENCE FOR STARK-INDUCED FREQUENCY SHIFTS IN THE FIR EMISSION

The first experimental evidence of a FIR frequency shift induced by Stark effect was reported by Stein *et al.* (1977) and Inguscio *et al.* (1977). In both cases a CH_3F laser at 496 μm was operated in presence of an electric field of the order of a few hundred V/cm. Stein *et al.* (1977) glued two strips of metal foil on the inside of a Pyrex hollow cylindrical waveguide; Inguscio *et al.* (1977) used a dielectric hollow waveguide with external Stark plates. The limitations of both the configurations have been discussed in Section IV.B. The frequency shifts were directly measured by beating the Stark laser output with a reference conventional laser (Stein *et al.*, 1977) or with the microwave power from a phase-locked klystron (Inguscio *et al.*, 1977). A maximum frequency shift of about 400 KHz was observed, corresponding to a linear Stark tuning coefficient of the order of 5 KHz/V cm^{-1}. The frequency shift could be observed only by slightly detuning the FIR resonator. These early experimental observations could be interpreted as a cavity pulling effect controlled by the electric field via Stark broadening of the gain curve. In presence of a moderate electric field, the gain curve of the active medium is symmetrically broadened, as derived in Sections III.A,B,C. As a consequence no shift direction is preferred. When the FIR resonator is slightly detuned, application of the electric field causes a net frequency shift. The actual laser frequency ν is determined by

$$(\nu - \nu_m)/\Delta\nu_m = (\nu_c - \nu)/\Delta\nu_c, \tag{29}$$

where ν_c is the resonant frequency of the empty laser cavity and $\Delta\nu_c$ its full width, ν_m is the frequency of the molecular transition and $\Delta\nu_m$ its full width. It is clear that an offset between ν_c and ν_m produces a frequency shift of the laser towards the cavity line center. The amount of shift depends on the cavity offset itself, as well as the magnitudes of $\Delta\nu_c$ and $\Delta\nu_m$. The last dependence is controlled by the Stark effect. Differentiation of the pulling equation with respect to the applied field E yields

$$\frac{d\nu}{dE} = \frac{\Delta\nu_c(\nu_c - \nu_m)}{(\Delta\nu_m + \Delta\nu_c)^2} \frac{d}{dE}(\Delta\nu_m). \tag{30}$$

In agreement with this equation the measured shifts were roughly proportional to the cavity offset and to $|E|$ and the direction of the shift was the same as the cavity detuning itself. The quantity $(d/dE)(\Delta\nu_m)$ in Eq. (30) can be estimated from Fig. 6 to be of the order of 20 KHz/V cm^{-1}. If $\Delta\nu_m$ is assumed to be 2 MHz and $\Delta\nu_c$ is estimated to be about 5 MHz, we obtain

$$\Delta\nu_c/(\Delta\nu_m + \Delta\nu_c)^2 \simeq 2 \times 10^{-1} \text{ (MHz}^{-1}). \tag{31}$$

The actual Stark coefficient depends on $\nu_c - \nu_m$. The observed shifts (a few kHz/V cm^{-1}) are consistent with reasonable values (a few MHz) assumed for $\nu_c - \nu_m$.

C. LARGE FREQUENCY TUNING BY STARK EFFECT ON FIR LINES OF CH$_3$F AND CH$_3$OH

The use of the hybrid metal–dielectric waveguide configuration allowed Evenson's group at NBS in Boulder to operate a CH$_3$F laser in presence of an electric field large enough to resolve the Stark components of the 496-μm line (Inguscio et al., 1979a; Strumia, 1978). The first high-efficiency operation of a Stark tuned CH$_3$OH laser was reported by Inguscio et al. (1978a): the strong line at 119 μm was observed to split onto a doublet. Large frequency tunings for the CH$_3$OH lines at 70.5 and 119 μm and for the CH$_3$F line at 496 μm were soon reported by Inguscio et al. (1979b). The scheme of a typical experimental apparatus is shown in Fig. 40.

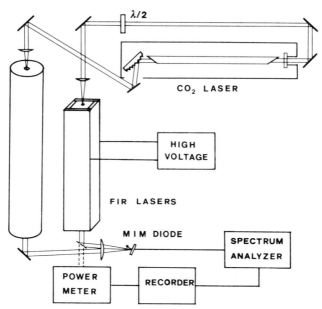

FIG. 40 General scheme of a typical experimental apparatus for observing the Stark effect on FIR lasers.

Resolution of the FIR Stark components is generally monitored with the interferogram obtained by scanning the cavity length. For small frequency shifts or for precise frequency measurements the output from the STIL (Stark-Tuned Infrared Laser) is mixed with the output of a FIR conventional laser or with a microwave signal.

1. CH_3F 496-μm Line

cw laser action was obtained for electric field intensity up to 2.8 kV/cm. Correspondingly, frequency tuning ranges were observed up to 60 MHz, i.e., much wider than the zero field gain curve width. In agreement with the theoretical results, the individual Stark components started to be resolved for electric fields of at least 500 V/cm, the precise value depending also on the length of the cavity actually used and on the pressure of the active gas. The observed Stark intensity pattern strongly depended on pump polarization orientation relative to the electric field. Typical results are shown in Fig. 41 for pump polarization, parallel and orthogonal to the electric field. The results are in agreement with the Stark patterns computed assuming no collisional M sublevels mixing in the involved levels (Figs. 8, 9, and 10). This experimental evidence of no mixing between the M sublevels was in agreement with results previously

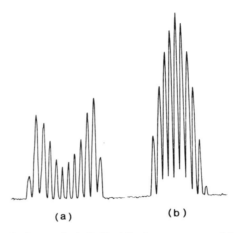

(a)   (b)

FIG. 41 Laser oscillation on the individual Stark components of the CH_3F 496-μm line. The electric field intensity was 1.9 kV/cm and CH_3F pressure 40 mtorr. The pump selection rules only were changed in the two cases. In (a), $\Delta M = 0$, all the 12 strongest components were detected; in (b), $\Delta M = \pm 1$, ten Stark components were observable.

obtained in collision-induced double resonance measurements (Shoemaker et al., 1974) and in collision induced Lamb dip laser Stark spectroscopy (Johns et al., 1975). In the collisions between polar molecules, dipole–dipole interaction dominates. As shown by Anderson (1949), the relevant matrix elements are the same as for the radiative transitions with selection rules $\Delta J = 0, \pm 1; \Delta K = 0; \Delta M = 0, \pm 1$. To first order, the transition rate for one of the collision partners, when J is fixed, is proportional to the matrix element squared:

$$|\langle J, K, M|V_{dd}|J, K, M \pm 1\rangle|^2 \propto \frac{K^2 (J M + 1)(J M)}{J^2 (J + 1)^2}, \qquad (32)$$

where V_{dd} is the dipole–dipole intermolecular potential. The transition rate strongly depends on the ratio between K and J. For $C^{12}H_3F$ ($J = 12$, $K = 2$), ΔM collisions can be considered weak as compared to ΔJ collisions. The absence of randomization causes the FIR Stark line shape to depend strongly on the pump selection rule.

Measurements on the CH_3F 496-μm line were also performed with pumping CO_2 radiation elliptically polarized so that both $\Delta M = 0$ and $\Delta M = \mp 1$ pump transitions were allowed. In general intermediate results between the two pure cases were observed; typical interferograms are shown in Fig. 42. The not exact tuning of the CO_2 frequency at the center of the CH_3F absorbing line could cause an asymmetry in the FIR Stark pattern, as discussed by Inguscio (1979) and as shown in Fig. 43.

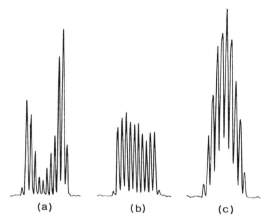

FIG. 42   Effect of the pump polarization on the output of the CH₃F Stark laser. Parallel and orthogonal polarizations are presented, respectively, in (a) and (c), while (b) is an intermediate case with elliptical polarization. The field strength was 1800 V/cm and the gas pressure was 56 mtorr (Inguscio *et al.*, 1979b).

3.   CH₃OH 118.8341-μm Line

Stark action on this popular line of methanol was observed in the presence of an electric field higher than 6 kV/cm. A large frequency tuning of the emitted line was observed and, in particular for electric field intensities

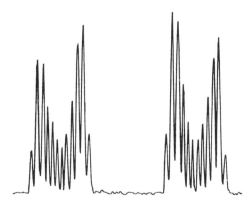

FIG. 43   Interferogram of the FIR output at 496 μm from a Stark CH₃F laser. The pump selection rule was $\Delta M = 0$, electric field strength 1500 V/cm, CH₃F pressure 40 mtorr, and the free-spectral range 98 MHz. The CO₂ laser was operated free running and during the registration the effect of frequency drift on the asymmetry of the Stark FIR pattern was clearly visible.

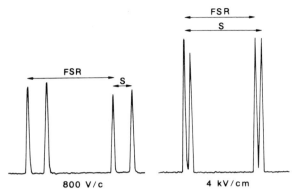

FIG. 44 Interferogram of the observed Stark splitting (S) of the CH$_3$OH 119-μm line. The free-spectral range (FSR) of the laser cavity is also indicated.

larger than 100 V/cm, the line splits into two components. The frequency separation between them increases linearly with electric field intensity. Typical high field results are shown in Fig. 44.

It is interesting to observe that at high electric fields the two components exhibit different intensities. The fact can be ascribed to a pump detuning introduced asymmetry (see Section V.E). The frequency splitting

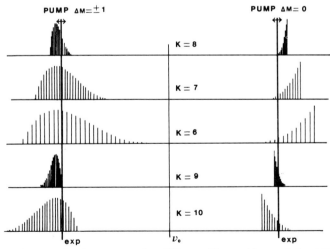

FIG. 45 Computed gain patterns for different M transitions of the 119-μm line of CH$_3$OH. The assumed J value is 16 and the patterns for different K values and both pump polarizations are displayed. It is assumed a Q branch pump transition. The continuous lines connecting the different diagrams are the experimental results ($E = 1$ kV/cm) (Inguscio *et al.*, 1979b).

as a function of the applied electric field is shown in Fig. 45 for $\Delta M = \mp 1$ pump selection rule.

An experimental Stark splitting coefficient of 26.5(5) MHz/kV cm^{-1} can be deduced. For $\Delta M = 0$, the observed splitting was a little larger. A high-resolution frequency measurement was performed (Inguscio et al., 1979b) by mixing the output of the Stark laser with that of a conventional cylindrical Pyrex waveguide laser. For an electric field of 300 V/cm the beat frequency for parallel pump polarization was about 300 KHz larger than the frequency obtained for orthogonal pump polarization. This corresponds to a Stark coefficient for $\Delta M = 0$ pump selection rule of 28.7 MHz/kV cm^{-1}. These results were confirmed by Henningsen (1980a). The difference in the frequency splitting of a line for the two possible pump selection rules $\Delta M = 0$ and $\Delta M = \mp 1$ is expected from the theory (see Section III.C). The amount and the sign of the difference is a function of K/J and its experimental value is particularly useful for assignment of the quantum number K. In Figure 46 is shown in the gain of a line with $J = 16$ and several values of $K(K = 6-10)$. It can be seen that the value of the splitting is practically independent of K when the pump is $\Delta M = \mp 1$, but, decreases with increasing K when the pump is $\Delta M = 0$. In the case of Fig. 46, a Q branch pump line ($\Delta J = 0$) has been assumed as for the 118-μm line ($J = 16$, $K = 8$).

3. Other CH$_3$OH Lines

Large Stark frequency tuning was observed on many CH$_3$OH laser lines. The Stark pattern resulted in a linear splitting into two components

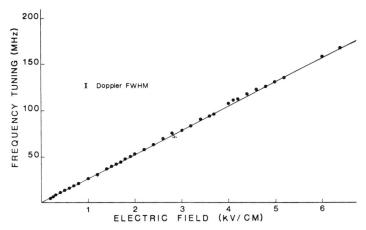

FIG. 46   Measured Stark frequency tuning of the CH$_3$OH 119-μm line (Bionducci et al., 1979).

(96.5, 99.3, 110.8, 193, 208, 261.5, 292 μm, in a multicomponents splitting (205 μm) and in a anomalous splitting into two components (37.9, 133.1 μm). Stark laser action on the 70.5-μm line [9–P(34) pump] was obtained for electric field intensities up to 1.4 kV cm^{-1}. The line was observed to split into two components with a linear Stark coefficient of 44.2 (10) MHz/kV cm^{-1} for $\Delta M = \mp 1$ pump selection rules. The result was confirmed by Henningsen (1980a) who observed also a small difference in the Stark coefficient for different pump selection rules.

With the same CO_2 pump line two new FIR laser lines at 208 and 205 μm appeared in presence of E field larger than ~ 1 kV cm^{-1} (Inguscio *et al.*, 1979b, f, g). The 208-μm line split by Stark effect into two components while the 205-μm line exhibited a multicomponents Stark pattern. The pumping of these off-resonant lines, the measurement of the Stark effect on the FIR emission and on the IR pump transitions by means of the TLD allowed a careful spectroscopic investigation of CH_3OH around the CO_2 9–P(34) line (Inguscio *et al.*, 1979g). The frequency tuning of the 208-μm line was the largest ever measured by Stark effect (see Table V). The measurements of the Stark effect on the 205-μm line lead to an assignment as (J, K): (6, 5) $\rightarrow$ (6, 5) $\rightarrow$ (5, 4). A TLD investigation on this line allowed the measurement of the Stark structure of the IR pump transition, which in conventional two-level high resolution spectroscopy is masked by the homogeneous linewidth (Inguscio *et al.*, 1980c; and see Section V.H).

The usual splitting into two components was reported on many lines: 96.5 μm (Inguscio *et al.*, 1979f), 99.3, 110.8, 261.5 μm (Inguscio *et al.*, 1980b) 193, 292 μm (Henningsen, 1980a). Whenever the assignment of the transitions is available, it turns out that the FIR selection rules are $\Delta J = \Delta K = \mp 1$, $K \sim \frac{1}{2}J$ in agreement with the predictions of Section III.C. An anomalous Stark splitting into two components was reported on the lines at 37.9 μm (Inguscio *et al.*, 1979f) and 133.12 μm (Henningsen, 1980b). Figure 47 shows the Stark effect observed for the 37.5-μm line pumped by 9–P(34). The experimental results can be interpreted by assuming a quadratic dependence on the electric field for one of the levels involved in the transition [Eq. (7)]. The high-field splitting values are consistent with a linear Stark coefficient of 59 MHz/kV cm^{-1}. The negative zero field intercept for the linear extrapolation suggests a splitting of the level of about 10 MHz. A list of the observed Stark effects on CH_3OH lines is reported in Table V. In the last column of the table is given the ratio between the observed Stark tuning and the Doppler FWHM. These figures can be considered as the number of lines needed to cover the same frequency range without the Stark effect.

TABLE V

Measured Stark Effects on CH_3OH FIR Laser Lines[a]

Line (μm)	CO_2 pump	Pump offset (MHz)	Tuning (MHz/kV cm⁻¹)	Stark lines	Tuning range (kV/cm)	Total tuning (MHz)	Tuning ratio
37.9	9–P(32)	−16	59	2	0–2	90	5
70.5	9–P(34)	+36	44.2	2	0–1.4	62	7
96.5	9–R(10)	0	12	2	0–3	36	5
99.3	9–P(36)	−89	29	2	0–1	29	4
110.8	9–P(36)	−80	24.6	2	0–1.3	32	6
118.8	9–P(36)	+29	26.4	2	0–6.5	172	30
133.1	9–P(24)	+6	101	2	0–0.5	50	8
193.1	9–P(38)	+13	34	2	0–0.35	12	3
205.6	9–P(34)	+120	35	4	0–2	70	20
208.3	9–P(34)	>130	135	2	0–2.7	364	120
261.5	9–P(12)	+85	18.3	2	0–2.2	40.3	16
292.2	9–P(38)	+13	20.5	2	0–0.2	4	2

[a] The pump offset values have been measured using the TLD technique.

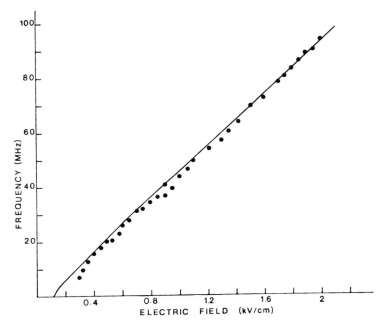

FIG. 47   Observed Stark frequency splitting of the 37.9-μm CH_3OH laser line ($\Delta M = \pm 1$).

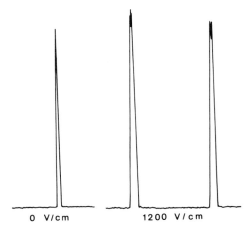

FIG. 48   Interferogram of the output of a Stark CH_3I laser at 508 μm. The free spectral range was 99 MHz.

D. SMALL FREQUENCY TUNING

As discussed in Sections III.A,C, there are transitions for which the Stark effect is not high enough to be observed directly in the FIR output interferogram as a set of resolved lines. This is for instance the case for high J transitions: the Stark broadening and lowering of the gain curve is too dramatic and laser action ceases before any Stark structure can be observed. However, as a general result, the Stark effect is effective in broadening and changing the intensity of the emitted lines, except for $K = 0$ in both the involved levels. A typical result for the 508-μm FIR laser line of CH_3I pumped with the CO_2 9–$P(34)$ line is shown in Fig. 48, where an

TABLE VI

FIR CH_3OH LASER LINES WITH SMALL
FREQUENCY TUNING[a]

Line	CO_2 pump	Pump offset	Reference
164.8	9–$R(10)$	0	c, d
232.9	9–$R(10)$	0	c, d
164.2	9–$P(12)$	+85	e
450.4	9–$P(12)$	+85	e
42.2	9–$P(32)$	−16	b
270	9–$P(32)$	−16	b
63.4	9–$P(34)$	+36	d
135.7	9–$P(36)$	−89	c
162.1	9–$P(36)$	−80	c
170.5	9–$P(36)$	+29	c, g
198.7	9–$P(38)$	+13	d, f
63.0	10–$R(16)$	−14	d
77.9	10–$R(16)$	−14	d
145.5	10–$R(32)$	<−45	b
242.8	10–$R(32)$	<−45	b
129.5	10–$R(34)$	−46	d
250.8	10–$R(34)$	−46	d
267.4	10–$R(34)$	−46	d
242.5	10–$R(34)$	−30	d
43.7	10–$R(36)$	+23	b
53.9	10–$R(36)$	+23	b
53	10–$R(46)$	+15	d
65	10–$R(46)$	+15	d
164	10–$R(48)$		b, d
286	10–$R(48)$		b, d

[a] Pump offsets are measured with the TLD technique.
[b] Inguscio et al. (1980d); [c] Inguscio et al. (1980a); [d] Inguscio et al. (1979f); [e] Inguscio et al. (1980b); [f] Henningsen (1980a); [g] Bionducci et al. (1979).

increase in the frequency emission bandwidth can be observed when the Stark field is applied to the active medium.

A list of lines for which a Stark effect on the intensity has been measured is reported in Table VI. It is worth noting that small frequency tunings or Stark fast control of the emitted frequency can be achieved also for these lines, as discussed in Sections IV.B and VI.B.

Stark frequency shifts were observed also on a HCOOH laser by De Marchi *et al.* (1979). Formic acid is an asymmetric molecule and the Stark effect is quadratic. By beating the signal from this FIR laser with a synthesized spectrally pure reference, the authors measured small frequency shifts on the three lines at 742.6, 432.7, and 393.6 μm. The largest shift was observed on the last one (80 KHz/100 V cm^{-1}).

Ammonia is another molecule with small FIR Stark frequency shifts. In this case the Stark effect in the levels of the excited vibrational state is quadratic and small, as a consequence of the large inversion levels' splitting. On the other hand, a large frequency tuning by Stark effect can be obtained in the vibrational transitions. As a consequence it is easy to bring new molecular absorptions into coincidence with pump CO_2 or N_2O laser lines (Section V.A).

However, because of the large tunability, this coincidence is obtained only over a limited range of Stark voltage and frequency tuning of the FIR laser is negligible.

E. IR–FIR TRANSFERRED LAMB DIP SPECTROSCOPY: STARK EFFECT IN THE PUMP TRANSITION

The saturation effects in the FIR output have been discussed in Section IV.E. The main result is that a dip in the FIR emission can be observed when the pump radiation is tuned exactly at the center of the absorbing transition. This effect can be easily observed for either Doppler or pressure broadened FIR transitions by frequency scanning the CO_2 pump line at fixed FIR cavity length. FIR cavity setting affects only the symmetry of the observed signal, but not the position of the dip which depends only on the pump detuning. The sensitivity and resolution of the technique allow to investigate the Stark effect on the molecular vibrational transitions involved in the laser action. As an example of linear Stark effect we discuss the results obtained for the CH_3OH line at 119 μm pumped by the $9-P(36)$ CO_2 laser radiation. As discussed in the previous section, in presence of an electric field this FIR laser line was observed to split into two components corresponding to $\Delta M = +1$ and $\Delta M = -1$ transition selection rules. The Stark effect on the pump transition can be computed according to the assignment by Danielewicz and Coleman (1977) and Henningsen (1977). The result is shown in Fig. 49. In case of $\Delta M = \pm 1$ pump selection rules, the absorbing transition itself is split into two multiplets. For low

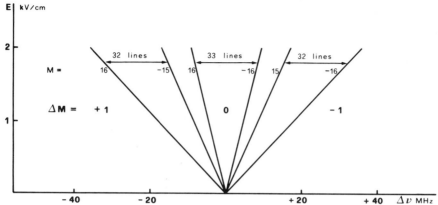

FIG. 49   Stark tunability of the CH_3OH pump transition (v, n, τ, J, K): $(1, 0, 1, 16, 8) \rightarrow$ $(0, 0, 1, 16, 8)$ yielding emission at 119 μm. $\Delta\nu_0 = 67.5$ MHz.

electric fields the frequency separation between each M component is low and two separate Lamb dips should be observed corresponding to the $\Delta M = +1$ and $\Delta M = -1$ pump transition components. This was actually observed as shown in Fig. 50.

The TLD was recorded with no electric field and in presence of an electric field of 700 V/cm. In the second case the Lamb dip corresponding to the $\Delta M = -1$ pump selection rule could be clearly seen. The dip corresponding to the $\Delta M = +1$ pump selection rule was less evident since it was close to the mode end of the CO_2 laser. The central crossover TLD

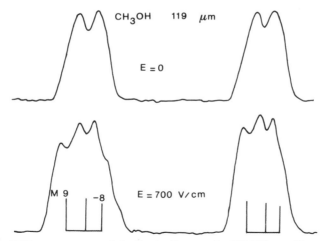

FIG. 50   TLD measurement of the Stark effect on the CH_3OH transition pumping the 119-μm line (Inguscio et al., 1979d).

could be also observed exactly at the same position as for zero electric field. At higher electric fields the two pump components were Stark broadened and the dip was difficult to observe. Moreover, the frequency splitting between the two components was so large that only one of them could be resonant with the pump radiation. A high electric field recording is shown in Fig. 51.

Only $\Delta M = -1$ pump transition components could be used. The measurement was performed at the particular electric field value (3.7 kV/cm) for which the frequency separation of the two FIR components was equal to the FIR resonator free spectral range, hence superimposed FIR action could be monitored for the same cavity setting. The computed gain is drawn below the experimental recording. The maximum gain on the line with pump $\Delta M = -1$ and FIR $\Delta M = -1$ selection rules is obtained for

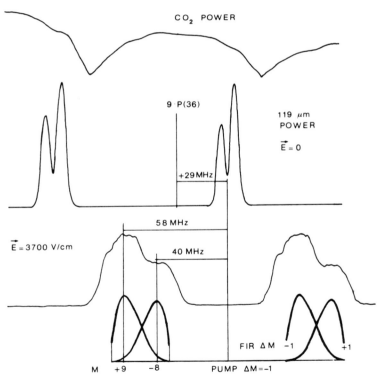

FIG. 51 Output power of the 119-μm CH$_3$OH laser line as a function of the CO$_2$ 9–P(36) pump frequency tuning (Inguscio et al., 1979d).

$M = +9$; while on the line with $\Delta M = -1$ pump and $\Delta M = +1$ FIR, it is obtained for $M = -8$. The calculated shifts from the unperturbed frequency are -58 and -40 MHz, respectively, in good agreement with the experimental results.

Stark detuning of the pump transition can limit high electric field operation of FIR lasers. A typical FIR power dependence on electric field intensity is shown in Fig. 52 for the 119-μm line pumped with $\Delta M = \mp 1$ selection rule. The large enhancement of the output power at very low electric fields depends on an increase in the infrared absorption of the gas and is discussed in Section V.F. A decrease in the output level is observable for Stark fields between 2 and 3 kV/cm. At these values the splitting between the $\Delta M = +1$ and $\Delta M = -1$ pump components is so large that the CO_2 laser can be frequency tuned to pump efficiently only one of them. As a consequence, the absorption of the gas is reduced by nearly a factor of 2. For electric fields larger than 6.5 kV/cm, the Stark effect on the pump transition is so large that the line frequency is nearly out of the tuning range of a conventional CO_2 pump laser (~ 90 MHz).

The TLD spectroscopy technique has also been successfully applied to investigate quadratic Stark effects in the IR pump transition. In this case a net frequency shift of the dip can be observable even at small electric fields. An example is given in Fig. 53. The frequency of the TLD on the FIR emission at 163 μm pumped with the 10-R(38) CO_2 laser line is shown

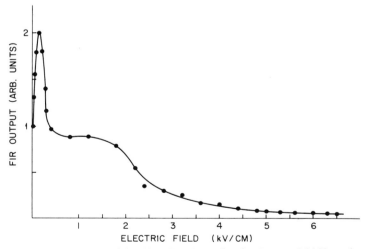

FIG. 52 Relative output of the CH_3OH 119-μm line for E up to 6.5 kV cm^{-1}: pressure, 40 mtorr; $\Delta M = \pm 1$. The CO_2 pump power was kept constant throughout. The cavity length was continuously tuned for the maximum output in presence of the electric field (Bionducci et al., 1979).

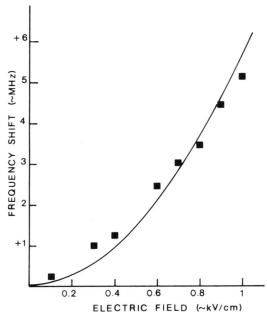

FIG. 53 Quadratic Stark effect on the IR pump transition of the CH_3OH yelding the 163-μm FIR laser line. The TLD technique was used. The continuous line shows the computed shift according to the assignment.

as a function of the electric field. The result is consistent with the presence of K splittings in both levels involved in the IR pump transition, as suggested by the assignment of Henningsen (1977) and measured in intracavity double resonance experiment by Arimondo *et al*. (1980).

F. ELECTRIC FIELD EFFECTS ON FIR OUTPUT LEVEL

A decrease in the power output was generally forecast for FIR lasers operated in presence of electric fields. In fact, the gain of the unperturbed line is divided between all the Stark components, according to the selection rule transition probabilities and saturation. On the contrary, small electric fields can cause a large enhancement of the FIR output, as first observed by Tobin and Jensen (1976a). Their result for the 119-μm line of the CH_3OH laser is shown in Fig. 54 (see also Fig. 52). Several authors reported similar results for other FIR lines (Tobin and Jensen, 1976b, 1977; Koo and Clapsy, 1979; Inguscio *et al*., 1979b, 1980a; Henningsen, 1980a; Bionducci *et al*., 1979). FIR line enhancement has been observed only for CO_2 pump polarization orthogonal to the electric field applied to the ac-

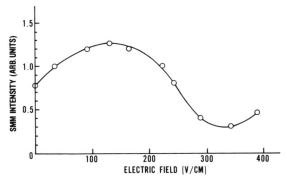

FIG. 54   Enhancement effect on a CH$_3$OH Stark FIR laser at 119 μm (Tobin and Jensen, 1976a).

tive medium. The relative output increase depends on the observed FIR line, on the FIR cavity losses, and in general on the threshold of the laser. The electric field at the maximum of the enhancement is of the order of 100 V/cm, the actual value depending also on the losses of the laser. Typical results for the CH$_3$F 496-μm line are shown in Fig. 55.

The CO$_2$ pump power and CH$_3$F gas pressure were kept constant during

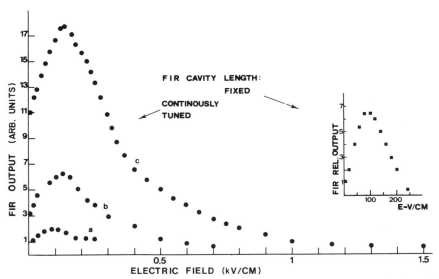

FIG. 55   Enhancement effect observed in the case of CO$_2$ 9–P(20) pump selection rule $\Delta M = \pm 1$ for the CH$_3$F 496-μm laser line (Inguscio *et al.*, 1980a, © 1980, IEEE).

the measurements. Frequency tuning of the laser emission was observed even at low electric field and the cavity length was always tuned for maximum FIR output at the different E values. The intensity of the strongest component was reported when the Stark structure of the line was resolved. By lowering the output coupling it was possible to measure the power emitted on three different transverse modes of the cavity (Curves a, b, c), hence with different losses. Changes in relative power enhancement and in the electric field at maximum output are evident. Measurements are also performed with cavity length optimized for zero Stark field. In this case, frequency detuning caused an apparent dramatic power cutoff even at relatively low electric fields, as shown in the insert of Fig. 55.

Enhancement measurements were also performed at different pressure values. Results for the CH_3OH 119-μm line are shown in Fig. 56. The maximum relative output increase depended on FIR laser threshold. The

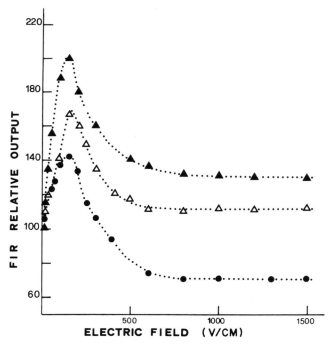

FIG. 56. Enhancement effect on 119-μm laser line for three different values of the CH_3OH (▲, 55; △, 90; and ●, 130 mtorr) pressure in the FIR cavity (CO_2 9–P(36) pump selection rule $\Delta M = \pm 1$) (Inguscio et al., 1980a, © 1980, IEEE).

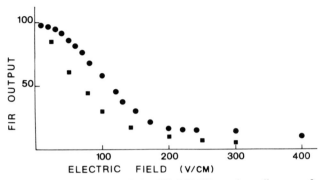

FIG. 57   Relative output power of the CH_3OH 170-μm laser line as a function of the Stark field (pump selection rules: ●, $\Delta M = \pm 1$; ■, $\Delta M = 0$) (Inguscio *et al.*, 1980a, © 1980, IEEE).

value of the electric field at the maximum and the power decrease after the maximum were not significantly affected by pressure.

Not all FIR lines exhibit this power enhancement feature. In Fig. 57 the power dependence on the electric field for the CH_3OH 170-μm line, is shown, a line that shares the upper level with the 119-μm line. No enhancement was observed for either orthogonal or parallel pump polarization. In the case of $\Delta M = 0$ pump selection rule, power decrease was

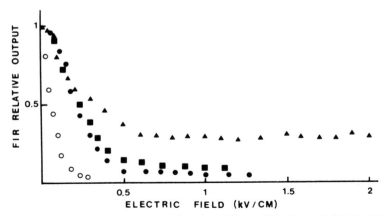

FIG. 58   FIR laser power output as a function of the applied electric field. The CO_2 laser radiation was linearly polarized orthogonal to the metal plates of the hybrid waveguide. (Pump selection rule $\Delta M = 0$. ○, 170; ▲, 119; and ■, 70 μm CH_3OH laser lines. ●, 496 CH_3F laser line.) (Inguscio *et al.*, 1980a, © 1980, IEEE).

stronger than for $\Delta M = \mp 1$ pump selection rule. Results on different lines with $\Delta M = 0$ pump selection rule are shown in Fig. 58. The electric field causes only a power decrease. It is worth noting that the power decrease at 170 μm is dramatic compared with that of other lines. The dependence of amplitude on the electric field was interpreted by Inguscio *et al.* (1980a) as an effect of increased pump absorption combined with Stark broadening and lowering of the gain curve as discussed in the Section V.G.

G. OPTOACOUSTIC STARK SPECTROSCOPY INVESTIGATION OF PUMP SATURATION

The enhancement effect in FIR emission is a consequence of the increase in the number of molecules in the excited level. Optically pumped FIR lasers are a three-level coupled system and within the Doppler profile of the IR transition only a fraction of molecules corresponding to the homogeneous linewidth is excited to the upper level. In typical FIR experimental arrangements, pump power saturates the absorbing molecular transition, as discussed in Section IV.E. As a consequence, absorption in the gas and the number of molecules in the excited level are increased if the Stark effect modifies the saturation condition. This effect is a molecular-Stark version of the stimulated Hanle effect, as discussed by Feld *et al.* (1974). Stark induced increases in the pump saturation parameter and hence in the infrared absorption of CH_3F and CH_3OH were investigated by Inguscio *et al.* (1979h, 1980a). Evidence of the effect was obtained by monitoring the multipass transmission at 10 μm of a Stark waveguide. A decrease in the signal was detected when an electric field orthogonal to the IR radiation polarization was applied on the absorbing gas. A typical result is shown in Fig. 59 for CH_3F (50 mtorr), absorbing CO_2 9–$P(20)$ radiation. The decrease of transmission was evident for field strengths larger than a few tens of V/cm. Recently evidence for Stark induced absorption increases was also reported by Koo and Clapsy (1979). A quantitative investigation could be performed (Inguscio *et al.*, 1980a) using a single-pass optoacoustic absorption cell. CO_2 frequency was PZT tuned for maximum optoacoustic signal at zero electric field. Increase in infrared absorption of the gas in presence of an orthogonal electric field caused a significant enhancement of the optoacoustic signal. Measurements were performed on various gases, with different CO_2 laser lines and pressures. Typical results for CH_3OH are showed in Fig. 60. Relative absorption increase was higher at lower pressures. A maximum increase of nearly 30% could be observed. These results are in agreement with the stimulated Hanle effect explanation since the saturation at fixed pump power intensity is higher at lower pressures. A detailed calculation of the absorption increase effect is rather complex, since many M sublevels are

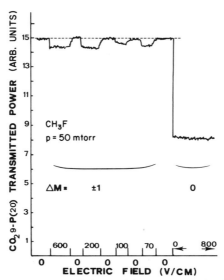

FIG. 59 Infrared CO_2 $9-P(20)$ radiation multipass transmission from the hybrid Stark waveguide resonator. With the CO_2 polarization parallel to the Stark field no absorption changes are observed for electric field up to 800 V/cm. On the contrary, an absorption increase can be observed when the CO_2 polarization is orthogonal to the Stark field (Inguscio et al., 1980a, © 1980, IEEE).

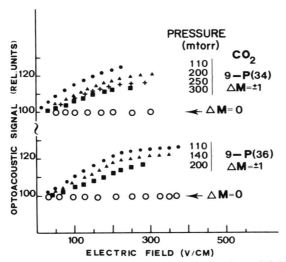

FIG. 60 Optoacoustic absorption measurements observed on CH_3OH for two CO_2 pump lines $9-P(34)$ and $9-P(36)$, respectively, as a function of the applied electric field. An absorption increase can be observed only when the CO_2 is polarized orthogonal to the Stark field (pump selection rule $\Delta M = \pm 1$) (Inguscio et al., 1980a, © 1980, IEEE).

197

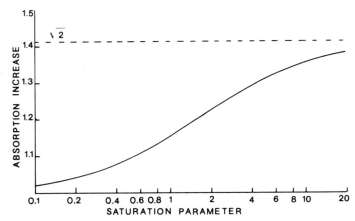

FIG. 61 Expected relative maximum absorption increase as a consequence of the non-linear zero field level crossing as a function of the saturation S in case of a J: $0 \rightarrow 1$ transition.

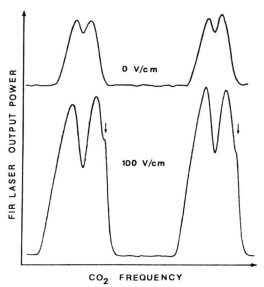

FIG. 62 Transferred Lamb dip (TLD) observed in the emission of the 119-μm FIR laser line as a function of the CO_2 radiation frequency tuning. Nearly 2 orders are scanned and the arrows indicate the position of the CO_2 mode change. The CH_3OH pressure was 35 mtorr. The CO_2 polarization was parallel to the Stark plates (Inguscio *et al.*, 1980a, © 1980, IEEE).

involved (Strumia *et al.*, 1981). In the case of J: $0 \to 1$ transitions the absorption increase can be easily calculated (Inguscio *et al.*, 1980a) using the theory of Feld *et al.* (1974). The computed maximum absorption relative increase is shown in Fig. 61 as a function of the saturation parameter $S = I_p/I_s$, where I_p is the pump intensity and I_s the saturation intensity.

Direct evidence of increase in homogeneous linewidth as a consequence of the Stark effect can be obtained using the transferred Lamb dip technique described in Sections IV.E and V.E. In Fig. 62 the FIR output at 119 μm is recorded as a function of pump frequency tuning. The two traces were recorded with the same pump power and with the FIR resonator tuned at the center of the gain curve. The upper trace was taken in absence of a Stark field while with the lower one a field of 100 V/cm was present. The emission enhancement was obtained for the whole CO_2 frequency tuning range. TLD width was nearly 1.5 times larger in presence of an electric field.

The absorption increases caused by the Stark effect can affect the FIR laser output. FIR output depends on the *pump rate* and on the *small signal gain curve*. The combined effects of the absorption increase and of the small signal gain curve decrease caused by the Stark effect explain the observed FIR enhancement features (see Fig. 63). As a consequence of the increase in the IR absorption, in case of $\Delta M = \mp 1$ selection rule, there is an increase of the pump rate for electric fields up to $100 \sim 200$ V/cm, as experimentally reported in Fig. 62 and schematically shown in Fig. 63a. A constant level is obtained for electric fields of about $300 \sim 500$ V/cm.

As for the small-signal gain curve, it is in any case lowered when an electric field is applied to the active medium, as a consequence of the Stark broadening and resolution. In order to combine this effect with the pump rate increase it is necessary to note that the gain decrease can be dramatic or not, depending on the number and relative intensity of the resolved components. Consider, for instance, the two extremal cases of the 119- and 170-μm FIR lines which share a common pump transition. The increase in the pump rate is, of course, common to the two lines, whereas the gain decrease for the 170-μm line is much larger than that of the 119-μm line. Experimental evidence for the different gain curve broadening and decrease can be inferred from the FIR power measurements with pump polarization parallel to the static electric field, as reported in Fig. 60 (in the case of $\Delta M = 0$ selection rule, the Stark tuning of the pump transition can be neglected and the FIR power decrease at low electric field can be mostly ascribed to the lowering of the gain curve). With an electric field of about 100 V/cm, the intensity at 119 μm (and also at 70 and 496 μm) is more than 90% the zero electric field value; on the

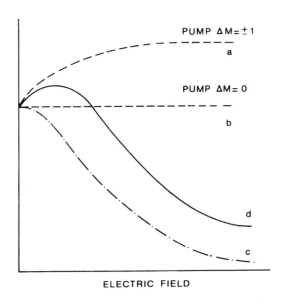

ELECTRIC FIELD

FIG. 63   Schematic explanation of the enhancement effects as a function of the applied electric field. Curves *a* and *b* are proportional to the upper level population for the two CO_2 polarizations. Curve *c* is proportional to the FIR laser gain, which is lowered as a consequence of the resolution of the Stark components. When the pump selection rule is $\Delta M = 0$, the net FIR gain is the product of curves *b* and *c*; when the pump is $\Delta M = \mp 1$, the net FIR gain is the product of curves *a* and *c* and an enhancement can be observed as schematically shown in curve *d* (Inguscio *et al.*, 1980a, © 1980, IEEE).

contrary, the intensity at 170 μm is dramatically reduced to about 30% (see also Fig. 57). This difference is caused by the different Stark patterns calculated in Fig. 64, according to the selection rules $\Delta M = \pm 1$. The Stark pattern of the 170-μm line consists of nearly 50 components, while for the 119-μm line there is a collapse into only two multiplets.

Considering again the FIR enhancement effect, it can be observed only for lines which have no dramatic decrease of the gain curve (Fig. 63c). The pumping rate increase predominates at low electric fields, while the gain decrease predominates once the constant level of the pump rate has been achieved. The resulting FIR enhancement curve is schematically shown in Fig. 63d. The relative importance of the two combined effects also depends on laser threshold.

FIR lines with a strong gain curve decrease exhibit in any case a power decrease. The low electric field pump rate increase only causes the FIR intensity decrease to be slower for $\Delta M = \pm 1$ than for $\Delta M = 0$ pump selection rules, as shown in Fig. 57.

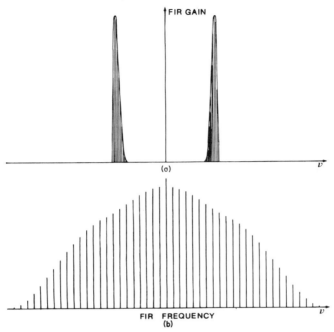

FIG. 64   Theoretical small signal gain Stark patterns of the CH_3OH (a) 119- and (b) 170-μm laser lines, respectively, assuming the pump selection rule $\Delta M = \pm 1$ and an electric field of 1 kV/cm (Inguscio *et al.*, 1980a, © 1980, IEEE).

H.   High-Resolution Spectroscopy in Stark-FIR Lasers: New Techniques

In Section IV.E we have illustrated the three-coupled level quantum-mechanical features of optically pumped FIR lasers. These features can lead to the development of new spectroscopic techniques yielding resolution inaccessible with conventional two-level schemes. Spectroscopy inside the homogeneous linewidth was recently reported by Inguscio *et al.* (1980c). The method can be understood by referring to Fig. 65. High-resolution optical spectroscopy of the radiative transitions between two energy levels, in general, allows resolution of the structures of the two states if they are larger than the energy indetermination. On the contrary, Fig. 65a shows a case in which the energy splittings are larger than the homogeneous width in both levels, although nearly equal; as a consequence, the photons associated with the $1 \rightarrow 2$ transition have nearly the same energy and cannot be resolved within the homogeneous linewidth. A three-coupled level version of the saturation spectroscopy technique is

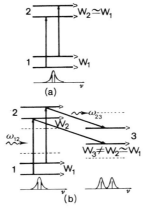

shown in Fig. 65b. The photons associated to the $1 \rightarrow 2$ transition, unresolved within the homogeneous linewidth, are coupled to the photons emitted in the $2 \rightarrow 3$ transition, in which the line structures are resolved. The saturation spectroscopy spectrum of the $1 \rightarrow 2$ transition can thus be transferred and resolved in the coupled $2 \rightarrow 3$ transitions, using the transferred Lamb dip technique. Referring to the principle of operation of an optically pumped FIR laser, 1 is a rotational level of the ground vibrational manifold and 2 and 3 are rotational levels of an excited vibrational manifold of a polar molecule. The $1 \rightarrow 2$ transition is the 10-μm pump transition and the $2 \rightarrow 3$ is the coupled FIR laser emission. Placing the active molecules in presence of a static electric field E allows Stark frequency tuning of the vibrational and rotational transitions. According to the theory developed in Section II, one shall have different Stark energies on the three levels involved in the pump cycle, W_1, W_2, and W_3. The absorbed IR or emitted FIR lines consist of different components, equally spaced by a quantity $|W_2 - W_1| E$ or $|W_3 - W_2| E$, respectively. When the frequency splittings are larger than the homogeneous linewidth, the Stark components can be resolved using conventional sub-Doppler techniques. In Stark spectroscopy of molecules a usual result is a large frequency tuning of the rotational FIR transition. In fact the changes in J and/or K values cause W_2 to be significantly different from W_3 [see Eq. (5)]. On the contrary, if the molecule is pumped through a ^qQ branch, the difference between W_1 and W_2 is due only to the μ value in the two vibrational states, which, in general, differ only by a few percent. As a consequence, the Stark structure of the infrared vibrational transition can be unresolved in spite of an energy sublevels splitting which can be much larger than the level width both in 1 and 2 states.

In Section V.C we have shown that in optically pumped molecular lasers collision-induced transitions between different M sublevels can be neglected. This means that in presence of an electric field the $1 \rightarrow 2 \rightarrow 3$ cycle is simply divided into many cycles, independent of each other, corresponding to the different M sublevels in the 1 state. The photon that originates from a given Stark component of the infrared transition $1 \rightarrow 2$ is labeled by the corresponding photon of the Stark component in the coupled FIR emission. If the latter is resolved, the structure of the IR transition too can be resolved by means of the IR–FIR transferred Lamb dip.

The technique is demonstrated in Fig. 66 for the FIR emission at 205 μm from CH_3OH pumped with the $9-P(34)$ CO_2 laser line. Following the scheme of Fig. 65b, for the 205-μm line, the three levels quantum numbers are (J, K): $(6, 5) \rightarrow (6, 5) \rightarrow (5, 4)$ (Inguscio et al., 1979g). Using Eq. (5) and the dipoles of Table I, the Stark splittings in the levels are given by $W_1 = 53.3$, $W_2 = 54.5$, and $W_3 = 61.0$ MHz/kV cm⁻¹.

The Stark splittings in the $1 \rightarrow 2$ infrared transition are given by

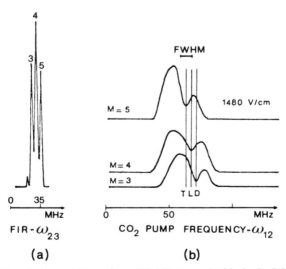

FIG. 66 FIR emission at 205 μm from CH_3OH pumped with the $9-P(34)$ CO_2 laser line in presence of an electric field of 1480 V/cm. (a) Resolution of the Stark components of the $2 \rightarrow 3$ transition. Interferogram obtained by scanning the FIR cavity at fixed pump frequency. (b) Transferred Lamb dips recorded on three different components of the FIR emission (Inguscio et al., 1980c).

$|W_2 - W_1| = 1.2$ MHz/kV cm^{-1}, while in the coupled $2 \to 3$ FIR transition they are $|W_3 - W_2| = 6.5$ MHz/kV cm^{-1}. The levels width γ is of the order of a few megahertz. As a consequence for electric fields of the order of 1 kV/cm, the Stark structure of the levels is resolved but cannot be directly observed in the $1 \to 2$ transition line since $|W_2 - W_1| < \gamma$. On the contrary with electric fields of the order of 1 kV/cm $|W_3 - W_2| > \gamma$ and the Stark structure of the $2 \to 3$ transition can be observed.

A typical recording is shown in Fig. 66a; FIR laser oscillation can be obtained on individual Stark components. By setting the FIR cavity length for oscillation on a given Stark component, and by scanning the CO_2 frequency, it is detected the Doppler free transferred Lamb dip corresponding to the saturation of the component of the IR pump transition to which the observed FIR component is coupled. A typical recording is shown in Fig. 66b. The transferred Lamb dips corresponding to different Stark FIR components are clearly observed at different positions during the CO_2 frequency scanning. The electric field intensity was 1480 V/cm, corresponding to a dip frequency separation $|W_2 - W_1| \simeq 1.8$ MHz. These dips should be unresolved in conventional sub-Doppler two-level spectroscopy (the homogeneous linewidth of the two-level transition can be estimated from the dips FWHM to be $\gamma \approx 5$ MHz).

Applications of the method can be foreseen for very high precision determination of the molecular excited dipoles, independently of a precise calibration of the electric field. In fact, the Stark structures recorded in Fig. 66a depend on the electric dipole moment of the excited level, while the frequency separation in the transferred Lamb dips recordings (Fig. 66b) depend on the difference of the dipole moments in the ground and excited states. In both cases the dependence on E is linear and the frequency splittings both at IR and FIR wavelengths can be easily determined with high accuracy by beating with reference lasers.

VI. Experimental Results—ac Electric Fields

Alternating electric fields can also be applied to the medium of optically pumped FIR lasers. The half width $\Delta\nu$ of the molecular levels is of the order of a few megahertz. The inverse of the half-width is a measure of the relaxation time of the molecules in the gas, or the time required for any transient phenomenon to disappear. As a consequence, if the frequency of the ac field is lower than $\Delta\nu$, the Stark field itself can be considered essentially static at any time and the Stark effect calculated accordingly. Effects on the amplitude and on the frequency of the lasers have been reported, as discussed below.

A. Effects on the Output Power

The application of an ac electric field allowed Fetterman *et al.* (1976) to reduce saturation owing to hole burning and hence pump the entire velocity distribution in the initial state. They observed an enhancement greater than 10 in the output from a NH_3 laser in presence of a 100 kHz ac field. ac electric field power enhancements were also observed for a CH_3OH FIR laser (Inguscio, 1979). An ac generator was capacitively coupled to the Stark plates. A sinusoidal function was used with a maximum of 32 V_{pp} and a maximum frequency of 1 MHz. A typical result for the CH_3OH 70.5-μm line is shown in Fig. 67.

The gas pressure was 160 mtorr, the laser cavity length was 154 cm, and the Stark plate separation was $d = 0.5$ cm. In 67a the ac frequency was kept constant at 500 kHz and the FIR amplitude was reported as a function of the applied ac field peak to peak amplitude. In 67b the peak to peak amplitude was kept constant at 24 V and the frequency of the ac field was changed. For the measurements reported in the upper curve only the ac electric field was applied. The lower curve was obtained by also applying a dc electric field of 60 V/cm. In this case the enhancement described in

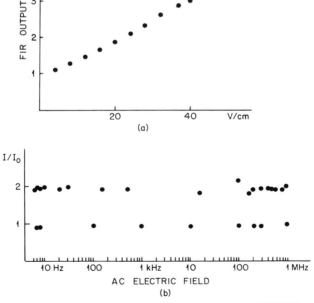

FIG. 67 Alternating electric field effects on the 70.5-μm CH_3OH FIR laser output level. (a) f = 500 kHz; DC, 0V.

Section V.F was observed and no further increase of the FIR output level was caused by the ac electric field.

Amplitude modulation of the FIR output by ac Stark fields have also been reported (Tobin and Jensen, 1976b; Koo and Clapsy, 1976).

B. HIGH-SPEED FM MODULATION

The ac Stark effect has been successfully used to generate high-speed frequency modulation of optically pumped FIR lasers. If the electric field is amplitude modulated, the molecular levels separation changes according to the applied field and the laser output is frequency modulated. In order to obtain a linear frequency modulation, one should choose transitions with linear Stark effect. An appropriate static electric field can also be applied to the active medium in order to obtain a weak dependence of the laser output power on the field strength. In this case, the FIR amplitude modulation is small and an almost pure FM can be obtained. High-speed frequency modulation was first reported by Stein et $al.$ (1977) on a 496-μm CH_3F laser. The experimental apparatus was essentially the same used for the observation of the frequency shifts induced by Stark effect (Section V.B). A maximum modulation frequency of 300 kHz was observed.

More recently modulation frequency up to 1 MHz was reported for the CH_3OH line at 119 μm (Inguscio et $al.$, 1978b, 1979b). The experimental apparatus was the same schematically shown in Fig. 40. The output from the Stark FIR laser was mixed in a MIM diode with the output of a conventional FIR laser. An ac voltage was applied to the Stark plates. The frequency of the modulating electric field was changed up to 1 MHz, while the amplitude was kept constant at 60 V_{pp}/cm. The signals shown in Fig. 68 were obtained using a spectrum analyzer connected to the amplitude heterodyne signal from the diode. Since only the Stark laser was modulated, the spectrum analyzer display represented the power density of this laser as a function of the frequency. A frequency modulated wave with sinusoidal modulation is by definition a wave in which the instantaneous angular velocity is varied according to the relation

$$\omega_i = \omega_c + 2\pi\Delta f \cos \omega_m t, \tag{33}$$

where ω_i is the instantaneous angular velocity, ω_c the angular velocity of the unmodulated wave (carrier wave), hence the average angular velocity; ω_m 2π times the modulating frequency f_m; and Δf the maximum deviation of instantaneous frequency from average.

The equation of the frequency modulated wave is commonly written in

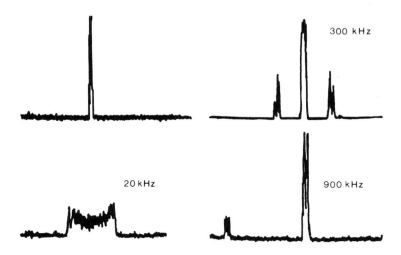

FIG. 68   Spectrum analyzer signals obtained by beating a frequency modulated CH_3OH laser at 119 μm with a reference laser.

the form

$$E = E_0 \sin(\omega_c t + m_f \sin \omega_m t), \qquad (34)$$

where $m_f = \Delta f/f_m$ is the modulation index of the frequency modulated wave. In the spectral analysis of the wave represented by Eq. (34), in addition to the carrier wave, frequency components spaced by the modulating frequency are present. The amplitude of the different frequency components depends upon the modulation index m_f. The character of the frequency spectra can be found in Terman (1955). When the modulation index exceeds unity, there are important higher order sideband components contained in the wave. This is the case of the spectrum obtained with low-frequency modulation and shown in Fig. 68. A modulation index of the order of 20 can be estimated. From the frequency modulation spectra reported in Fig. 68 for $f_m = 300$ kHz and $f_m = 900$ kHz, a modulation index of 1.2 and 0.4, respectively, can be estimated. The results are consistent with the static electric field tuning sensitivity, as discussed by Inguscio (1979).

These results demonstrate that optically pumped FIR lasers can be electronically high-speed frequency modulated. The method is generally applicable to any lasing medium. It can be successfully also used for FIR lines which do not exhibit a net frequency shift but only a broadening.

In this case the cavity pulling effect controlled by the Stark broadening of the gain curve can be used as discussed in Section V.B and experimentally demonstrated by Stein *et al.* (1977). Another appreciable feature of the method is that no introduction of lossy material into the laser is required. The maximum modulation speed that one can actually achieve also depends on the modulation index. Results even better than those described in this section can be obtained increasing the ac signal amplitude or choosing FIR lines with larger Stark coefficient. Stein and Van der Stadt (1977) and Stein *et al.* (1978) used this method to phase lock a CH_3F laser at 496 μm.

High-speed frequency modulation of FIR lasers is needed, for instance, in order to be able to phase lock such lasers with no loss of precision to frequencies multiplied upward from the radio-frequency region.

It is worth noting that the frequency noise of free-running FIR systems is of the order of 1 kHz and as a consequence a band width of only 10–20 kHz should be sufficient to phase lock the optically pumped lasers.

VII.   Conclusion

Optically pumped FIR molecular lasers have proved to be a powerful tool for spectroscopic investigation. Laser system have achieved output levels of many mW cw, power fluctuations less than a few percent and free-running frequency stability as good as 1 part to 10^9.

Stark operation of optically pumped FIR lasers has been observed for electric fields applied to the active molecules as high as several tens of kV/cm.

As a first result, *FIR operation on new lines* has been made possible by Stark tuning into resonance infrared pump transitions which are close to CO_2 or N_2O laser lines.

As a more impressive result the possibility of *wide-frequency tuning* has been demonstrated as well as *fast-frequency control* of the FIR emissions. Use of these results can be foreseen in applied physics: frequency locking of FIR oscillators in frequency multiplications chains; electronic phase control of FIR sources used in interferometry for plasma diagnostic. Beside the use in applied physics, Stark FIR lasers are powerful tools for investigation in *fundamental physics*.

Direct frequency measurement of the FIR output allows the determination of the energy of highly excited rotovibrational levels of the active molecules. As a result it has been possible to investigate small effects on the molecular constants that could not be determined by conventional microwave spectroscopy. This leads to a better understanding of the structure of key molecules widely used in other fields of the modern molecular spectroscopy, such as the laser-Stark or the microwave-laser spectroscopy.

The three-coupled level features of optically pumped FIR lasers made it possible to develop new *high-resolution spectroscopy* techniques.

The transferred Lamb dip (TLD) technique allows the transfer of infrared molecular saturation spectra to the coupled FIR transitions. Owing to the nonlinear response of the laser, the sensitivity and the resolution of the conventional saturation spectroscopy can be significantly increased. For instance, it has been demonstrated that in the IR–FIR Stark spectroscopy of molecules it is possible to obtain spectral resolutions which, in conventional IR Stark saturation spectroscopy, are inaccessible inside the homogeneous width of the lines. The understanding of some important features of the molecule–molecule collision interactions has been made possible by the Stark operation of FIR lasers. The application of the electric field and the resolution of the Stark structure of the laser line allow a direct visualization of the nonrandomizing effect of the collisions on the M levels of the molecule. Also the FIR laser output power can be affected by the application of an electric field on the active medium. Under proper experimental conditions it was possible to observe large power enhancements. This effect is owing to an increase in the infrared absorption caused by small electric fields. It can be understood in terms of nonlinear Hanle effect occurring in a quantum-mechanical system in presence of energy level degeneracy and saturating laser radiation. The observation of this effect allows a generalization to a multilevel model of results previously obtained in simpler physical schemes.

The new techniques so far recalled can be a vaulable help for the understanding of the basic physical mechanism underlying optically pumped laser action.

Moreover the experimental results obtained by using these new techniques are very sensitive to the quantum numbers and to the selection rules of the molecular transitions involved in the pump-laser action.

The molecular spectra can be much complicated by the presence of even simple physical interactions, so that even the spectra of simple molecules may become practically ununderstandable. The new techniques based on the Stark operation of FIR lasers provide new experimental tools which allow to tackle even very complex situations.

ACKNOWLEDGMENTS

Research on optically pumped submillimeter wave lasers by the authors was and is supported by the Gruppo Nazionale di Struttura della Materia of the National Research Council and the Italian Ministry of Education.

The results described in this work would not have been possible without the help and support of many people. Special thanks are extended to our friend Dr. A. Moretti for his enthusiasm during many hours spent in laboratory. We wish to thank for active help in our experimental and theoretical studies to Drs. G. Bionducci, N. Ioli, G. Moruzzi, M. Pellegrino. The skillful technical assistance of Mrs. M. Francesconi and M. Montanari is acknowledged.

For many valuable discussions in the course of the research and writing this review we are greatly indebted to: E. Arimondo, Pisa; K. J. Button, M.I.T.; T. Y. Chang, Bell; E. D. Danielewicz, Aerospace Corp.; K. M. Evenson, NBS; M. S. Feld, M.I.T.; H. R. Fetterman, M.I.T.; J. O. Henningsen, Copenhagen; D. T. Hodges, Aerospace Corp.; J. Jimenez, Orsay; T. Oka, Ottawa; J. Sattler, Harry Diamond; M. S. Tobin, Harry Diamond; C. O. Weiss, PTB.

Last, but not least, the work would never have been completed without the support, love, and understanding of our wives Rossana and Giovanna.

REFERENCES

Abrams, L. R. (1979). *In* "Laser Handbook" (M. L. Stitch, ed.), Vol. 3, pp. 41–88. North-Holland Publ., Amsterdam.

Adam, B., and Kneubühl, F. (1975). *Appl. Phys.* **8**, 281–291.

Anderson, P. W. (1949). *Phys. Rev.* **76**, 647–661.

Arimondo, E., and Inguscio, M. (1979). *J. Mol. Spectrosc.* **75**, 81–86.

Arimondo, E., Inguscio, M., Moretti, A., Pellegrino, M., and Strumia, F. (1980). *Opt. Lett.* **5**, 496–498.

Bava, E. *et al.* (1977). *Opt. Commun.* **21**, 46–48.

Bedwell, D. J., Duxbury, G., Herman, H., and Orengo, C. A. (1978). *Infrared Phys.* **18**, 453–460.

Benedetti, R., Di Lieto, A., Inguscio, M., Minguzzi, P., Strumia, F., and Tonelli, M. (1977). *Proc. Symp. Frequency Control, 31st, Atlantic City, New Jersey* pp. 605–611.

Bionducci, G., Inguscio, M., Moretti, A., and Strumia, F. (1979). *Infrared Phys.* **19**, 297–308.

Birnbaum, G. (1967). *J. Chem. Phys.* **46**, 2455–2460.

Bluyssen, H. J. A., Van Etteger, A., Maan, J. C., and Wyder, P. (1980). *IEEE J. Quantum Electron.* **QE-16** 1347–1351.

Brewer, R. G. (1973). *In* "Fundamental and Applied Laser Physics" (M. S. Feld, A. Javan, and N. A. Kurnit, eds.) (*Proc. Esfahan Symp.*), pp. 421–436. Wiley, New York.

Burkhard, D. G., and Dennison, D. M. (1951). *Phys. Rev.* **84**, 408–417.

Chang, T. Y. (1974). *IEEE Trans. Microwave Theory Tech.* **MTT-22**, 983–988.

Chang, T. Y. (1977). *In* "Topics in Applied Physics" (Y. R. Shen, ed.), Vol. 16, pp. 215–272. Springer Verlag, Berlin and New York.

Chang, T. Y., and Bridges, T. J. (1970). *Opt. Commun.* **1**, 423–426.

Chantry, G. W., and Duxbury, G. (1974). *Methods Exp. Phys.* **3A**, 302–394.

Coleman, P. D. (1973). *IEEE J. Quantum Electron.* **QE-9**, 130–138.

Cosmovici, C. B., Inguscio, M., Strafella, F., and Strumia, F. (1979). *Astron. Space Sci.* **60**, 475–491.

Danielewicz, E. J., and Coleman, P. D. (1977). *IEEE J. Quantum Electron.* **QE-13**, 485–490.

Danielewicz, E. J., Plant, T. K., and De Temple, T. A. (1975). *Opt. Commun.* **13**, 366–369.

Degnan, J. J. (1976). *Appl. Phys.* **11**, 1–33.

De Marchi, A., Godone, A., and Bava, E. (1979). *Proc. 33rd Conf. on Freq. Control,* 504–510.

De Temple, T. A., and Danielewicz, E. J. (1976). *IEEE J. Quantum Electron.* **QE-12**, 40–47.

Duxbury, G., and Herman, H. (1978). *J. Phys. B* **11**, 935–949.

Evans, D. E., Prutney, S. L., and Sexton, M. C. (1980). *Infrared Phys.* **20**, 21–27.

Evenson, K. M., Broida, H. P., Wells, J. S., Mahler, R. J., and Mizushima, M. (1968). *Phys. Rev. Lett.* **21**, 1038–1040.

Evenson, K. M. *et al.* (1977). *IEEE J. Quantum Electron.* **QE-13**, 442–444.

Feld, M. S., Sanchez, A., Javan, A., and Feldman, B. J. (1974). *Proc. Colloq. Int. CNRS* No. 217, pp. 87–104.

Feldman, B. J., and Feld, M. S. (1972). *Phys. Rev. A* **5**, 899–918.

Feldman, B. J., and Feld, M. S. (1975). *Phys. Rev. A* **12**, 1013–1018.

Fesenko, L. D., and Dyubko, S. F. (1976). *Sov. J. Quantum Electron.* **6**, 839–843.

Fetterman, H. R., Schlossberg, H. R., and Parker, C. D. (1973). *Appl. Phys.* **23**, 684–686.

Fetterman, H. R., Parker, C. D., and Tannenwald, P. E. (1976). *Opt. Commun.* **18**, 10–12.

Forber, R. A., Tenenbaum, J., and Feld, M. S. (1980), *Int. J. IR MM Waves* **1**, 527–560.

Freund, S. M., Duxbury, G., Romheld, M., Tiedje, J. T., and Oka, T. (1974). *J. Mol. Spectrosc.* **52**, 38–57.

Garmire, E. (1976). *Appl. Opt.* **15**, 3037–2039.

Garmire, E., McMahon, T., and Bass, M. (1976). *Appl. Opt.* **15**, 145–150.

Gordy, W., and Cook, R. L. (1970). "Microwave Molecular Spectra." Wiley (Interscience), New York.

Henningsen, J. O. (1977). *IEEE J. Quantum Electron.* **QE-13**, 435–441.

Henningsen, J. O. (1980a). *J. Mol. Spectrosc.* **83**, 70–93.

Henningsen, J. O. (1980b). *Proc. Int. Conf. IR MM Waves, 5th, Wurzburg* pp. 373–374.

Henningsen, J. O., and Jensen, H. G. (1975). *IEEE J. Quantum Electron.* **QE-11**, 248–252.

Heppner, J., and Weiss, C. O. (1977). *Opt. Commun.* **21**, 324–326.

Hodges, D. T., and Hartwick, T. S. (1973). *Phys. Lett.* **23**, 252–253.

Hodges, D. T., Tucker, J. R., and Hartwick, T. S. (1976). *Infrared Phys.* **16**, 175–182.

Hodges, D. T., Foote, F. B., and Reel, R. D. (1977). *IEEE J. Quantum Electron.* **QE-13**, 491–494.

Hodges, D. T. (1978). *Infrared Phys.* **18**, 375–384.

Inguscio, M. (1979). Ph. D. Thesis, Scuola Normale Superiore, Pisa. Univ. Microfilms International, Diss-Abstr. C 14/4644.

Inguscio, M., Minguzzi, P., and Strumia, F. (1975). *Proc. Int. Conf. Infrared Phys., 1st, Zurich* pp. 229–230.

Inguscio, M., Minguzzi, P., and Strumia, F. (1976). *Infrared Phys.* **16**, 453–456.

Inguscio, M., Minguzzi, P., and Tonelli, M. (1977). *Opt. Commun.* **21**, 208–210.

Inguscio, M., Minguzzi, P., Strumia, F., and Tonelli, M. (1978a). *Proc. Int. Conf. Submm. Waves, 3rd, Guildford,* pp. 144–145.

Inguscio, M., Minguzzi, P., Strumia, F., and Tonelli, M. (1978b). *Proc. Int. Conf. Submm. Waves, 3rd, Guildford,* pp. 187–188.

Inguscio, M., Strumia, F., Evenson, K. M., Jennings, D. A., Scalabrin, A., and Stein, S. R. (1979a). *Opt. Lett.* **4**, 9–11.

Inguscio, M., Minguzzi, P., Moretti, A., Strumia, F., and Tonelli, M. (1979b). *Appl. Phys.* **18**, 261–270.

Inguscio, M., Moretti, A., and Strumia, F. (1979c). *Proc. Int. Conf. Infrared Phys., 2nd, Zurich* pp. 199–200.

Inguscio, M., Moretti, A., and Strumia, F. (1979d). *Optics Commun.* **30**, 355–360.

Inguscio, M., Moretti, A., and Strumia, F. (1979e). *In* "Springer Series in Optical Sciences" (H. Walther and K. W. Rothe, eds.), Vol. 21, pp. 634–636. Springer Verlag, Berlin and New York.

Inguscio, M., Moretti, A., Moruzzi, G., and Strumia, F. (1979f). *Proc. Int. Conf. IR MM Waves Their Appl., 4th IEEE* Catalogue No. 79 CH 1384-7 MTT, pp. 201–202; and *Int. J. IR MM Waves* (1981), Vol. 2, No. 5.

Inguscio, M., Moretti, A., and Strumia, F. (1979g). *Proc. Int. Conf. IR MM Waves Their Appl., 4th* IEEE Catalogue No. 79 CH 1384-7 MTT, pp. 205–206.

Inguscio, M., Moretti, A., and Strumia, F., (1979h). *Proc. Int. Conf. Infrared Phys, 2nd, Zurich* pp. 211–213.

Inguscio, M., Moretti, A., and Strumia, F. (1980a). *IEEE J. Quantum Electron.* **QE-16**, 955–963.

Inguscio, M., Ioli, N., Moretti, A., Moruzzi, G., and Strumia, F. (1980b). *Proc. Int. Conf. IR MM Waves, 5th* 61–62; *Opt. Commun.* **37**, 211–216.

Inguscio, M., Moretti, A., and Strumia, F. (1980c). *Opt. Commun.* **35**, 64–68.

Inguscio, M., Moretti, A., and Strumia, F. (1980d). *Opt. Commun.* **32**, 87–90.

Inguscio, M., Moretti, A., and Strumia, F. (1980e). *In* "Lasers '79" (V. J. Corcoran, ed.), pp. 678–685. STS Press, McLean, Virginia.

Ioli, N., Moruzzi, G., and Strumia, F. (1980). *Lett. Nuovo Cimento* **28**, 257–264.

Ivash, E. V., and Dennison, D. M., (1953). *J. Chem. Phys.* **21**, 1804–1816.

Izatt, J. R., Bernard, P., and Mathieu, P. (1979). *Proc. Int. Conf. IR MM Waves Their Appl., 4th* IEEE Catalogue No. 79 CH 1384-7 MTT, pp. 165–166; *Int. J. IR MM Waves* **2**, 65–70.

Jimenez, J. J. (1979). *Radio Sci.* **14**, 541–560.

Johns, J. W. C., Mc Kellar, A. R. W., Oka, T., and Romheld, M. (1975). *J. Chem. Phys.* **62**, 1488–1496.

Kneubühl, F. K. (1977). *J. Opt. Soc. Am.* **67**, 959–963.

Kneubühl, F. K., and Affolter, E. (1979). *In* "Infrared and Submillimeter Waves" (K. J. Button, ed.), Vol. I, pp. 235–278. Academic Press, New York.

Kneubühl, F. K., and Sturzenegger, Ch. (1980). *In* "Infrared and Submillimeter Waves" (K. J. Button, ed.), Vol. 3. pp. 219–274. Academic Press, New York.

Koepf, G. A., and Smith, K. (1978). *IEEE J. Quantum Electron.* **QE-14**, 333–338.

Koo, K. P., and Clapsy, P. C. (1976). *Proc. Int. Conf. Submm. Waves Their Appl., 2nd* IEEE Catalogue No. 76 CH 1152-8 MTT, pp. 171–172.

Koo, K. P., and Clapsy, P. C. (1979). *Appl. Opt.* **18**, 1314–1321.

Kramer, G., and Weiss, C. O. (1976). *Appl. Phys.* **10**, 187–188.

Kwan, Y. Y., and Dennison, D. M. (1972). *J. Mol. Spectrosc.* **43**, 291–319.

Lees, R. M. (1972). *J. Chem. Phys.* **57**, 2249–2252.

Lees, R. M., and Baker, J. G. (1968). *J. Chem. Phys.* **48**, 5299–5318.

Lourtioz, J. M., Adde, R., Bouchon, D., and Pontnau, J. (1979). *Rev. Phys. Appl.* **14**, 323–330.

Lund, M. W., Cogan, J. N., and Davis, J. A. (1979). *Rev. Sci. Instrum.* **50**, 791–792.

Malk, E. G., Niesen, J. W., Parson, D. F., and Coleman, P. D. (1978). *IEEE J. Quantum Electron.* **QE-14**, 544–550.

Marcatili, E. A. J., and Schmeltzer, R. A. (1964). *Bell Syst. Tech. J.* **43**, 1783–1809.

Marshall, M. D., and Muenter, J. S. (1980). *J. Mol. Spectrosc.* **83**, 279–282.

Oka, T. (1973). *Adv. At. Mol. Phys.* **9**, 124–206.

Paschen, F. (1889). *Wied. Ann.* **37**, 69.

Plainchamp, P. M. (1979). *IEEE J. Quantum Electron.* **QE-15**, 860–864.

Radford, H. E., Peterson, F. R., Jennings, D. A., and Mucha, J. A. (1977). *IEEE J. Quantum Electron.* **QE-13**, 92–94; correction, **QE-13**, 881.

Redon, M., Gastaud, C., and Fourrier, M. (1979a). *IEEE J. Quantum Electron.* **QE-15,** 412–414.

Redon, M., Gastaud, C., and Fourrier, M. (1979b). *Opt. Commun.* **30,** 95–98.

Redon, M., Gastaud, C., and Fourrier, M. (1980). *Infrared Phys.* **20,** 93–98.

Schmidt, J., Berman, P. R., and Brewer, R. G. (1973). *Phys. Rev. Lett.* **31,** 1103–1106.

Seligson, D., Ducloy, M., Leite, J. R. R., Sanchez, A., and Feld, M. S. (1977). *IEEE J. Quantum Electron.* **QE-13,** 468–472.

Shimizu, F. (1969). *J. Chem. Phys.* **51,** 2754–2755.

Shoemaker, R. L., Stenholm, S., and Brewer, R. G. (1974). *Phys. Rev.* **10A,** 2037–2050.

Stein, S. R., and Van de Stadt, H. (1977). *Proc. Int. Conf. Frequency Control, 31st* 601–604.

Stein, S. R., Risley, A. S., Van de Stadt, H., and Strumia, F. (1977). *Appl. Opt.* **16,** 1893–1896.

Stein, S. R., Risley, A. S., Van de Stadt, H., and Strumia, F. (1978). *Optical Spectra* **10,** 36–37.

Strumia, F. (1978). *In* "Coherence in Spectroscopy and Modern Physics" (F. T. Arecchi, R. Bonifacio, and M. O. Scully, eds.), pp. 381–395. Plenum Press, New York.

Strumia, F., Inguscio, M., and Moretti, F. (1981). *In* "Laser Spectroscopy V" (T. Oka and B. P. Stoicheff, eds.), Vol. 30, Springer-Verlag, Berlin and New York.

Temkin, R. J. (1977). *IEEE J. Quantum Electron.* **QE-13,** 450–454.

Temkin, R. J., and Cohn, D. R. (1976). *Opt. Commun.* **16,** 213–217.

Terman, F. E. (1955). "Electronic and Radio Engineering." McGraw-Hill, New York.

Tobin, M. S., and Jensen, R. E. (1976a). *Appl. Opt.* **15,** 2023–2024.

Tobin, M. S., and Jensen, R. E. (1976b). *Proc. Int. Conf. Submm. Waves Their Applications, 2nd* IEEE Catalogue No. 76 CH 1152-8 MTT, pp. 167–168.

Tobin, M. S., and Jensen, R. E. (1977). *IEEE J. Quantum Electron.* **QE-13,** 481–484.

Tobin, M. S., and Koepf, G. A. (1980). *Proc. Int. Conf. IR MM Waves, 5th* pp. 306–307.

Townes, C. H., and Schawlow, A. L. (1955). "Microwave Spectroscopy." McGraw-Hill, New York.

Tucker, J. R. (1974). *IEEE Trans. Microwave Theory Tech.* **MTT-22,** 1117.

Uehara, K., Shimizu, T., and Shimoda, K. (1968). *IEEE J. Quantum Electron.* **QE-4,** 728–732.

Von Engel, A. (1965). "Ionized Gases." Oxford Univ. Press (Clarendon), London and New York.

Weiss, C. O. (1976). *IEEE J. Quantum Electron.* **QE-12,** 580–584.

Weitz, E., and Flynn, G. (1971). *J. Appl. Phys.* **42,** 5187–5190.

Weitz, E., Flynn, G., and Ronn, A. M. (1972). *J. Chem. Phys.* **56,** 6060–6067.

Wollrab, J. E. (1961). "Rotational Spectra and Molecular Structure." Academic Press, New York.

Yamanaka, M. (1977). *J. Opt. Soc. Am.* **67,** 952–958.

INFRARED AND MILLIMETER WAVES, VOL. 5

CHAPTER 4

The GaAs TUNNETT Diodes

Jun-ichi Nishizawa

Research Institute of Electrical Communication
Tohoku University
Sendai, Japan

I.   INTRODUCTION   215
     A. *Dispersion Effects in Avalanche Phenomena*   217
     B. *Time Constant in Tunnel Junction*   228
     C. *Temperature Dependence*   229
     D. *Efficiency of Carrier Generation*   230
     E. *Displacement Current in Transit Time Effect*   232
II. TUNNETT DIODE   238
     A. *Design of TUNNETT Diode*   238
     B. *Comparisons among Materials*   243
     C. *Preparation of TUNNETT Diode*   244
     D. *Results of the Measurements*   251
     E. *Noise in TUNNETT Diode*   258
III. FUTURE OF TUNNETT   260
     A. *High-Power TUNNETT Diode*   260
     B. *GUNNETT Diode*   261
     C. *Before and after TUNNETT*   263
VI. CONCLUSION   264
     REFERENCES   265

I. Introduction

The avalanche multiplication phenomenon in a semiconductor has been found by the author (Watanabe and Nishizawa, 1952), McKay and McAfee (1953), and Gunn (1956). The transit time negative resistance diode was proposed independently by the author (Nishizawa and Watanabe, 1953) and by Shockley (1954) as a solid-state microwave source. The operation principle of the avalanching negative resistance diode, which is called the IMPATT (*impa*ct ionization *t*ransit *t*ime) diode, is a combination of the avalanche injection and the transit time effect of injected carriers. The carrier injection is usually quenched by the field modulation induced by the space charge of injected carriers, released with the transit time as time constant as in the case of the Geiger–Müller

215

counter. The TUNNETT (*tunnel* injection *transit time*) diode was proposed by the author (Nishizawa and Watanabe, 1958) as the result of analysis of the avalanching negative resistance diode.

In this paper the avalanche phenomenon is analyzed, and it is concluded that the avalanche buildup does not follow the ac voltage when the transit angle of the avalanche region cannot be neglected at higher frequencies. The avalanche buildup can finish after so many repetitions of the rushing of electrons forward and the rushing of holes backward, increasing carriers by ionization through the avalanche region. The tunnel injection at the reverse biased junction will be useful as an injection source of the transit time negative resistance diode in place of the avalanche injection at high frequencies.

Despite the importance of the buildup time and the spatial distribution of the avalanche injection (Nishizawa and Watanabe, 1958), the conventional treatment of the avalanche injection has long been quite oversimplified. Such a treatment is partly owing to the fact that the ionization coefficient of carriers is a strong function of the electric field intensity and its analytical calculation is very difficult. McKay and McAfee (1953) assumed that electrons and holes of the avalanching carriers would be represented by the electrons, which would drift in the same direction, and also that two electrons would be created as the result of the ionization of the electron. Since the assumption disregards the distributed duplex feedback mechanism in both directions in the avalanche injection, it leads to an erroneous conclusion that the current never be infinity and that the time constant is the transit time of electrons through the avalanche layer of the avalanche diode. In the actual avalanche injection, the generated holes in the avalanche region drift in the opposite direction to that of the generated electrons, and the electrons generated by holes, even at the injected point of electrons, drastically increase the current multiplication and the time constant of the avalanche buildup. This oversimplified approximation was corrected by Misawa (1966), but he did not give any physical explanation about a large effect of the time constant in the avalanche injection. The superiority of the TUNNETT diode to the IMPATT diode by virtue of the high frequency, low noise, and low applied voltage has experimentally been confirmed on GaAs $p^{+}-n$ diodes by the authors' research group in 1968 (Okabe *et al.*, 1968).

At that time the TUNNETT and IMPATT modes and a hybrid of the TUNNETT and IMPATT modes had been obtained in the course of study (Okabe *et al.*, 1968). Recently the performance limit of the IMPATT diode at high frequency has been recognized by other workers (Elta and Haddad, 1979) and the hybrid mode diode, which is named MITATT, has been developed as a cw source at 150 GHz by Fetterman *et al.* (Elta *et al.*, 1980), which is the first successful achievement of cw operation.

The TUNNETT diode has an estimated high-frequency capability up to 1000 GHz, the low noise property and small dc bias voltage of less than about 10 V compared with those of the IMPATT diode, and will be a useful device for the millimeter to submillimeter wave region as oscillators, amplifiers, etc.

A. DISPERSION EFFECTS IN AVALANCHE PHENOMENA (Nishizawa *et al.*, 1974)

In this section we shall discuss the dispersion effect in the avalanche phenomenon. For many years, the avalanche injection has been thought of as only a time delay phenomenon. However, this is not correct and should be considered as the dispersion phenomena. Figures 1a and 1b show sche-

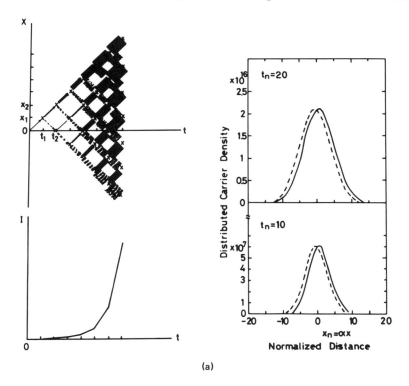

(a)

FIG. 1a Schematic illustration of the avalanche buildup process, where flowing current can be described as

$$J_s \frac{(\alpha - \beta) \exp[(\alpha - \beta)W_i]}{\alpha - \beta \exp[(\alpha - \beta)W_i]},$$

when $\alpha \gg \beta$, $J_s \exp(\alpha W_i)$, and when $\alpha = \beta$, $J_s/(1 - \alpha W_i)$, where J_s is a saturation current before ionization.

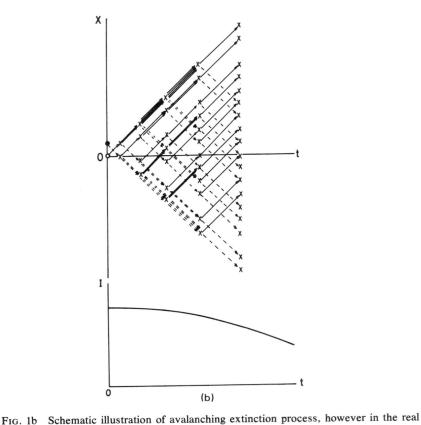

FIG. 1b   Schematic illustration of avalanching extinction process, however in the real phenomenon the number of ionization should be quite small. The figure is only to illustrate the fact that the ionization is spatially distributed to resist quick quench in time W_i/v_n.

matic illustrations of the avalanche buildup process and the extinction process, respectively, under the assumption that the ionization coefficients of an electron (α) and hole (β) are the same and that an electron and a hole have the same saturation velocity $v_n = v_p$. In the real space, $\alpha/\beta = \exp[(\alpha - \beta)W_i]$, when $\alpha = \beta$, $\alpha W_i = 1$, gives the condition for the infinite current, and W_i is the thickness of the avalanche region. Therefore, the number of ionization, to give infinite current, for a certain electron penetrating through the region with the thickness W_i is less than 10, and, when $\alpha = \beta$, it is unity, which is quite small. Then the situation of avalanching by both electrons and holes is an extraordinarily difficult phenomenon to imagine, and the figure gives only an idea for a better understanding of the avalanche injection because the coefficients α and β are shown having much larger value over current infinity. Symbol x is the position in the impact ionization region. At first an electron travels toward the positive x direction gaining the energy from the electric field as $E = qE \, \Delta x$. When the electron gains sufficient energy ($\sim 2E_g$), impact ionization occurs and the electron-hole pair is created at $t = t_1$. Then the electron again travels toward the positive x direction and the hole travels to the negative x direction. Since the next impact ionization occurs at $t = t_2$, new electrons and holes appear at $x = x_2$ and at $x = 0$ at $t = t_2$. The succession of these processes yields a large amount of electrons and holes. It is seen from Fig. 1a that the carriers are increased and the distribution of electrons and holes is spatially dispersed; for example, the effect of the ionization at $x > 0$ influences the carrier density, even at $x < 0$, through the generated holes after a certain time interval, which drastically increases the time constant for the stabilization of the flowing current. In the quenching phase, with a slightly decreased field intensity, the carriers begin to attenuate at $t = 0$; however, there is a possibility for these carriers to ionize, then a longer period of time is required for the decrease of carriers because of the spatially distributed ionization (as is schematically shown in Fig. 1b). If the large amplitude operation is considered, the sudden decrease of the field intensity lower than the threshold of impact ionization can suppress carrier generation instantaneously and the carriers flow from the ionizing region at least in time W_i/v_n: In this case, the phase inversion of the applied voltage before the injection current buildup increases not only the time delay but also the effective resistance of the avalanche injection region, which means that the carrier dispersion occurs in the buildup process of the avalanche multiplication current. This effect is called the *a*valanche *i*nduced *d*ispersion (AID) effect, which has been analyzed by the author (Nishizawa *et al.*, 1974). The AID effect plays an important role in the IMPATT diode because it deteriorates the device performance at a higher frequency. The AID effect appears most strongly under the condition that $\alpha = \beta$ and $v_n = v_p$, because the avalanche buildup time is longest. By contrast, the case of the carrier multiplication by single carriers such as the gas discharge, $\alpha \gg \beta$, shows the fastest buildup time of W_i/v_n.

The thickness of the ionization region for the sufficient avalanche process buildup also imposes the performance limit on the IMPATT diode since the ionization coefficient saturates from several hundred kV/cm and the buildup time is also limited and with the increase of the frequency thickness of the transit time region becomes thinner. Finally, the whole region of the IMPATT diode becomes the avalanche region and the transit time effect of the injected avalanche carriers disappears, which corresponds to the Misawa-type diode. Some results of calculations are described below.

The IMPATT diode with the $n^+-p-p^+-p-p^+$ (Nishizawa and Watanabe, 1958) structure as shown in Fig. 2 is considered. The high

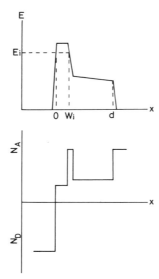

FIG. 2   IMPATT diode.

constant electric field at $x = 0$ to W_i is the avalanche region. The basic equations are

$$\frac{\partial p}{\partial t} = -\frac{1}{q}\frac{\partial J_p}{\partial x} + \beta(E)v_p p + \alpha(E)v_n n, \tag{1}$$

$$\frac{\partial n}{\partial t} = \frac{1}{q}\frac{\partial J_n}{\partial x} + \beta(E)v_p p + \alpha(E)v_n n, \tag{2}$$

$$J = J_p + J_n + \epsilon\frac{\partial E}{\partial t}, \tag{3}$$

$$J_p = q(v_p p - D_p \text{ grad } p), \tag{4}$$

and

$$J_n = q(v_n n + D_n \text{ grad } n). \tag{5}$$

Poisson's equation is given by

$$-\frac{\partial^2 V}{\partial x^2} = \frac{\partial E}{\partial x} = \frac{q}{\epsilon}(p - n - N_A^-) \approx -\frac{q}{\epsilon}N_A^-. \tag{6}$$

When the condition of $E \sim$ constant, the space charge effect is ignored and $v_p \gg D_p\beta$, $v_n \gg D_n\alpha$ is assumed. The diffusion current in the ava-

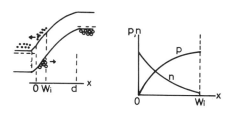

FIG. 3 Avalanche of carriers (Juñ-ichi Nishizawa and Ken Suto, 1980).

lanche region is also neglected. Then Eqs. (1) and (2) can be reduced to the simple form

$$\frac{1}{v_p}\frac{\partial p}{\partial t} + \frac{\partial p}{\partial x} = \beta(p + r\sqrt{b}\,n), \tag{1'}$$

$$\frac{1}{v_p}\frac{\partial n}{\partial t} - \sqrt{b}\,\frac{\partial n}{\partial x} = \beta(p + r\sqrt{b}\,n), \tag{2'}$$

where $r = \alpha/\beta < 1$ and $\sqrt{b} = v_n/v_p$.

At the dc steady state, if the $p = 0$ at $x = 0$ is assumed, J equals $qv_p(p + \sqrt{b}\,n)$, neglecting of the diffusion current.

The hole and electron distribution is obtained from Eqs. (1') and (2'),

$$p(x) = \frac{Jr}{qv_p(1-r)}\{\exp[\beta(1-r)x] - 1\}, \tag{7}$$

$$n(x) = \frac{1}{\sqrt{b}}\frac{J}{qv_p(1-r)}\{1 - r\exp[\beta(1-r)x]\}. \tag{8}$$

This is illustrated in Fig. 3. The electron density increases from $n(W_i)$ toward the $x = 0$ direction. The total current J is expressed, using $n(W_i)$, as

$$J = \frac{q(1-r)v_n n(W_i)}{1 - r\exp[\beta(1-r)W_i]}. \tag{9}$$

The ac solution is made using Eqs. (1') and (2'). If $\omega/v_p \ll \beta(E)(1 - r)$ is fulfilled, the time derivative term can be neglected. The ionization coefficient is expressed as

$$\beta(E) = \beta(E_0) + \frac{\partial \beta}{\partial E}\tilde{E}e^{j\omega t}, \tag{10}$$

where $\beta(E_0)$ is the dc value of β, and $\tilde{E}e^{j\omega t}$ is the ac perturbation term. By combining Eqs. (9) and (10), the ac component of the current is derived as

$$\tilde{J} = \frac{q v_n n(W_i) r (1 - r)^2 W_i \exp[\beta(E_0)(1 - r)W_i]}{\{1 - r \exp[\beta(E_0)(1 - r)W_i]\}^2} \left(\frac{\partial \beta}{\partial E}\right) \tilde{E}$$

$$= \frac{-J^2 r}{q v_n n(W_i)} \exp[\beta(E_0)(1 - r)W_i] \left(\frac{\partial \beta}{\partial E}\right) \tilde{V}(W_i). \qquad (11)$$

The admittance per unit area of the avalanche region is expressed in addition to the displacement current through the depletion layer capacitance as

$$g_i = \frac{J^2 r}{q v_n n(W_i)} \exp[\beta(E_0)(1 - r)W_i] \left(\frac{\partial \beta}{\partial E}\right),$$

$$(12)$$

$$b_i = \omega \frac{\epsilon}{W_i} = \omega c_i \qquad [\text{at } \frac{\omega}{v_p} W_i \ll \beta(1 - r)W_i].$$

This result is given under the assumption of $\omega/v_p \ll \beta(1 - r)$. Under a high-frequency condition that $\omega/v_p \gg \beta(E)(1 - r)$, Eqs. (1') and (2') can be reduced to

$$\frac{j\omega \tilde{p}}{v_p} = \frac{j\omega \tilde{n}}{v_p} = \beta(\tilde{p} + r\sqrt{b}\,\tilde{n}) + \frac{\partial \beta}{\partial E} \tilde{E}(p_0 + r\sqrt{b}\,n_0)$$

$$\approx \frac{\partial \beta}{\partial E}(p_0 + r\sqrt{b}\,n_0)\tilde{E}, \qquad (13)$$

neglecting the spatial derivative terms, which are small compared with the time derivative terms. The above equation means that the phase of the ac current owing to carrier has a time lag of $\pi/2$ radian with respect to the ac voltage and the inductance occurs in the equivalent circuit. Equation (13) holds for the case of the phase angle of the avalanche region ($\theta = \omega(W_i/v_p) \gg \beta(E)(1 - r)W_i$) is greater than 1 because $\beta(E)W_i \approx 1$. At higher frequencies the avalanche injection region and the drift region is not separated, so the delay time effect of the avalanching carriers should be taken into account at high frequencies.

The equivalent circuit of the avalanche region at a small phase angle is shown in Fig. 4. Now the avalanche injection of the p^+-i-n^+ diode has been analyzed in more detail (Nishizawa et al., 1978c).

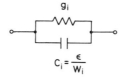

FIG. 4   Equivalent circuit of the avalanche region in a small transit angle.

Neglecting the diffusion of the carriers, $J_n(x)$ is given by

$$J_n = \frac{K}{(1 - r)\alpha_0} \{\exp[(1 - r)\alpha_0 x]$$

$$- \frac{x}{W_i} (\text{esp}[(1 - r)\alpha_0 W_i] - 1) - 1\}$$

$$+ \frac{x}{W_i} (J_0 - (J_{n0} + J_{p0})) + J_{n0} \qquad 0 < x < W_i \qquad (14)$$

and $J_p(x) = J_0 - J_n(x)$, where Eqs. (1), (2), and (6) and the assumption of $v_n = v_p = v$, $\beta = r\alpha$, $\alpha_0 = \alpha(E = E_0 \text{ const})$ are used.

Hence K is an integral constant given by

$$K = \frac{\alpha_0 J_{n0}(1 - r) + r\alpha_0 J_0 - [J_0 - (J_{n0} + J_{p0})]/W_i}{1 + \{1 - \exp[(1 - r)\alpha_0 W_i]\}/(1 - r)\alpha_0 W_i}, \qquad (15)$$

and J_0 is the total dc current density. The dc electron current distribution in a uniformly avalanching diode is shown in Fig. 5 as a function of a typical value of current density (1 kA/cm²). It is evident that the distribution of a sort of carriers for $r = 1$ is almost linear while that for $r = 0$ is extremely sharp. Thus, the thickness of the avalanche region to generate sufficient carriers is maximum for $\alpha = \beta$ and decreases as r deviates from unity.

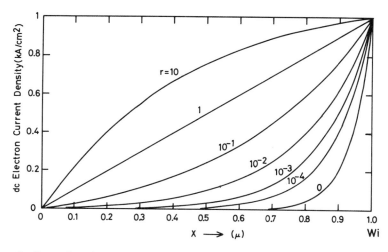

FIG. 5 Shown here is dc electron current distribution in a uniformly avalanching diode. $J_0 = 1$ kA/cm², $W_i = 1$ μm. (J. Nishizawa, T. Ohmi, and M. S. Niranjian, Avalanche induced dispersion in IMPATT diodes, *Solid-State Electronics* **21**. Copyright 1978, Pergamon Press, Ltd.)

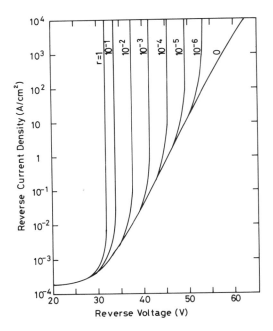

FIG. 6 $I-V$ characteristic of a uniformly avalanching diode. $W_i = 1$ μm. (J. Nishizawa, T. Ohmi, and M. S. Niranjian, Avalanche induced dispersion in IMPATT diodes, *Solid-State Electronics* **21**. Copyright 1978, Pergamon Press, Ltd.)

Figure 6 shows the current voltage characteristic as a function of the ratio r for a uniformly avalanching diode neglecting space charge with a 1-μm thick depletion layer. The sharp $I-V$ characteristics of $r = 1$ are seen. The voltage for a given current density, as well as the incremental resistance, is minimum for $r = 1$ and maximum for $r = 0$ (or ∞). This is understandable from the fact that the regenerative feedback present in the avalanching carrier is totally absent when either electrons or holes are capable of ionization, while in the case of $r = 1$ it becomes maximum. The difference becomes noticeable when the feedback exceeds a critical value as can be seen in the $I-V$ curves for the lower values of r in Fig. 6, the curves branch off from the non-feedback ($r = 0$) curve at the correspondingly increased values of total current. This behavior is found to have a considerable effect on the small signal ac behavior.

Figure 7 shows the small signal admittance of a uniformly avalanche diode at a constant dc current level of $J_0 = 1$ kA/cm^2 and $W_i = 1$ μm. The negative conductance for $r = 1$, which has the lowest field intensity, varies very little over a wide range of frequencies and has a maximum value which is always lower than that for any other r values and finally tends to the positive resistance. In every case, the negative resistance dis-

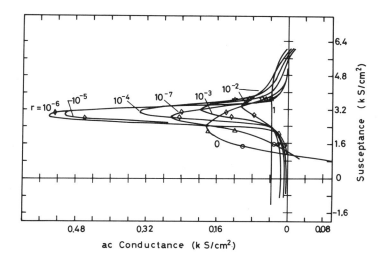

FIG. 7 Small signal admittance of a uniformly avalanching diode. $J_0 = 1$ kA/cm^2, $W_i = 1$ μm. (J. Nishizawa, T. Ohmi, and M. S. Niranjian, Avalanche induced dispersion in IMPATT diodes, *Solid-State Electronics* **21**. Copyright 1978, Pergamon Press, Ltd.)

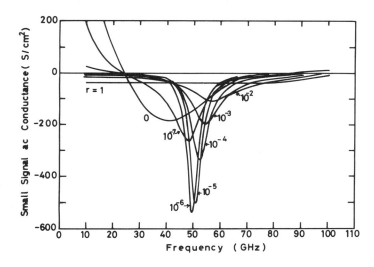

FIG. 8 Conductance versus frequency for a uniformly avalanching diode. $J_0 = 1$ kA/cm^2, $W_i = 1$ μm. (J. Nishizawa, T. Ohmi, and M. S. Niranjian, Avalanche induced dispersion in IMPATT diodes, *Solid-State Electronics* **21**. Copyright 1978, Pergamon Press, Ltd.)

appears at a frequency lower than 100 GHz.

Figure 8 shows the small signal ac conductance versus frequency for a uniformly avalanching diode in which $J_0 = 1$ kA/cm^2 and $W_i = 1$ μm. The negative conductance of $r = 1$ is very low and varies very little over the wide frequency ranges. Figure 9 shows the small signal ac conductance versus frequency for a diode with a triangle field profile in which $J_0 = 1$ kA/cm^2 and $W_i = 1$ μm.

Figures 8 and 9 suggest that the resistance of the avalanching region start to be increased after dispersion at a frequency lower than 100 GHz. As the ionization rate ratio deviates from unity, the buildup time effect tends to play an important part in the small signal operation even in a diode having a uniform field profile because of the localization of the avalanche region. The maximum negative conductance in an optimum ionization rate ratio is two or three times larger than that of the conventional IMPATT diodes having a nearly triangle field profile.

For higher frequencies, the buildup current in the avalanche region is one of the marginal sides. Impedance of the avalanche injection region becomes larger and inductive. This corresponds to the dispersion and not to the simple time delay in the avalanche region. For this reason, the thickness of the avalanche region for a higher frequency should be designed to be sufficiently thin to reduce the buildup time delay. Also in order to keep a constant injection current, the intensity of applied electric field should be increased. The reduction of the thickness of the avalanche

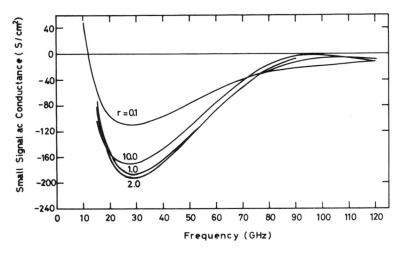

FIG. 9   Conductance versus frequency for a diode having a triangle field profile. $J_0 = 1$ kA/cm^2, $W_i = 1$ μm. (J. Nishizawa, T. Ohmi, and M. S. Niranjian, Avalanche induced dispersion in IMPATT diodes, *Solid-State Electronics* **21**. Copyright 1978, Pergamon Press, Ltd.)

region to avoid the dispersion increases the impedance because of the lower ionization in the region, which compensates the negative resistance of the transit region and decreases the output power of the diode. If the applied voltage in the avalanche region is increased and finally exceeds the tunneling injection, the diode is automatically changed into a TUN-NETT diode.

A more simplified calculation for the buildup shows that the front of distributed carriers caused from the avalanche moves at the avalanche induced drift velocity (v_a) and spread out with the avalanche induced diffusion constant (D_a) as given by

$$v_a = (\alpha v_n^2 - \beta v_p^2)/(\alpha v_n + \beta v_p) \qquad (16)$$

and

$$D_a = \alpha \beta v_n v_p (v_n + v_p)^2/(\alpha v_n + \beta v_p)^3. \qquad (17)$$

The center of the group of distributed avalanching carriers moves with the value of αv_n^2 and βv_p^2 (Nishizawa et al., 1974; Ohmi, 1977).

The following three situations are given by

(a) $\alpha v_n^2 > \beta v_p^2$. In this case the group of avalanching carriers moves in the same direction of electrons.

(b) $\alpha v_n^2 = \beta v_p^2$. In this case the shape of distribution of avalanching carriers does not seem to move.

(c) $\alpha v_n^2 < \beta v_p^2$. In this case the shape of distribution of avalanching carriers seems to move toward the same direction of holes. This is shown in Fig. 10.

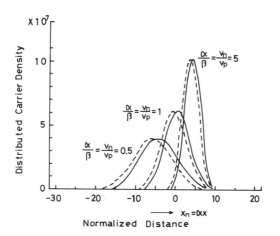

FIG. 10 Avalanche buildup process of distributed carriers ($t_n = 10$). ———, electron; ---, hole.

B. TIME CONSTANT IN TUNNEL JUNCTION

As discussed in the preceding section, the buildup time of the avalanche region of the IMPATT diode does not follow the ac voltage at high frequency owing to the AID effect.

In this Section the time constant of the tunnel junction is described. Figure 11 shows schematic illustrations of the $p^+-n^+-i-n^+$ structure, the electric field profile and the band diagram of the TUNNETT diode. In the $p^+-n^+-i-n^+$ structure, the electron tunnels from the p^+ side valence band to the n^+ side conduction band.

The minimum thickness of the tunnel junction is given approximately by

$$\Delta x = \mathrm{E_g}/-qE \le W_t, \tag{18}$$

where $\mathrm{E_g}$ is band-gap energy of the semiconductor, q is the electron charge, E is the electric field intensity, and the minimum voltage for V_t is equal to $\mathrm{E_g}$. Applied voltage can be reduced because of the diffusion potential, as is well known in the case of the Esaki diode where tunneling can occur without applied voltage. When $\mathrm{E_g} = 1$ eV and $E = 1000$ kV/cm, Δx is 100 Å (10^{-6} cm) and is about 1 order of magnitude smaller than that of the avalanche injection region width of the IMPATT diode operating over 100 GHz (Ishibashi *et al.*, 1977; Kuno, 1979). To generate sufficient injecting carriers, Δx is decreased inversely proportional to the electric field intensity of the p^+-n^+ junction, while the avalanche injection region does not decrease with increase in electric field intensity because of the saturation of the ionization coefficients of an electron and a hole.

The current density of the tunneling junction is given by

$$J_t = \frac{1}{a^3} \frac{qEa}{h} \exp\left(-\frac{\pi^2(2m_e^*)^{1/2}\mathrm{E_g^{3/2}}}{2hqE}\right), \tag{19}$$

where a is the lattice constant, E is the electric field intensity, m_e^* is the effective mass of an electron, $\mathrm{E_g}$ is the band-gap energy, q is the electron charge, and h is Planck's constant. The transit time of the carrier in tun-

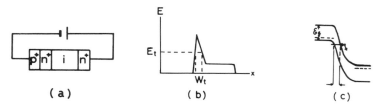

FIG. 11   TUNNETT diode (a) structure, (b) electric field profile, and (c) band diagram: $\Delta x = (\mathrm{E_g}/-qE) < W_t$ (Jun-ichi Nishizawa and Ken Suto, 1980).

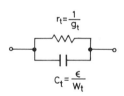

FIG. 12 Equivalent circuit of the tunnel injection region.

neling is neglected since the tunneling occurs by the quantum-mechanical transition of an electron from the valence band to the conduction band.

The above equation is valid as far as the modulation of the field intensity at the injection point by the transit space charge is followed by the change of the electric field intensity at the tunneling junction at high frequencies.

The small signal ac current is given by differentiating equation (19),

$$\tilde{J}_t \simeq (\pi^2 (2m_e^*)^{1/2} E_g^{3/2} / 2hqE^2) \tilde{E} J_t = A J_t \tilde{E}, \tag{20}$$

where A is the constant inversely proportional to the square of the field intensity, and $\tilde{J}_t$ and $\tilde{E}$ are the ac components of J_t and E, respectively.

The tunneling junction can be represented by the conductance given by Eq. (20). Since the displacement current flows in the depletion layer capacitance, the equivalent circuit of the tunnel junction is given by the parallel of the resistance and the capacitance as shown in Fig. 12. Then the capacitance increases inversely proportional to W_t and the conductance increases with J_t. Hence J_t is a strong function of the electric field intensity and thus the time constant of the tunnel junction is shortened with an increase of the electric field intensity. This point is essentially different from that of the avalanche injection so that the tunnel injection will act as a useful injection source of the transit time negative resistance diode up to 1000 GHz.

C. Temperature Dependence

The difference of the tunnel injection and the avalanche injection have been discussed in Sections A and B; the temperature dependence of the tunnel injection and the avalanche injection will now be discussed. As can be seen in Eq. (19), the tunneling can be increased with the decrease of the band-gap energy and the increase of the electric field intensity. The band-gap energy slightly decreases with increasing temperature, so the tunnel injection increases as temperature increases as shown in Fig. 13.

On the contrary, the avalanche injection decreases with increasing junction temperature, because electrons or holes lose the energy gained drifting through the high electric field region by the lattice scattering,

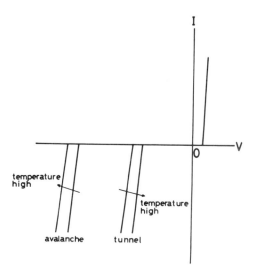

FIG. 13   Difference of the temperature dependence of the tunnel and the avalanche injection.

especially optical phonon emission. The temperature dependence of the tunnel injection and avalanche injection affects the breakdown voltage temperature dependence as shown in Fig. 13. The temperature dependence of the $I-V$ characteristic is used so as to determine by which the injection mechanism occurs mainly in the junction.

D.   EFFICIENCY OF CARRIER GENERATION

Smaller dc input power consumption in the injection is needed as the injection source of the transit time negative resistance diode. From this point of view, the tunnel injection is superior to that of the avalanche injection. The transition voltage from the avalanche to the tunnel injection differs depending on the material such as Ge, Si (Tokuyama, 1962; Tyagi, 1968; Sze and Gibbons, 1966), GaAs (Okabe *et al.*, 1968), and InP.

The electric field dependence of the injection process is important. The small signal ac increment of the ionization coefficient of an electron is given similarly to Eq. (10) as

$$\alpha(E) = \alpha(E_0) + \frac{\partial \alpha}{\partial E} \tilde{E} e^{j\omega t}.$$

The quantity $\partial \alpha / \partial E$ of GaAs is calculated as shown in Fig. 14. The $\partial \alpha / \partial E$ increases with the electric field intensity, then saturates and finally decreases.

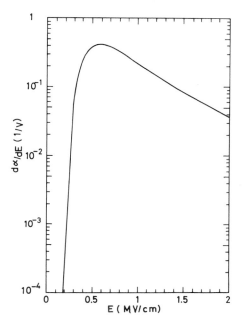

FIG. 14 $\partial\alpha/\partial E$ of GaAs. $\alpha = 3.5 \times 10^5 \exp -[(6.85 \times 10^5)/E]^2$.

The detailed ac component of the avalanche region is given by Eq. (11) as

$$\tilde{J} = \frac{qv_n n(W_i)r(1 - r)^2 W_i \exp[\alpha(E_0)(1 - r)W_i]}{\{1 - r \exp[\alpha(E_0)(1 - r)W_i]\}^2} \left(\frac{\partial\alpha}{\partial E}\right) \tilde{E}.$$

The conduction current in the avalanche region is determined by the $\partial\alpha/\partial E$. Hence the $\tilde{J}$ has a maximum at a certain electric field intensity as shown in Fig. 14. On the contrary, the small signal tunnel current is given by Eq. (20) as

$$\tilde{J}_t \simeq AJ_t\tilde{E}.$$

Since J_t is a strong function of the electric field intensity, the tunnel injection is superior to that of the avalanche injection at higher electric field intensity. With the smaller dc bias voltage and the higher electric field intensity dependence of the tunnel injection, the efficiency of the carrier generation is higher than that of the avalanche injection.

E.   Displacement Current in Transit Time Effect

The analysis of the modulated carrier drifting space (Nishizawa and Watanabe, 1958) shall now be described with regard to drifting carriers between the base and the collector.

The transit time effect of carriers of the transit time negative resistance diode is accounted for prior to the consideration of the displacement current. After the injection of carriers owing to the avalanche or tunnel injection the carriers flow into the drift region where both drifting and diffusion of injected carriers should be considered. The induced current flows into the external circuit outside the diode owing to the space charge of the drifted hole or electron. The induced current is solved using Poisson's equation given by

$$d\tilde{E}/dx = q\tilde{n}/\epsilon \tag{6'}$$

and the total ac current $\tilde{J}$ given by

$$\tilde{J} = \tilde{J}_p + \tilde{J}_n + j\omega\epsilon\tilde{E}. \tag{3'}$$

The diode under consideration is the IMPATT diode as shown in Fig. 2 and Fig. 3. the injected carriers of electron shall be considered.

Since the current continuity is maintained, $\tilde{J}$ is equal to the induced current having flowed to the external circuit. The hole is neglected in the analysis of the drift space. The distribution of the drifting electrons is solved using Eqs. (4) and (5) as

$$\tilde{n}(x) = \tilde{n}(W_i) \exp\left(\left\{\frac{v_n}{D_n} - \left[\left(\frac{v_n}{D_n}\right)^2 + 4j\frac{\omega}{D_n}\right]^{1/2}\right\}\frac{x - W_i}{2}\right). \tag{21}$$

The $\tilde{E}$ is solved from Eqs. (6') and (14):

$$\tilde{E}(x) = \frac{2\tilde{n}(W_i)q}{\epsilon} \frac{\exp\left(\left\{\frac{v_n}{D_n} - \left[\left(\frac{v_n}{D_n}\right)^2 + 4\frac{j\omega}{D_n}\right]^{1/2}\right\}\frac{x - W_i}{2}\right) - 1}{\frac{v_n}{D_n} - \left[\left(\frac{v_n}{D_n}\right)^2 + 4\frac{j\omega}{D_n}\right]^{1/2}}$$

$$+ \tilde{E}(W_i) = -\frac{q\tilde{n}(W_i)D_n}{2\epsilon j\omega}\left\{\frac{v_n}{D_n} + \left[\left(\frac{v_n}{D_n}\right)^2 + 4\frac{j\omega}{D_n}\right]^{1/2}\right\}$$

$$\times \left[\exp\left(\left\{\frac{v_n}{D_n} - \left[\left(\frac{v_n}{D_n}\right)^2 + 4\frac{j\omega}{D_n}\right]^{1/2}\right\}\frac{x - W_i}{2}\right) - 1\right]$$

$$+ \tilde{E}(W_i). \tag{22}$$

The ac voltage of the drift region is obtained from the integration of Eq. (15) as follows:

$$\tilde{V}(x) - \tilde{V}(W_i) = \frac{q\tilde{n}(W_i)D_n^2}{4\epsilon\omega^2} \left\{ \frac{v_n}{D_n} + \left[\left(\frac{v_n}{D_n} \right)^2 + \frac{4j\omega}{D_n} \right]^{1/2} \right\}^2$$

$$\times \left[\exp \left(\left\{ \frac{v_n}{D_n} - \left[\left(\frac{v_n}{D_n} \right)^2 + \frac{4j\omega}{D_n} \right]^{1/2} \right\} \frac{x - W_i}{2} \right) - 1 \right]$$

$$- \frac{q\tilde{n}(W_i)}{2\epsilon j\omega} [v_n + (v_n^2 + 4j\omega D_n)^{1/2}](x - W_i)$$

$$- \tilde{E}(W_i)(x - W_i). \tag{23}$$

The induced current $\tilde{J}$ is constant in any place, so the $\tilde{J}$ is obtained from Eq. (3') at $x = W_i$ as

$$\tilde{J} = \frac{q\tilde{n}(W_i)[v_n + (v_n^2 + 4j\omega D_n)^{1/2}]}{2} + j\omega\epsilon\tilde{E}(W_i). \tag{24}$$

$\tilde{E}(W_i)$ is obtained from $[\tilde{V}(W_i) - \tilde{V}(d)]$ at Eq. (16), then the induced current $\tilde{J}$ is obtained as,

$$\tilde{J} = \frac{jq\tilde{n}(W_i)}{4\omega(d - W_i)} [v_n + (v_n^2 + 4j\omega D_n)^{1/2}]^2 \left[\exp \left(\left\{ \frac{v_n}{D_n} \right. \right. \right.$$

$$\left. \left. - \left[\left(\frac{v_n}{D_n} \right)^2 + 4 \frac{j\omega}{D_n} \right]^{1/2} \right\} \frac{d - W_i}{2} \right) - 1 \right]$$

$$+ j\omega\epsilon \frac{\tilde{V}(W_i) - \tilde{V}(d)}{d - W_i}, \tag{25}$$

where the first term corresponds to the admittance owing to the induced current and the second term is the flowing displacement current in the depletion layer capacitance of the drift region.

The admittance of the drift region y_s is derived from the first term of Eq. (18) divided by the drift region voltage $\tilde{V}(W_i) - \tilde{V}(d)$:

$$y_s = - \frac{jq\tilde{n}(W_i)}{4\omega(d - W_i)} [v_n + (v_n^2 + 4j\omega D_n)^{1/2}]^2\beta_t \left(\tilde{E}(W_i)(d - W_i) \right.$$

$$+ \frac{q\tilde{n}(W_i)}{2j\epsilon\omega} [v_n + (v_n^2 + 4j\omega D_n)^{1/2}](d - W_i) + \frac{q\tilde{n}(W_i)}{4\epsilon\omega^2}$$

$$\left. \times [v_n^2 + (v_n^2 + 4j\omega D_n)^{1/2}]^2\beta_t \right)^{-1},$$

where

$$\beta_t = 1 - \exp \left(\left\{ \frac{v_n}{D_n} \right) - \left[\left(\frac{v_n}{D_n} \right)^2 + 4j \frac{\omega}{D_n} \right]^{1/2} \right\} \frac{d - W_i}{2} \right). \tag{26}$$

Here β_t is the factor of the phase rotation of the drift region and the attenuation of the carrier density due to diffusion.

At the ordinary frequency $v_n^2/D_p \gg \omega$ is fulfilled. So we obtain the next approximation:

$$\left[\left(\frac{v_n}{D_n}\right)^2 + 4\frac{j\omega}{D_n}\right]^{1/2} \simeq \frac{v_n}{D_n} + 2j\frac{\omega}{v_n} + 2\frac{\omega^2 D_n}{v_n^3},$$

Then y_s is approximated as

$$y_s \simeq -\frac{jq\tilde{n}(W_i)}{\omega(d - W_i)}\left(v_n + \frac{\omega^2 D_n^2}{v_n^3} + j\frac{\omega D_n}{v_n}\right)^2 \beta_t' \left[\tilde{E}(W_i)(d - W_i)\right.$$

$$+ \frac{q\tilde{n}(W_i)}{j\omega\epsilon}\left(v_n + \frac{\omega^2 D_n^2}{v_n^3} + j\frac{\omega D_n}{v_n}\right)(d - W_i) + \frac{q\tilde{n}(W_i)}{\omega^2\epsilon}$$

$$\times \left(v_n + \frac{\omega^2 D_n^2}{v_n^3} + j\frac{\omega D_n}{v_n}\right)^2 \beta_t'\right]^{-1}, \tag{27}$$

where

$$\beta_t' \simeq 1 - \exp\left[-\left(\frac{\omega^2 D_n}{v_n^3} + j\frac{\omega}{v_n}\right)(d - W_i)\right]. \tag{28}$$

Here, β_t' is the approximated factor of the phase rotation and the attenuation owing to diffusion of the drifted electrons in the drift region. If the transit angle $\omega(d - W_i)/v_n$ is smaller than π radian, Eq. (20) is calculated as

$$y_s \simeq \frac{q\tilde{n}(W_i)\left[v_n + \frac{\omega^2 D_n^2}{v_n^3} + j\omega\left(\frac{D_n}{v_n} - \frac{d - W_i}{2}\right)\right]}{\tilde{E}(W_i)(d - W_i) + \frac{q\tilde{n}(W_i)}{2\epsilon}(d - W_i)^2} \tag{29}$$

with the extension of the second-order terms of the exponential function of β_t'. It is shown that y_s is dependent on the admittance of the avalanche region because y_s is related to the ratio of $\tilde{n}(W_i)$ to $\tilde{E}(W_i)$.

At a low frequency such as $\omega/v_n \ll \alpha(E)(1 - r)$, the relation between $E(W_i)$ and $n(W_i)$ is given by Eq. (7):

$$\tilde{n}(x) = \frac{Jr}{qv_n}\exp[\alpha(E_0)(1 - r)W_i]\left(\frac{\partial\alpha}{\partial E}\right)\tilde{E}(W_i)W_i. \tag{30}$$

Since $n(x)$ is in phase with the ac voltage $\tilde{E}(W_i)W_i$, Eq. (22) is given as

$$y_s \simeq \frac{\frac{JrW_i}{v_n}\exp[\alpha(E_0)(1 - r)W_i]\left(\frac{\partial\alpha}{\partial E}\right)}{(d - W_i) + \frac{Jr}{2\epsilon}\frac{W_i}{v_n}\exp[\alpha(E_0)(1 - r)W_i]\left(\frac{\partial\alpha}{\partial E}\right)}$$

$$\times \left[v_n + \frac{\omega^2 D_n^2}{v_n^3} + j\omega\left(\frac{D_n}{v_n} - \frac{(d - W_i)}{2}\right)\right]. \tag{31}$$

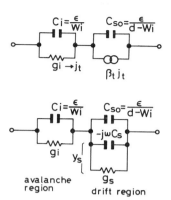

FIG. 15 Equivalent circuit of the avalanche injection region and its approximation at a small transit angle.

The transit region length $d - W_i$ is larger than D_n/v_n. Then Eq. (24) at a small transit angle shows that y_s is represented by the parallel of the positive conductance and the negative capacitance, so the negative conductance does not appear as shown in Fig. 15.

Equations (26) and (27) do not express simple approximations in a large transit angle region. However the relation between $\bar{n}(W_i)$ and $\tilde{E}(W_i)$ is given by Eq. (30) under the condition that $\omega/v_n \ll \alpha(E)(1 - r)$. If the current density is small and the voltage drop owing to the space charge is neglected, y_s is almost the same as $-j\beta_t'$. β_t or β_t' is illustrated as shown in Fig. 16. $-j\beta_t'$ is most negative when the transit angle $\theta = \omega(d - W_i)/v_n$ is nearly equal to the $\frac{3}{2}\pi$ radian. Then the maximum negative conductance is obtained at the $\frac{3}{2}\pi$ radian.

The impedance of the drift region is given by

$$Z_s \equiv \frac{1}{y_s} = \frac{j}{\beta_t}\left[\frac{4\omega\tilde{E}(W_i)(d - W_i)^2}{q\bar{n}(W_i)[v_n + (v_n^2 + 4j\omega D_n)^{1/2}]^2}\right.$$
$$\left. - j\frac{2(d - W_i)^2}{\epsilon[v_n + (v_n^2 + 4j\omega D_n)^{1/2}]}\right]$$
$$- \frac{(d - W_i)}{j\omega\epsilon} \simeq \frac{j}{\beta_t'}\left[\frac{\omega\tilde{E}(W_i)(d - W_i)^2}{q\bar{n}(W_i)\left(v_n + \frac{\omega^2 D_n^2}{v_n^3} + j\omega\frac{D_n}{v_n}\right)^2}\right.$$
$$\left. - j\frac{(d - W_i)^2}{\epsilon\left(v_n + \frac{\omega^2 D_n^2}{v_n^3} + j\frac{\omega D_n}{v_n}\right)}\right] - \frac{d - W_i}{j\omega\epsilon}. \qquad (32)$$

Here β_t' has a negative phase $-j\theta_1$ when the transit angle is near the $\frac{3}{2}\pi$

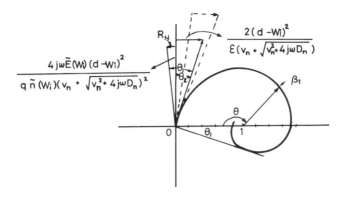

FIG. 16   Transport function diagram of the drift region β_t (or β'_t).

radian and the term in the bracket has also a negative phase as $-j\theta_2$ as shown in Fig. 16. If $\theta_1 > \theta_2$, the phase angle of the first term is $\frac{1}{2}\pi + \theta_1 - \theta_2$ and has a value between the $\frac{1}{2}\pi$ and π radians.

Hence the first term of Z_s is represented by a series of the negative resistance $-R_N$ and the inductance $j\omega L$. The second term is the negative capacitance $-C'_s = -(d - W_i)/\epsilon$. So the impedance of the transit region is given by the parallel circuit of Z_s and the depletion layer capacitance as shown in Fig. 17 with the avalanche injection region. The above equivalent circuit is valid only when the transit angle of the drift region is nearly the $\frac{3}{2}\pi$ radian and the transit angle of the avalanche region is very small.

Next the admittance of the TUNNETT diode is considered below.

The electron tunneling from the p^+ valance band into the n^+ conduction band as shown in Fig. 11 is assumed and the tunneling region from 0 to W_i is assumed. Then the ac density of electrons at $x = W_i$ is given by

$$\tilde{n}(W_i) = \frac{2AJ_t}{q[v_n + (v_n^2 + 4j\omega D_n)^{1/2}]} \tilde{E}(W_i).$$

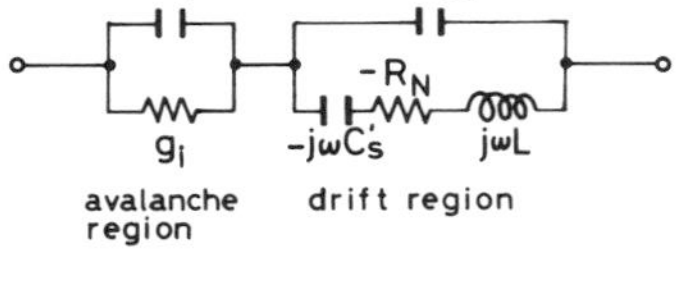

FIG. 17   The equivalent circuit of the IMPATT diode when the transit angle of the drift region is near at $\frac{3}{2}\pi$ radian.

The $\bar{n}(W_i)$ and the $\tilde{E}(W_i)$ are nearly in phase. Then the admittance of the drift region is given in a way similar to Eq. (27) by

$$
y_s \simeq \frac{-\dfrac{jAJ_t}{2\omega(d-W_i)}\left(v_n + \dfrac{\omega^2 D_n^2}{v_n^2} + j\omega\,\dfrac{D_n}{v_n}\right)\beta'_t}{(d-W_i) + \dfrac{AJ_t}{j\omega\epsilon}(d-W_i) - \dfrac{AJ_t}{2\epsilon\omega^2}\left(v_n + \dfrac{\omega^2 D_n^2}{v_n^3} + j\,\dfrac{\omega D_n}{v_n}\right)\beta'_t}. \tag{33}
$$

The variation of the phase of y_s is similar to $-j\beta'_t$ when the second and third terms of the denominator are neglected on account of their smallness compared to the first term.

The negative conductance is maximum when the transit angle of the drift region is near the $\frac{3}{2}\pi$ radian. Then the equivalent circuit of the TUN-NETT diode in which the transit angle is at near the $\frac{3}{2}\pi$ radian is expressed the same as shown in Fig. 17. This equivalent circuit is valid up to a frequency which is considerably higher than that of the IMPATT diode since the tunnel injection region of the TUNNETT diode is very thin.

The ac current is the sum of the conduction current $\tilde{J}_n + \tilde{J}_p$ and the displacement current $j\omega\varepsilon\tilde{E}$. The displacement current increases in proportion to the dielectric constant and the frequency. If the absolute value of $|\omega\varepsilon\tilde{E}|$ is greater than the conduction current $|\tilde{J}_n + \tilde{J}_p|$, the negative resistance owing to the transit time effect of the drift region of the transit time negative resistance diode will disappear progressively.

The dielectric constant of the semiconductor is 16 for Ge, 11.8 for Si, and 10.9 for GaAs. Hence the semimetal will be more useful for the TUN-NETT diode at high frequencies than the semiconductor (Nishizawa, 1975).

In the future the Misawa type or double drift type of the TUNNETT should be analyzed in the same manner. As a whole, the minimum applied voltage for the TUNNETT is the sum of the voltages for tunneling and drift reduced by the diffusion voltage; V_{diff}, and is given by

$$
V_a = V_t + V_d - V_{\text{diff}}.
$$

The minimum value of V_t equals E_g. The drift field intensity for the drift space should be minimum and is proposed to keep the saturation velocity for the drifting carriers to keep the high-frequency operation. Then in the case of electrons in GaAs which operates at 300 GHz, the minimum voltage for drift space is 0.08 V. If the width of the drift space is thin sufficiently small compared to the Debye length, the field can be established only by the diffusion potential, and the total minimum applied voltage is quite small.

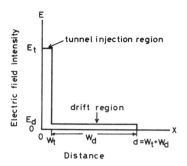

FIG. 18   Ideal electric field profile of the TUNNETT diode.

II.   TUNNETT Diode

A.   DESIGN OF TUNNETT DIODE

The ideal electric field profile of the TUNNETT diode is shown in Fig. 18. The field intensity E_t of the tunnel injection region should be higher than about 1000 kV/cm, and the thickness W_t should be less than about 100 Å in order to enhance the tunnel injection. As a result, the voltage for the tunnel injection is much lower than that for the avalanche injection. If the capacitive current through the depletion layer for the tunnel junction is neglected, the maximum tunnel injection occurs at the $\frac{1}{2}\pi$ radian. Hence the width of the drift region W_d should be chosen as

$$W_d = \frac{3v_s}{4f}, \tag{34}$$

where f is the oscillation frequency and v_s is the saturation velocity of carriers. The $p^+-n^+-i(\nu)-n^+$ structure as shown in Fig. 19a is suitable for

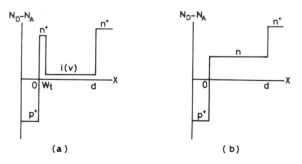

(a)   (b)

FIG. 19   Doping profile of the TUNNETT diode (a) $p^+-n^+-i(\nu)-n^+$ and (b) p^+-n-n^+ structure.

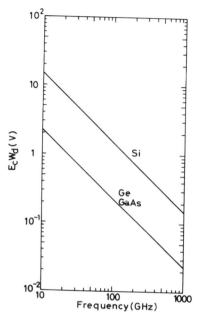

FIG. 20   Relation between f and $E_c W_d$.

realizing an approximately ideal electric field profile compared with the p^+-n-n^+ structure as shown in Fig. 19b. If v_s is 1×10^7 cm/sec and $f = 300$ GHz, W_d is calculated to be 0.25 μm. The minimum voltage for the drift region is calculated to be as $W_d E_c$, where E_c is the critical field for the saturation velocity. The relation between $W_d E_c$ and the frequency of several materials is shown in Fig. 20. The tunnel injection occurs in a very thin region, so the determination of W_d in the TUNNETT diode is valid as compared with that of the IMPATT diode in which most of the bias voltage is consumed in the avalanche injection region at high frequencies.

The efficiency of the TUNNETT diode is defined as

$$\eta = \frac{P_{out}}{P_{in}} = \frac{(\cos \varphi) I_{rf} V_{rf}}{I V_a} = \frac{(\cos \varphi) I_{rf} V_{rf}}{I(V_t + V_d - V_{diff})}, \qquad (35)$$

where I and V_a are the dc bias current and voltage, V_t and V_d are the voltages applied to the tunnel injection region and the drift region, and I_{rf} and V_{rf} are the rf current and the voltage, respectively. The reduction of V_t of the dc bias voltage is necessary to increase the diode efficiency. V_t is calculated to be less than 2 V under the condition that $E_t = 2000$ kV/cm and $W_t = 100$ Å in Fig. 18, however, it can be enough even with diffusion

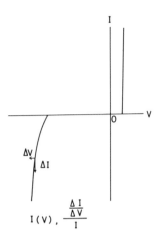

FIG. 21 $I-V$ characteristics of $p-n$ diode (Jun-ichi Nishizawa, Kaoru Motoya, and Yasuo Okuno, 1978).

potential like in the case of Esaki diode. On the contrary, for the avalanche injection, the avalanche voltage is larger than 5.3 V and the voltage for V_d is 0.08 V at 300 GHz, so the total voltage is larger than about 5.4 V in the case of GaAs. The following quantity of γ is defined to estimate the $I-V$ characteristics of the tunnel junction (Nishizawa, 1975).

$$\gamma = \frac{\dfrac{dI}{dV}}{I} = \frac{\dfrac{d(J_t S)}{d(E_t W_t)}}{I} = \frac{G}{I}, \tag{36}$$

where I and V are the dc bias current and the voltage, respectively. This quantity is the injection efficiency by the feedback field intensity from the drift region per units of flowing dc current, which corresponds to the current ratio for rf and dc. The γ is the quantity of the differential conductance of the injection region as presented in Eq. (20) divided by the dc current as shown in Fig. 21, so γ should be higher in order to increase the efficiency of the TUNNETT diode.

Various structures for tunnel injection as a $p-n$ diode, the Schottky barrier diode and the Metal Insulator Semiconductor (MIS) diode are compared. The $p-n$ diode is thought to be the most promising one. The direct tunnel current density in a $p-n$ junction is given by the following equation (Kane, 1961; Moll, 1964).

$$J_t = \frac{\sqrt{2}\, q^3 m^* V_a E}{4\pi^3 \hbar^2 E_g^{1/2}} \exp\left(-\frac{\pi m^{*1/2} E_g^{3/2}}{2\sqrt{2}\, q E \hbar}\right), \tag{37}$$

where E is the electric field intensity in the junction, E_g is the band-gap en-

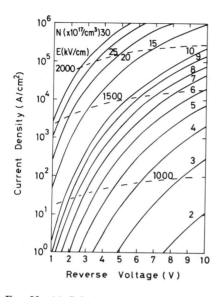
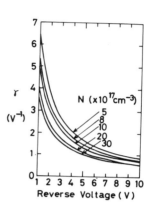

FIG. 22 (a) Calculated result of $I-V$ characteristic of the p^+-n diode. ($E_g = 1.43$ eV, $m_e^* = 0.068\ m_0$, $m_{lh}^* = 0.12\ m_0$, $\varepsilon_s = 10.9$.) (b) Calculated result of γ of (a) the p^+-n diode (Jun-ichi Nishizawa, Kaoru Motoya, and Yasuo Okuno, 1978).

ergy, V_a is the applied voltage, and m^* is the reduced effective mass of the form (Butcher *et al.*, 1962) of

$$m^* = [(1/m_e^*) + (1/m_{lh}^*)]^{-1}, \tag{38}$$

where m_e^* is the electron effective mass and m_{lh}^* is the effective light hole mass. Given E by the peak field of the p^+-n (abrupt) junction, E is given by

$$E = \left(\frac{2qN_D(V_a + V_b)}{\varepsilon_s}\right)^{1/2}, \tag{39}$$

where V_b is the built-in voltage of the junction a part of the total diffusion voltage V_{diff}, N_D is the donor density in the n layer, and ε_s is a dielectric constant of the material.

Substituting Eq. (39) into Eq. (37) yields (Nishizawa *et al.*, 1978a)

$$J_t = B_1\sqrt{V_a + V_b}\,V_a \exp\left(-\frac{B_2}{\sqrt{V_a + V_b}}\right), \tag{40}$$

where

$$B_1 = \frac{q^3}{2\pi^3\hbar^2}\left(\frac{qm^*N_D}{\varepsilon E_g}\right)^{1/2} \tag{41}$$

and

$$B_2 = \frac{\pi E_g}{4q\hbar} \left(\frac{\varepsilon E_g m^*}{qN_D}\right)^{1/2}.$$

(42)

The results of calculations are shown in Fig. 22a and in Fig. 22b. The material parameter of GaAs is chosen for the calculations. The results of calculations from Eq. (40) show that a considerable amount of the tunneling current over 10^4 A/cm^2 will flow into the diode with $N_D > 6 \times 10^{17}$ cm^{-3} at a bias voltage lower than 10 V. The tunnel current of the reverse biased Schottky barrier diode is given by the equation (Padovani and Stratton, 1966),

$$J_{SB} = \frac{A^* \pi E_{00} \exp[-2V_b^{3/2}/3E_{00}(V_a + V_b)^{1/2}]}{kT[V_b/(V_b + V_a)]^{1/2} \sin\{\pi kT[V_b/(V_b + V_a)]^{1/2}/V_{00}\}},$$

(43)

where A^* is a modified Richardson constant in which the electron mass is replaced by the effective electron mass in a semiconductor, V_b is the built-in voltage, and V_a is the bias voltage, respectively.

The results of calculations of I–V and γ of the Schottky barrier diode in which the material parameter is chosen for GaAs are shown in Fig. 23a, b. The tunnel current and γ of the Schottky barrier diode are much smaller than those of the p–n diode. These results suggest superiority of the p–n junction diode to the Schottky barrier diode for the TUNNETT diode. Moreover, the temperature dependence of the I–V characteristic is considerably larger in the case of the Schottky barrier.

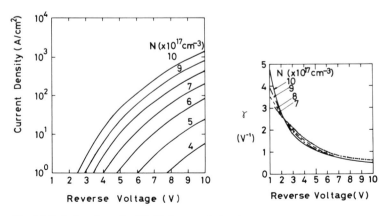

FIG. 23   (a) Calculated result of I–V characteristic of the Schottky barrier diode. (b) Calculated result of γ of (a) the Schottky barrier diode (Jun-ichi Nishizawa, Kaoru Motoya, and Yasuo Okuno, 1978).

B. COMPARISONS AMONG MATERIALS

The material in which the tunnel injection occurs at a smaller voltage that is effective in increasing the voltage efficiency and also the larger γ material that is effective in increasing the current efficiency are suitable for the fabrication of the TUNNETT diode. The small band-gap material such as Ge is suitable for this purpose, and the direct tunneling is higher in efficiency than the indirect tunneling. The small ionization constant α is desirable, but this is not the serious point. Because the tunneling occurs even in a narrow space, the avalanche injection cannot be induced. However, the occurrence of avalanche injection, which generates much noise yet can compensate temperature coefficient, can be avoided with the introduction of the $p^+-n^+-i(\nu)-n^+$ diode.

The higher saturation velocity of carriers is more desirable in order to raise the oscillation frequency and the feasibility of the fabrication of the drift region which becomes shorter at a higher frequency. The smaller diffusion coefficient of carriers is required in order to prevent the spatial dispersion of injected carriers. The measured diffusion coefficient is shown in Fig. 24 (Okamoto et al., 1965). The reliable literature values of the saturation velocity and the diffusion coefficient of carriers at high electric field intensity are not available at present, so the investigation of these physical parameters is necessary for the design of the TUNNETT diode.

The input power density of the TUNNETT diode is of the order of 10^5 W/cm² in the case of $P_{in} = 1$ W and the junction area $= 1 \times 10^{-5}$ cm². The high thermal conductivity of material is needed for the heat convection. The material parameters of the several semiconductors are tabulated

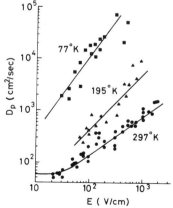

FIG. 24 Electric field dependence of the diffusion coefficient of Ge. There are many possibilities to tend to zero after the velocity saturation because electron energy is absorbed by the optical lattice vibration (K. Okamoto, J. Nishizawa, and K. Takahashi, 1965).

TABLE I

MATERIAL PARAMETERS OF Ge, Si, AND GaAs

Properties	Ge	Si	GaAs
$\mathscr{E}_g$ (eV) (300°K)	0.803	1.12	1.43
Effective mass m^*/m_0 electron	$m_l^*=1.6$ $m_t^*=0.082$	$m_l^*=0.97$ $m_t^*=0.19$	0.068
hole	$m_{lh}^*=0.04$ $m_{hh}^*=0.3$	$m_{lh}^*=0.16$ $m_{hh}^*=0.5$	0.12 , 0.5
Saturation velocity (cm/sec) electron	6×10^6	10^7	10^7
hole	6×10^6	6×10^6	10^7
Thermal conductivity at 300°K (W/cm°C)	0.64	1.45	0.46

in Table I. Ge will be a promising material with its narrower band-gap energy than that of GaAs, however, lower diffusion potential.

C. PREPARATION OF TUNNETT DIODE

The first TUNNETT oscillation was realized experimentally in 1968 from the GaAs p^+-n diode fabricated by conventional slow cooling liquid phase epitaxy (Okabe *et al.*, 1968). Recently several types of TUNNETT diodes such as $p^+-n^+-i(\nu)-n^+$, p^+-n-n^+, and p^+-n diodes have been fabricated by the use of the author's new liquid phase epitaxial growth method, which is named the temperature difference method under controlled vapor pressure (TDM CVP) (Nishizawa *et al.*, 1975b). In this method the deviation from stoichiometric composition of GaAs is compensated by the application of the arsenic vapor pressure. The near perfect crystals of GaAs and GaP have successfully been obtained (Nishizawa *et al.*, 1977a), and the excellent LEDs (*light emitting diodes*) have also been achieved (Nishizawa *et al.*, 1977a).

The epitaxial growth apparatus for TDM CVP is shown in Fig. 25. The sliding carbon boat is used to multilayer successive epitaxial growth. The temperature difference in Ga melt is settled by the tungsten heater attached locally to the outside of the Ga melt container, and the GaAs is deposited onto the substrate in which temperature is kept constant.

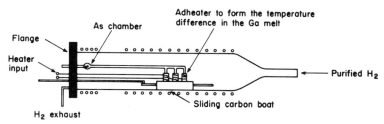

FIG. 25   Growing appratus for the TDM CVP liquid phase epitaxy.

1. p^+-n-n^+ and p^+-n Diode†

The p^+-n-n^+ and p^+-n diodes are fabricated by the following process:

(1)   The $n-n^+$ or n layers are grown on the p^+ (100) Zn-doped substrate ($\rho : 4 \sim 6 \times 10^{-3}$ Ω cm). The Te-doped polycrystal is used as a source in the Ga melt and the doping density is controlled by changing the growth temperature T_g;

(2)   Formation of the n ohmic contact;

(3)   Thinning the wafer by the lapping and the chemical etching to a thickness of about 10 μm;

(4)   Formation of the p^+ ohmic contact;

(5)   Dicing;

(6)   Bonding on the gold plated copper stem; and

(7)   Adjusting the junction area with the chemical etchant.

The cross section of the diode structure is illustrated in Fig. 26. The $C-V$ and $I-V$ characteristics have been measured. The $1/C^2$ versus V plot of the p^+-n-n^+ diode is shown in Fig. 27. It is clear that a near-ideal

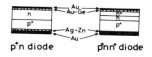

Diode structure	Growth temperature (°C)	Doping density of n layer (cm⁻³)	Doping density of n⁺ layer (cm⁻³)	Thickness of n layer (μm)
p^+n	600 – 670	5–10 X 10¹⁷	———	1 – 2
p^+nn^+	640 – 650	1 X 10¹⁸	>3 X 10¹⁸	0.7 – 3

FIG. 26   Structure of the p^+-n-n^+ diode and the properties of the epitaxially grown wafers. (Jun-ichi Nishizawa, Kaoru Motoya, and Yasuo Okuno, GaAs TUNNETT diodes, *IEEE Transactions on Microwave Theory and Techniques* **26** (No. 12) 1031. © 1978 IEEE.)

† See Okabe *et al.* 1968; Nishizawa *et al.* 1976b, 1978a, 1978b.

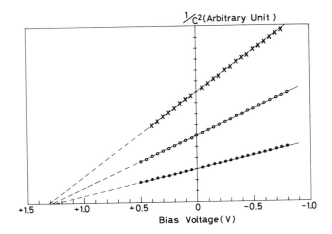

FIG. 27 $1/C^2-V$ plots of the p^+-n-n^+ diodes fabricated from three different wafers. ●, A; ×, B; and ○, C. (Jun-ichi Nishizawa, Kaoru Motoya, and Yasuo Okuno, GaAs TUNNETT diodes, *IEEE Transactions on Microwave Theory and Techniques* **26** (No. 12) 1031. © 1978 IEEE.)

abrupt junction is formed because T_g is relatively low to prevent the Zn diffusion into the n layer during the LPE (*liquid phase epitaxy*). The $I-V$ characteristics of the p^+-n diodes with the doping density of the n layer (N_D) are shown in Fig. 28 with the calculated $I-V$ lines. The flowing current is increased with the ambient temperature of the diode of $N_D = 8 \times 10^{17}$ cm^{-3}. In our experiments the p^+-n diode in which $N_D \geq 8 \times 10^{17}$ cm^{-3} is the tunnel diode. The current of the p^+-n diode in which $N_D = 5 \times 10^{17}$ cm^{-3} is suddenly increased at about 6 V and $J = 200$ A/cm^2. In the p^+-n diode, the temperature coefficient changes the sign at the diode with an increased voltage of 5.3 V as shown in Fig. 29.

2. $p^+-n^+-i(\nu)-n^+$ Diode†

The $p^+-n^+-i(\nu)-n^+$ diode has been fabricated in order to raise the efficiency of the TUNNETT diode, since the $p^+-n^+-i(\nu)-n^+$ diode is the most promising structure by virtue of its thin tunnel region and lower doping drift region as predicted by the original device structure of the TUNNETT diode, because the highly doped tunnel junction can generate much injecting carriers with smaller voltage V_t and at the same time the time constant C_t/G_t is much smaller than that of the lightly doped tunnel junction. Several attempts have been made to fabricate the $p^+-n^+-i(\nu)-n^+$ diode on the p^+ substrate. However, the control of the thin n^+ layer has not been realized. This is partly owing to the instability

† See Nishizawa *et al.*, 1980.

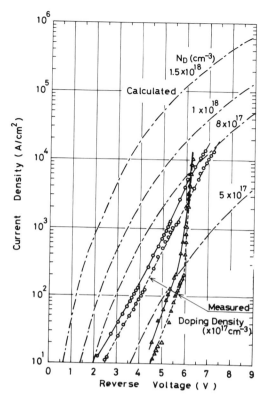

FIG. 28 *I–V* characteristics of the GaAs $p^{+}-n$ diode. Calculated lines (—·—) are also indicated. ($E_{\mathrm{g}} = 1.43$ eV, $m_{e}^{*} = 0.068\ m_{0}, m_{\mathrm{lh}}^{*} = 0.12\ m_{0}, \varepsilon_{\mathrm{s}} = 10.9$.) ———, 353 K; ---, 293 K; ○, 8; △, 5. (Jun-ichi Nishizawa, Kaoru Motoya, and Yasuo Okuno, GaAs Tunnett diodes, *IEEE Transactions on Microwave Theory and Techniques* **26** (No. 12) 1031. © 1978 IEEE.)

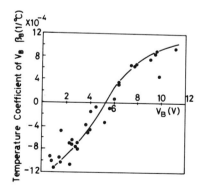

FIG. 29 Relation between V_{B} and β_{B} in the $p^{+}-n$ diode. $J = 100$ A/cm².

247

of the beginning of epitaxial growth on the chemically etched substrate.

The new technique has been developed to form a thin n^+ layer by the introduction of the growth diffusion TDM CVP. The successive growth onto the n^+ Si doped (ρ ; 1×10^{-3} Ω cm) substrates have been carried out. At first the n^+ layer is grown on the n^+ substrate after the slight melt-back process; and then the second $i(\nu)$ layer and the third p^+ layer are grown. Since the p^+ Ga melt for the third layer growth contains the Ge as a p type impurity and S as a minor n type dopant, then the n^+ diffused layer into $i(\nu)$ layer is formed during the p^+ layer growth because sulfur, which is contained less than germanium, has much larger diffusion constant. T_g is about 750 °C.

The cleavaged cross-sectional view of the epitaxially grown $p^+-n^+-i(\nu)-n^+$ wafer is shown in Fig. 30. The doping profile measured is also shown in Fig. 31. The total thickness of the $n^+-i(\nu)$ layer is approximately 0.4 μm. The p^+-n^+ interface was measured by the $C-V$ and the $i(\nu)-n^+$ transition region was estimated from the doping profile of the $p^+-i(\nu)-n^+$ diode. The fabrication processes of the TUNNETT diode after epitaxial growth was nearly the same as the above-mentioned p^+-n-n^+ diode. The processes are (a) p^+ ohmic contact formation, (b) wafer thinning process, (c) n^+ ohmic contact formation, (d) dicing into about 100 μm $\times$ 100 μm, (e) bonding on the gold-plated copper stem, and finally (f) adjusting the junction area with chemical etchant.

The cross section of the $p^+-n^+-i(\nu)-n^+$ diode is illustrated in Fig. 32 and the $I-V$ characteristics are shown in Fig. 33. This curve shows that the diode has a mixed mode with avalanche because the breakdown volt-

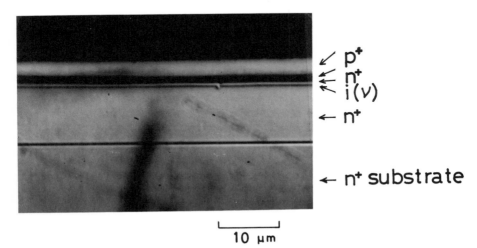

FIG. 30   Cross sectional view of the epitaxially grown wafer ($p^+-n^+-i(\nu)-n^+A$) (J. Nishizawa, K. Motoya, and K. Suzuki, 1980).

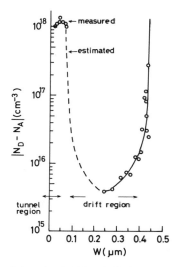

FIG. 31 Doping profile of the $p^+-n^+-i(\nu)-n^+$ diode A (J. Nishizawa, K. Motoya, and K. Suzuki, 1980).

age is as high as 7 V and should be less than a few volts and also the efficiency and the frequency should be much higher.

3. Schottky Barrier Diode†

The Schottky barrier diode whose structure is $Pt-n^+-n-n^+$ and $Pt-n-n^+$ has been fabricated by use of the conventional slow cooling method. The n or $n-n^+$ layers were grown on the Te doped n^+ substrate $(\rho ; 1 \times 10^{-3} \ \Omega$ cm).

The fabrication processes are the almost the same as the $p-n$ diode except for the Schottky barrier metal evaporation. The measured barrier height is 0.94 eV. The measured $I-V$ characteristics of the Schottky barrier diode $Pt-n-n^+$ are shown in Fig. 34. The $I-V$ expressed as $J = J_t$ $\exp(V/V_t)$ form. The temperature coefficient of $V_B(\beta_B)$ is measured by the $I-V$ characteristics by varying the temperature. Temperature coefficient

FIG. 32 Cross section of the $p^+-n^+-i(\nu)-n^+$ TUNNETT diode.

† See Ohmi and Motoya, 1976.

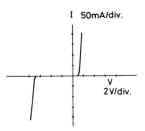

FIG. 33 I–V characteristic of the p^+–n^+–$i(\nu)$–n^+ diode A ($S = 7 \times 10^{-6}$ cm^2).

β_B of the diode in which the doping density of the n layer (N_D) is 3.5×10^{17} cm^{-3} is zero, and β_B changes negative with increasing doping density of the n layer $>3 \times 10^{17}$ cm^{-3}. The temperature dependence of the I–V characteristics of the diode ($N_D \simeq 5.9 \times 10^{17}$ cm^{-3}) are shown in Fig. 35. The current of the diode increases with increasing temperature.

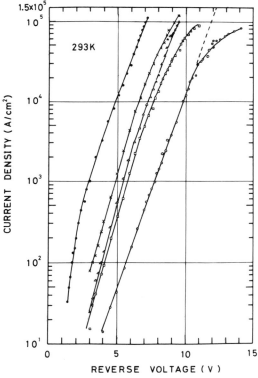

FIG. 34 I–V characteristics of the Pt–n–n^+ diodes with changing the doping density of n layer. Doping density (cm^{-3}): ●, 1×10^{18}; ×, 6.3×10^{17}; △, 5.9×10^{17}; □, 5×10^{17}; ○, 3.5×10^{17}.

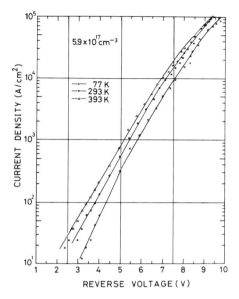

FIG. 35 Temperature dependence of $I-V$ characteristics of the Pt$-n-n^+$ diode ($N_D = 5.9 \times 10^{17}$ cm^{-3}). $\times$, 77 K; $\bullet$, 293 K; $\triangle$, 393 K (T. Ohmi and K. Motoya, 1976).

D. RESULTS OF THE MEASUREMENTS†

The oscillation experiment of the fabricated diodes have been performed using the circuit as shown in Fig. 36a and the Y band (170–260 GHz) TUNNETT oscillator is shown in Fig. 36b.

The diode with quartz stand-off structure is biased through the insulated bias post. The oscillation outpower is adjusted by the short plunger behind the diode and the bias post in order to obtain the maximum output power. The several rectangular waveguide circuits from the R band

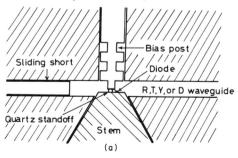

(a)

FIG. 36 (a) Cross section of the oscillator circuits (J. Nishizawa, K. Motoya, and Y. Okuno, 1979).

† See Okabe *et al.*, 1968; Nishizawa *et al.* 1974, 1977b, 1978a,b, 1979, 1980c.

FIG. 36   (b) Y band TUNNETT diode oscillator. The scale is 1 mm/div. (K. Motoya, Y. Okuno, and J. Nishizawa, 1978).

(75–110 GHz), T band (110–170 GHz), Y band (170–260 GHz), and D band (220–325 GHz) have been used as shown in Fig. 37. The output power was detected by the Si point contact diode. The detector was calibrated by a 190–210 GHz multiplier in which the GaAs Schottky barrier diode operates as a doubler of the 100 GHz crystron output power and the thin film thermocouple type power meter (Anritsu MP85B). The oscillation frequency is measured by the detection of voltage of the standing wave in the waveguide. The diode is biased pulsively with 100 nsec to 3 μsec in width and the repetition rate of 100 Hz to 10 kHz.

Figure 38 shows the relation between the breakdown voltage V_B and the

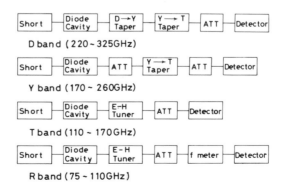

FIG. 37   Waveguide circuits for the oscillation experiments. Output power of D and Y bands are detected by the T band detector through the tapered waveguide sections.

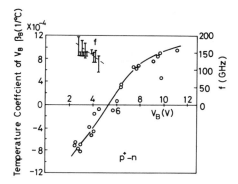

FIG. 38 Relation between the breakdown voltage (V_B), its temperature coefficient (β_B) and the oscillation frequency of the GaAs p^+-n diode.

oscillation frequency of the p^+-n diode. (Okabe *et al.*, 1968). It is clear that the diodes in which β_B is negative oscillates higher frquency than those of the diodes in which β_B is positive, which is the characteristic of the TUNNETT oscillation. The oscillation frequency of the p^+-n diodes have increased as V_B decreased.

The oscillation characteristics of the p^+-n and p^+-n-n^+ diodes are

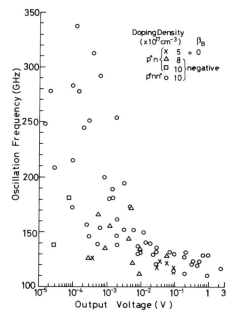

FIG. 39 Relation between the oscillation frequency and the output voltage of the detector of the p^+-n and p^+-n-n^+ diodes (J. Nishizawa, K. Motoya, and Y. Okuno, 1979).

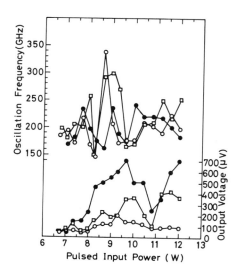

FIG. 40 Oscillation characteristic of submillimeter wave GaAs p^+-n-n^+ diode. $S = 7.8 \times 10^{-6}$ cm². Bias post (mm$^\phi$): ●, 0.36; □, 0.5; ○, 0.56 (J. Nishizawa, K. Motoya, and Y. Okuno, 1979).

shown in Fig. 39. The maximum oscillation frequency (f_{max}) of the p^+-n diode of $N_D = 5 \times 10^{17}$ cm^{-3} is 130 GHz and the diodes of $N_D \gtrsim 8 \times 10^{17}$ cm^{-3} oscillate above 130 GHz. With increasing N_D, the oscillation frequency of the p^+-n and p^+-n-n^+ diodes increases with the enhancement

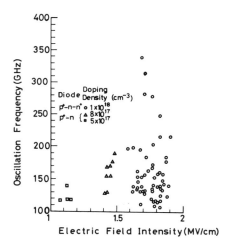

FIG. 41 Relation between the oscillation frequency and the electric field intensity of the p^+-n and p^+-n-n^+ diodes.

of the tunnel injection and the submillimeter fundamental oscillation frequency of 338 GHz (λ = 0.89 mm) has been obtained from the p^+-n-n^+ diode in which N_D is 1×10^{18} cm^{-3} as shown in Fig. 40. The peak pulsed output power has been estimated to be 10 mW.

The relation between the maximum electric field intensity (E_m) of the p^+-n junction and the oscillation frequency is shown in Fig. 41. The E_m over 1400–1500 kV/cm seems to be needed for the p^+-n GaAs TUNNETT diodes from the experimental results.

The threshold current density (J_{th}) for the oscillations of the p^+-n and p^+-n-n^+ diodes is in the range $8-10 \times 10^4$ A/cm^2 considerably higher than that for the cw operation. Most of the bias voltage of the p^+-n and p^+-n-n^+ diodes under experiments is only consumed to extend the depletion layer with regard to the injection region and not for the drift region; consequently, the relatively larger bias current is needed to establish the high-field intensity in the drift region for the oscillation. This lower efficiency caused by the lower resistivity in the drift region will be improved with the introduction of the $p^+-n^+-i(\nu)-n^+$ structure. Threshold applied voltage has been also still much higher than that estimated value mentioned above.

The experimental results of the $p^+-n^+-i(\nu)-n^+$ diodes are described below. The relation between the oscillation frequency and the current density of the $p^+-n^+-i(\nu)-n^+$ diodes is shown in Fig. 42 in comparison with the p^+-n-n^+ diode. The J_{th} of 3×10^4 A/cm^2 has been obtained from the $p^+-n^+-i(\nu)-n^+$ diode and this value of J_{th} is one-half or smaller than those of the p^+-n-n^+ diodes.

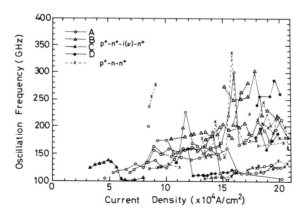

FIG. 42 Relation between the oscillation frequency and the current density of the $p^+-n^+-i(\nu)-n^+$ diodes A, B, C, and D and p^+-n-n^+ diodes (J. Nishizawa, K. Motoya, and K. Suzuki, 1980).

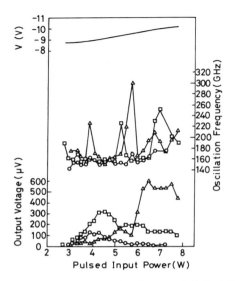

FIG. 43   Oscillation characteristics of the $p^+-n^+-i(\nu)-n^+$ diode A ($S = 3.75 \times 10^{-6}\,\text{cm}^2$) mounted in the Y band cavity. $f_{max} = 301$ GHz. Bias post (mm$^\phi$): $\bigcirc$, 0.58; $\triangle$, 0.5; $\square$, 0.46 (J. Nishizawa, K. Motoya, and K. Suzuki, 1980).

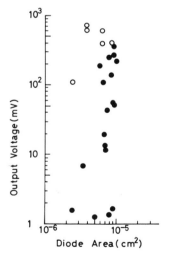

FIG. 44   Comparison of the output voltage of the detector at T band between the $p^+-n^+-i(\nu)-n^+$ diode and the p^+-n-n^+ diodes. $\bigcirc$, $p^+-n^+-i(\nu)-n^+$; $\bullet$, p^+-n-n^+ (J. Nishizawa, K. Motoya, and K. Suzuki, 1980).

The maximum fundamental oscillation frequency of up to 301 GHz have been obtained from $p^+-n^+-i(\nu)-n^+$ diode at a lower input power level than that of the p^+-n-n^+ diode as shown in Fig. 43. The active layer thickness of this diode is about 0.4 μm. The standing wave in the waveguide at 301 GHz oscillation is very fine and the low frequency oscillation has not been observed. A comparison of the output power in the T band between the $p^+-n^+-i(\nu)-n^+$ and p^+-n-n^+ diode is shown in Fig. 44. It is seen that the higher output power is obtained from $p^+-n^+-i(\nu)-n^+$ diodes, then the efficiency of the $p^+-n^+-i(\nu)-n^+$ diode is superior to that of the p^+-n-n^+ diodes.

The oscillation characteristics at the Y band between the $p^+-n^+-i(\nu)-n^+$ and p^+-n-n^+ diodes in which the junction area is nearly the same are shown in Fig. 45. The higher output power has been obtained from the $p^+-n^+-i(\nu)-n^+$ diode at a lower input power level. These results are considered to be the evidence of the higher efficiency of the $p^+-n^+-i(\nu)-n^+$ diode over the p^+-n-n^+ and p^+-n diodes.

The temperature dependence of the $I-V$ characteristics have been investigated in the preparation of the GaAs TUNNETT diode. This property affects the oscillation characteristics of the TUNNETT diode as shown in Fig. 46. With an increase of the junction temperature varied by increasing the pulse width the output power of the oscillator is increased. This property is considered to provide an effective means to check whether the diode is oscillated by the tunnel injection.

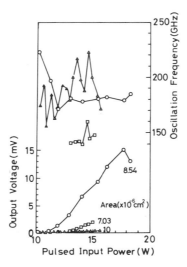

FIG. 45 Comparison of the oscillation characteristics between the $p^+-n^+-i(\nu)-n^+$ diode and p^+-n-n^+ diodes. $\bigcirc$, $p^+-n^+-i(\nu)-n^+$ diode A; $\triangle$, $\square$, p^+-n-n^+ (J. Nishizawa, K. Motoya, and K. Suzuki, 1980).

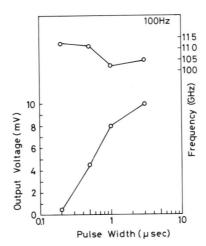

FIG. 46   Pulsed oscillation characteristics of the $p^+-n^+-i(\nu)-n^+$ diode A ($S = 1.6 \times 10^{-5}$ cm²) (Jun-ichi Nishizawa, 1980).

The oscillation characteristics of the Pt–n GaAs diode are shown in Fig. 47. As mentioned in Section II.A, the Schottky barrier diode is considered inferior to the p–n diode. The f_{max} of 154 GHz has been obtained in our experiment and the device performance of the Schottky barrier diode as the TUNNETT is not good in comparison to the $p^+-n^+-i(\nu)-n^+$, p^+-n-n^+, and p^+-n diodes as expected.

E.   NOISE IN TUNNETT DIODE

The low noise of the tunnel injection of the TUNNETT diode is an

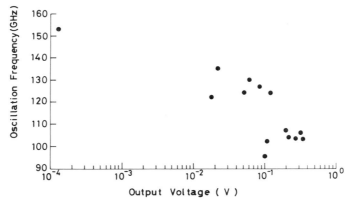

FIG. 47   Relation between the output voltage of the detector and the oscillation frequency of the Schottky barrier TUNNETT (Pt–GaAs) diode (J. Nishizawa, K. Motoya, and Y. Okuno, 1979).

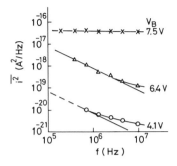

FIG. 48 Mean square noise current versus frequency of the GaAs p^+-n diodes. $I = 2$ mA. $\times$, $V_B = 7.5$ V; $\triangle$, $V_B = 6.4$ V; $\bigcirc$, $V_B = 4.1$ V (T. Okabe and J. Nishizawa, 1969).

attractive character for application such as oscillators and amplifiers in the millimeter to submillimeter wave region. The mean square noise current $\overline{i^2}$ (A²/Hz) of the dc biased p^+-n diode has been measured as shown in Fig. 48 (Okabe and Nishizawa, 1969). There are a slight $1/f$ noise overlapped on the constant noise independent of the frequency. Several diodes with different V_B have been measured. With decrease in V_B the noise is greatly decreased and finally the shot noise of $\overline{i^2} = 2qI$ has been confirmed as shown in Fig. 49. The $\overline{i^2}$ of the diode of $V_B = 4.1$ V is 4 orders of magnitude smaller than that of the diode with $V_B = 7.5$ V. The noise in the avalanche diode of $V_B = 7.5$ V is nearly constant, independent of the frequency.

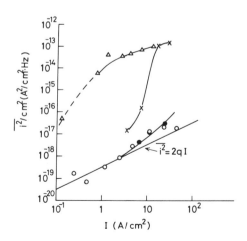

FIG. 49 Mean square noise current versus current density of GaAs p^+-n diode. $\triangle$, $V_B = 7.5$ V; $\times$, $V_B = 6.4$ V; $\bigcirc\bullet$, $V_B = 4.1$ V. (T. Okabe and J. Nishizawa, 1969).

The reduction of the J_{th} of the TUNNETT diode is important for increasing the efficiency and lowering the noise level in the device performance.

III. Future of TUNNETT

A. High-Power TUNNETT Diode

The noise in the TUNNETT diode is the shot noise only if $1/f$ noise is successively reduced. Then the output power level of the TUNNETT diode can be much smaller than that of the IMPATT diode.

The output power of the TUNNETT diode has attained 1 mW 150 GHz cw as reported by Fetterman et al. (Elta et al., 1980). The diode prepared in the author's laboratory by Motoya is very close to 100 mW in pulse operation at 300 GHz. Moreover, the structure of the diode is rather unsatisfactory in that it needs about 7 V, which is very large compared with the theoretical value and very close to the IMPATT mode, though the temperature coefficient is positive.

Based on the theoretical analysis, the applied voltage for the TUN-NETT can be as low as 1 V for 300 GHz and is much smaller than that of the IMPATT. And the flowing current can be much higher, because the energy dissipation in the injection region is much smaller than that in the IMPATT, and the available power is expected to be larger.

Since there is no minimum current limitation on the flowing current for the oscillation, continuous oscillation should be easily realized in the TUNNETT diode. The problem involved is the control of the distributed impurity of the TUNNETT diode

$$\frac{q}{\epsilon} \int N_d \, dx = E_t - E_d, \tag{44}$$

between the drift region and the tunneling region, which corresponds to the balance between the tunneling region and the transit time drift region.

The voltage (V_{tr}) required to the transition region (W_{tr}) between the

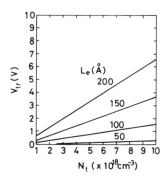

FIG. 50.   The relation between V_{tr} and the impurity density (N_t) of the tunneling region as a parameter of Le. $N_d = 10^{16}$ cm^{-3}.

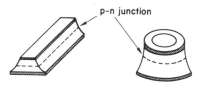

FIG. 51 Structure of the distributed type diode. Ring structure has no standing wave when output circuit is distributedly coupled, and the efficiency becomes maximum. It also has memory action in the sense of direction of propagating oscillating wave (Watanabe *et al.*, 1960).

tunneling region and the drift region is calculated. The impurity profile is given as follows (Shimizu, 1961):

$$N = N_t \exp\left(-\frac{x - W_t}{Le}\right) + N_d, \tag{45}$$

where N_t is the doping density of the tunneling region, N_d is the doping density of the drift region, W_t is the thickness of the tunneling region and Le is the diffusion length of the impurity in the tunneling region. The calculated V_{tr} is shown in Fig. 50 as a parameter of Le.

V_{tr} is increased with the increase of Le and N_t. The small V_{tr} is needed to increase the efficiency of the Tunnett diode. The TUNNETT is expected to operate even with high dc current for both higher power generation and continuous operation.

These should be solved earlier as well as the reduction of reduced impurity contents in drift region.

For the higher power realization the distributed structure can be introduced as shown in Fig. 51, which is applied to the Esaki diode and also is suggested to be applicable to the avalanche diode and the Gunn diode (Watanabe *et al.*, 1960). Ono and Mizuno realized the distributed structure of the Gunn diode (Aishima *et al.*, 1977). This is a very useful technique because the junction capacitance can be neglected and the area is increased without the decrease of the impedance.

B. GUNNETT DIODE

The GUNNETT diode and the GUNPATT diode have been proposed by the author (Nishizawa, 1971, 1974). The GUNPATT diode has a combined operational mode with the GUNN effect (transferred electron effect) and the IMPATT can prevent the diffusion attenuation of the density wave through the drift region. The operation principle of the GUNNETT diode is the use of the GUNN effect of the drift region of the TUNNETT diode.

As discussed in Section I.E, the phase of the injected carriers in the drift region rotates and also attenuated owing to the diffusion. In the GUNNETT diode and the GUNPATT diode the attenuation of the

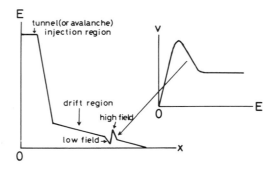

Fig. 52   Schematic illustration of the GUNNETT diode.

drifting carriers is prevented by the use of the negative $v-E$ characteristic of the GUNN effect.

This is schematically illustrated in Fig. 52. The lightly doped drift region is formed in order to set the drift field intensity in the negative $v-E$ region. The electric field intensity of the front of injected carriers is higher than that of the rear of injected carriers owing to the space charge effect. Then the velocity of the front part of electrons is slower than that of the rear of electrons. Hence the shape of electrons becomes very thin. This phenomenon of carriers acts as the prevention of the diffusion of the injected carriers. As explained above, this phenomenon is useful for both of the TUNNETT and IMPATT diode.

The high power and high efficiency of the GaAs X band IMPATT in the literature is said to be the mode of the GUNPATT diode (Hirachi *et al.*, 1976). Considering the $v-E$ characteristic and the injection mechanism,

TABLE II

SEVERAL OPERATION MODE OF THE TRANSIT TIME
NEGATIVE RESISTANCE DIODE

Injection mechanism	E_d	Mode of operation	Avalanche multiplication
Tunnel	I	GUNNETT	small or none
	II , III	TUNNETT	
Avalanche starts by the tunnel injection	I	Hybrid	large
	II , III		
Mixed of avalanche and tunnel	I		
	II , III		
Avalanche	I	GUNPATT	
	II , III	IMPATT	

several operation modes of the transit time negative resistance diode are tabulated in Table II. The material, which has a satellite valley in the conduction band such as GaAs, InP, etc., will be able to be the GUNNETT diode. The high power and higher frequency from the GUNNETT diode will be expected. In this case, the resonator structure for the acoustic oscillation seems very important for the operation.

C. Before and after TUNNETT

Field effect transistors are extending the frequency range and the power. But the static induction transistor, which corresponds to the short channel punching through the pre-space charge conduction FET (*field effect transistor*), has a much smaller time constant than that of the FET because of the short channel and the short gate. Therefore, the SIT (*static induction transistor*) (Nishizawa *et al.*, 1975a) is expected to operate at a much higher frequency than the FET. However, both have a common operation mechanism, control of the channel width. On the other hand, the potential barrier in SIT is very similar to the base punching through the bipolar transistor which operates prior to the attainment of space charge conduction. The main difference between SIT and BPT (*bipolar transistor*) is the mechanism of the control of the potential barrier height; in the SIT, it is controlled through the depletion layer by the base voltage even in the MOS type, and in the BPT, it is controlled through the base resistance. Therefore, with increasing frequency, control of the potential barrier height becomes difficult in the BPT and is easier in the SIT. Based on the difference of the operation mechanisms, the SIT can operate in a manner similar to the BPT even with the MIS structure, which has an extremely small storage effect, even not with the MOS structure, the SIT is principally a unipolar device and has no minority carrier storage effect. Therefore the SIT is much superior to the BPT.

As mentioned already, if the thickness of the potential barrier is reduced in the SIT, which can be realized only in the SIT but not in the BPT because of the drastic increase of base resistance, and finally it follows the thermionic emission equation and the α cutoff frequency can be determined from the distribution of the initial velocity. And this situation is named the Ideal SIT by the author (Nishizawa *et al.*, 1975a; Nishizawa, 1978, 1979, 1980) and the Ballistic transistor very recently by Eastman *et al.* (Shur and Eastman, 1979). The Ballistic SIT is expected to be capable of operation up to 718 GHz by the author (Nishizawa, 1978) and up to 800 GHz by Bozler (Bozler and Alley, 1980). The permeable base transistor (Bozler and Alley, 1980) is the buried Schottky gate type SIT which was described even in the first paper of SIT (Nishizawa *et al.*, 1975a). The Ballistic transistor which corresponds to SIT with the very thin potential barrier was also described even in the first paper of SIT, and in that case $I–V$

characteristics was estimated to show thermionic emission equation (Nishizawa et al., 1975a, Nishizawa, 1978, 1979, 1980) and α cutoff frequency was estimated to be determined by the distribution of the initial velocity of the injected carriers (Nishizawa, 1978). And if the injection source region is replaced by the tunneling junction, the transistor promises to be particularly effective in reducing the capacitance between source and gate and increasing the transfer conductance. In the tunneling junction, the dependence of the injecting carrier on the field intensity is so large that the transfer conductance Gm is much higher than the conventional SIT. The figure of merit of the SIT can be represented by Gm/C, so the highest frequency of operation becomes also very high and is easily estimated to be in the range of THz. This kind of SIT is named the Static Induced Tunnel Transistor (SIT^2).

If the transit time effect in the depletion layer between the intrinsic gate and drain to increase the thickness of the layer, this kind of transistor can be called the Static Induced Transit Time Transistor (SIT^3). When both mechanisms are used, the transistor can be the Static Induced Tunnel Transit Time Transistor (SIT^4).

Anyhow, three kinds of transistor may be applied for the range of THz, competing with the TUNNETT, which is useful in the same range. All of these are low noise devices.

After THz range, Raman (Pidgeon et al., 1971) and Brillouin (Nishizawa and Suto, 1980) semiconductor lasers are expected to be most useful devices in the future as reported by the author (Nishizawa, 1963), up to 100 THz of the field of the conventional semiconductor laser diode.

IV. Conclusion

It is concluded that the TUNNETT is the most promising device intermediate between the static induction transistor and the Raman and Brillouin semiconductor lasers.

The much higher power output also can be expected as the results of the realization of optimum field distribution. It is also certain that the TUNNETT will be capable of much higher efficiency cw operation in the very near future. The frequency limit is expected to be as high as about a few terahertz though it depends on the fabrication technique and on the ballistic effect. The ballistic effect seems to induce a very fast speed for carriers in the transit region and the design rule should be largely changed, partly there are some results which coincides much better for such theoretical estimation expecting ballistic effect rather than the classical equation. A new design rule to be conceived will enable the TUNNETT diode to have larger dimensions which makes easier its fabrication. These are the future problems to be solved.

REFERENCES

Aishima, A., Suzuki, S., Yokoo, K. and Ono, S. (1977). 5 ~ 20 GHz wide band GUNN diode oscillator, *Conv. Record Semicond. Sect. J.E.C.E.* pp. 6–159, (in Japanese).

Bozler, C. O., and Alley, G. D. (1980). Fabrication and numerical simulation of the permeable base transistor, *IEEE Trans. Electron Devices* **ED-27** (6), 1128–1141.

Butcher, P. N., Hulme, K. F., and Morgan, J. R. (1962). Dependence of peak current density on acceptor concentration in germanium tunnel diodes, *Solid-State Electron.* **5,** 358–360.

Elta, M. E., and Haddad, G. I. (1979). High frequency limitation if Impatt, Mitatt, and Tunnett mode devices, *IEEE Trans. Microwave Theory Tech.* **MTT-27** 442–449.

Elta, M. E., Fetterman, H. R., Macropoulos, W. V., and Lambert, J. J. (1980). 150 GHz GaAs MITTAT source, *IEEE Electron Device Lett.* **EDL-1** (6), 115–116.

Gunn, J. B. (1956). Avalanche injection in semiconductors, *Proc. Phys. Soc. London Ser. B.* **69,** 781–790.

Hirachi, Y., Kobayashi, K., Ogasawara, K., Hisatsugu, T., and Toyama, Y. (1976). A new operation mode surfing mode in high-low-type GaAs IMPATT's, *Tech. Dig. IEDM* pp. 102–105.

Ishibashi, T., Nakamura, T., and Ohmori, M. (1977). Submillimeter wave silicon impatt diodes, *IECE Jpn. Tech. Group Meeting* MW76-137 (in Japanese).

Kane, E. O. (1961). Theory of tunneling, *J. Appl. Phys.* **32** (1), 83–91.

Kuno, H. J. (1979). IMPATT devices for generation of millimeter waves, "Infrared and Millimeter Waves," Vol. 1, Source of Radiation, Chapter 2. Academic Press, New York.

McKay, K. G., and McAfee, K. B. (1953). Electron multiplication in silicon and germanium, *Phys. Rev.* **91** (5), 1079–1084.

Misawa, T. (1966). Negative resistance on p-n junction under avalanche breakdown conditions, Part I and II, *IEEE Trans. Electron Devices* **ED-13** 137–151.

Moll, J. L. (1964). "Physics of Semiconductors." McGraw-Hill, New York.

Nishizawa, J. (1963). Future of semiconductor laser, *Denshi Kagaku* **13** (4), 17–20, 30–31 (in Japanese).

Nishizawa, J. (1971). Panel discussion of several problems on semiconductor oscillators, *Joint Conv. Record Four Inst. Elec. Eng., Jpn.* pp. 121–144.

Nishizawa, J. (1974). Progress of compound semiconductor devices, *Denshi-Zairyo* 18–22 (in Japanese).

Nishizawa, J. (1978). Panel discussion on SIT, *Semicond. Electron. (Handotai-Kenkyu)* **15,** 319–358.

Nishizawa, J. (1979). Recent progress and potential of SIT, *Digest Tech. Papers, Int. Conf. Solid State Devices, 11th, August* A-0-1.

Nishizawa, J. (1980). Recent progress and potential of SIT, *Jpn. J. Appl. Phys. Suppl. 19–1* **19,** 3–11.

Nishizawa, J. and Suto, K. (1980). Semiconductor Raman laser, *J. Appl. Phys.* **51** (5), 2429–2431.

Nishizawa, J., and Watanabe, Y. (1953). The contract research report to the Nippon Telegraph and Telephone Public Corporation.

Nishizawa, J., and Watanabe, Y. (1958). High frequency properties of the avalanching negative resistance diode, *Sci. Rep. Res. Inst. Tohoku Univ.* **10** (2), 91–108.

Nishizawa, J., and Yamamoto, K. (1978). High-frequency high-power static induction transistor, *IEEE Trans. Electron Devices,* **ED-25,** 314–322.

Nishizawa, J., Ohmi, T., and Sakai, T. (1974). Millimeter-wave oscillation from Tunnett diode, *Proc. Eur. Microwave Conf.* pp. 449–453.

Nishizawa, J., Terasaki, T., and Shibata, J. (1975a). Field-effect transistor versus analog transistor (Static induction transistor), *IEEE Trans. Electron Devices* **ED-22,** 185–197.

Nishizawa, J., Okuno, Y., and Tadano, H. (1975b). Nearly perfect crystal growth of III-V compounds by the temperature difference method under controlled vapor pressure, *J. Crystal Growth* **31**, 215–222.

Nishizawa, J., Suto, K., and Teshima, T. (1977a). Minority-carrier lifetime measurements of efficient GaAlAs p-n heterojunctions, *J. Appl. Phys.* **48** (8), 3484–3495.

Nishizawa, J., Motoya, K., and Okuno, Y. (1977b). 200 GHz tunnett diodes, *Proc. Conf. Solid State Devices, 9th August* C-2-2.

Nishizawa, J., Motoya, K., and Okuno, Y. (1978a). GaAs tunnett diodes, *IEEE Trans. Microwave Theory Tech.* **MTT-26** (12), 1029–1035.

Nishizawa, J., Motoya, K., and Okuno, Y. (1978b). *Jpn. J. Appl. Phys. Suppl. 17-1* **17**, 167–172.

Nishizawa, J., Ohmi, T., and Niranjian, M. S. (1978c). Avalanche induced dispersion in Impatt diodes, *Solid-State Electron.* **21**, 847–858.

Nishizawa, J., Motoya, K., and Okuno, Y. (1979). Submillimeter wave oscillation from GaAs TUNNETT diodes, *Proc. Eur. Microwave Conf.* pp. 463–467.

Nishizawa, J., Motoya, K., and Suzuki, K. (1980). GaAs $p^+-n^+-i(\nu)-n^+$ tunnett diodes, *Proc. Eur. Microwave Conf., 10th* pp. 667–671.

Nishizawa, J. (1975). Tunnett diode, *Oyo Butsuri* **44**, 821–825 (in Japanese).

Ohmi, T., (1977). "Semiconductor Electronics" (Handotai-Kenkyu), Vol. 13, Chapter 4. Kogyo-chosakai.

Ohmi, T., and Motoya, K. (1976). Millimeter wave oscillations from tunnett diodes, *IECE Jpn. Tech. Group Meeting* **ED-75-71** (in Japanese).

Okabe, T., and Nishizawa, J. (1969). Some consideration of tunnel injection transit-time (tunnett) diode oscillator, *IECE Jpn. Tech. Group Meeting* **ED-69-19.**

Okabe, T., Takamiya, S., Okamoto, K., and Nishizawa, J. (1968). Bulk oscillation by tunnel injection, presented at *IEEE Int. Electron Devices Meeting, December.* Also, RIEC Technical Rep. **TR-31,** Tohoku Univ.

Okamoto, K., Nishizawa, J., and Takahashi, K. (1965). Measurement of hot carrier diffusion constant in semiconductors, *J. Appl. Phys.* **36** (12), 3716–3722.

Padovani, F. A., and Stratton, R. (1966). Field and thermionic-field emission in Schottky barriers, *Solid-State Electron.* **9** (7), 695–707.

Pidgeon, C. R., Lax, B., Aggarwal, R. L., and Chase, C. E. (1971). Tunable coherent radiation source in the 5 μ region, *Appl. Phys. Lett.* **19** (19).

Shimizu, A., and Nishizawa, J. (1961). Alloy-diffused variable capacitance diode with large figure-of-merit, *I. R. E. Trans. Electron Devices* **ED-8,** 370–377.

Shockley, W. (1954). Negative resistance arising from transit time in semiconductor diodes, *Bell Syst. Tech. J.* **33**, 799.

Shur, M. S., and Eastman, L. F. (1979). Ballistic transport in semiconductor at low temperature for low-power high-speed logic, *IEEE Trans. Electron Devices* **ED-26** (11), 1677–1683.

Sze, S. M., and Gibbons, G. (1966). Avalanche breakdown voltages of abrupt and linearly graded p-n junctions in Ge, Si, GaAs and GaP, *Appl. Phys. Lett.* **8,** 111 (1966).

Tokuyama, T. (1962). Zener breakdown in alloyed germanium p^+-n junctions, *Solid-State Electron.* **5**, 161–169.

Tyagi, M. S. (1968). Zener and avalanche breakdown in silicon alloyed p-n junctions, *Solid-State Electron.* **11**, 95–115.

Watanabe, Y., and Nishizawa, J. (1952). Reverse characteristic of the semiconductor rectifier, *Record Elec. Commun. Eng. Conversazione, Tohoku Univ.* **21** (3).

Watanabe, Y., Nishizawa, J., Yamamoto, T. and Shimizu, A. (1960). Wide band parametric amplification of the distributed semiconductor diode, *IECE Jpn., Tech. Committee Meeting, January.*

CHAPTER 5

Measured Performance of Gyrotron Oscillators and Amplifers*

V. L. Granatstein, M. E. Read, and L. R. Barnett

Naval Research Laboratory
Washington, D.C.

I.	INTRODUCTION	267
II.	GYROTRON TRAVELING-WAVE TUBE AMPLIFIERS	272
	A. *Gyro-TWT Theory*	273
	B. *Gyro-TWT Experiments*	275
	C. *Summary of Performance and Prospects*	280
III.	GYROMONOTRON OSCILLATORS	281
	A. *Review of the Theory*	283
	B. *Qualitative Features of Oscillator Performance*	286
	C. *Parametric Dependence of Oscillator Performance*	290
	D. *Efficiency Enhancement*	293
	E. *Stability and Coherence*	295
	F. *High-Average Power Performance*	296
	G. *Summary of Performance and Prospects*	296
APPENDIX.	MEASUREMENT METHODS	298
	A. *Cold Tests*	298
	B. *Electron Beam Performance*	299
	C. *Tube Performance (Power, Frequency, Mode)*	300
	REFERENCES	301

I. Introduction

In the first volume of this series on "Infrared and Millimeter Waves," there appeared an excellent review paper on the subject of "Gyrotrons" written by Hirshfield (1979). That paper correctly predicted that "Achievements over the new few years can be expected to be rapid." The present paper reports on experimental studies of gyrotron physics and technology that have taken place in the two-year span since the Hirshfield review was published; the stress is on work in the United States. Gyrotron experimental research at the Institute of Applied Physics,

* Supported in part by the Naval Electronic Systems Command and by the Department of Energy.

Gorky, USSR, was extensively covered in the Hirshfield review and in a number of other recent review papers (Flyagin *et al.*, 1977; Hirshfield and Granatstein 1977; Andronov *et al.*, 1978; Gapanov *et al.*, 1980; Symons and Jory, 1981); this work on high-power, efficient, gyrotron oscillators at millimeter wavelengths had set the pace in the field ever since the mid 1960s; however, to our knowledge, little new has been published by the gyrotron group at Gorky in the last two years. Most of the newly published experimental results have been reported by two groups working in the U.S. at the Naval Research Laboratory (NRL) in Washington, D.C. and at Varian Associates, Palo Alto, California. The research of both these groups shall be described in the present paper, but the work at NRL will be stressed because of the authors' greater familiarity with its content.

Experimental studies in gyrotron physics have also recently been initiated at least a dozen new research institutions in the United States, the United Kingdom, France, Israel, Japan, and the Peoples Republic of China. An accelerating flow of research accomplishment is anticipated. Such a rapid growth in activity is owing to the breakthrough that gyro-

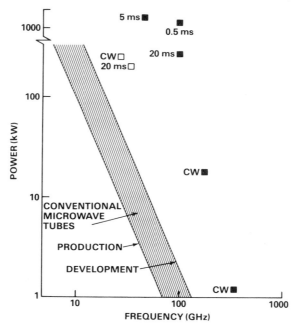

FIG. 1   Continuous-wave and long pulse microwave tube state of the art 1980. (■, Gyrotrons developed at The Institute of Applied Physics, Gorky, USSR; □, gyrotrons operated in the USA.)

trons have achieved in power levels at millimeter wavelengths and to requirements for high-power millimeter wave sources for heating of plasmas in controlled thermonuclear fusion research.

In Fig. 1, the reported power levels in long pulse (60.1 msec) gyrotron oscillators are plotted as a function of frequency. It is clear that power levels orders of magnitude above the capabilities of conventional microwave tubes have been achieved throughout the millimeter wave band (1–10 mm). All the data points in Fig. 1 come from experiments at the Institute of Applied Physics in Gorky except for the two open square data points. The data point at 28 GHz represents the achievement by Varian Associates of 212-kW cw power (Symons and Jory, 1981), by far the largest cw power achieved to date with gyrotrons or for that matter with any tube type at frequencies near the millimeter wave band.

The open square data point at 35 GHz represents a 150-kW gyrotron oscillator operated at NRL with 20 msec pulse duration (Read *et al.*, 1980). Measurements made of this tube's performance will be described in detail in Section III. This gyrotron has been applied to electron cyclotron resonance heating (ECRH) of the plasma in a large Tokamak device at the Oak Ridge National Laboratory (Gilgenbach *et al.*, 1980). Results of the heating experiment were impressive as shown in Fig. 2. Approximately 60% of the gyrotron power was absorbed by the plasma, with projection of virtually complete absorption in a tokamak of reactor size.

Gyrotrons emit coherent radiation at the electron cyclotron frequency. Basically, they have been capable of unusually large power levels at short wavelengths because the resonant wavelength is fixed by the strength of an externally applied magnetic field rather than by the scale of a circuit structure as in most conventional tubes. Hence, in principle, one can more successfully employ circuits with larger dimensions which are capable of handling higher power levels. The gyrotron circuits moreover are uncomplicated and inexpensive, usually consisting simply of an unloaded circular waveguide or of a cavity formed by using a length of such a waveguide with end reflectors.

The new complications in gyrotron technology involve the requirement for both a stronger magnet (usually superconducting) to provide the resonant axial field and a novel type of electron gun. The electron guns employed in gyrotrons typically provide a hollow beam of spiraling electrons which have a large fraction of their energy in velocity components transverse to the tube axis since only transverse energy is converted into electromagnetic radiation. Magnetron injection guns (Tsimring, 1972; Gol'denberg and Petelin, 1973; Avdoshin *et al.*, 1973, Avdoshin and Gol'denberg, 1973, Seftor *et al.*, 1979a). have thus far been used most successfully with their cathodes operated in a temperature limited mode.

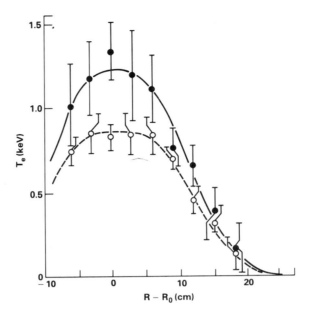

F$_{IG}$. 2   Results of electron cyclotron heating (ECH) experiment in the ISX-B tokamak using a 35-GHz gyrotron oscillator. Curves show electron temperature profiles in the tokamak plasma. Dashed line denotes data taken before ECH; solid line represents data at the end of ECH. Line average electron density was ~ 10^{13} cm^{-3}. Data are for 80-kW microwave injection from a 35-GHz gyrotron oscillator having a 10 msec pulse duration (from Gilgenbach *et al.*, 1980).

Temperature limited operation has been necessary in order to minimize velocity spread in the electron beam. Excessive velocity spread degrades gyrotron performance especially when the interaction region is long as in gyrotron amplifiers. As will be described below, considerable effort is being devoted to developing gyrotron electron guns of improved design that will be capable of producing electron beams with reduced velocity spread.

Large electron guns capable of driving gyrotrons with considerably greater power capabilities than those represented by the data in Fig. 1 are also an area of active research and development. It is projected that in thermonuclear reactors, highly efficient gyrotrons capable of multimegawatt power levels (cw) at frequencies of 100 GHz or greater will be required. This projection is driving exploration of new configurations and mechanisms so that the present state of art may be surpassed. As gyrotron circuits are designed for higher powers at shorter wavelengths, more attention must be paid to the conflicting problems of mode competition and thermal loading; design tradeoffs and operating limits are encountered. Thus gyrotron oscillator research has many challenges still to over-come and is being actively pursued.

While gyrotron oscillators are the preferred gyrotron configuration for energetic applications such as plasma heating, there has also been interest in developing gyrotron amplifers because of the potentially superior char-acteristics in information-carrying systems such as communication links and radars. Among the amplifier advantages are substantial instantaneous bandwidth, ability to generate complex waveforms, flexibility in signal modulation, and ability to control the phase and frequency of the signal. However, gyrotron amplifier experiments are in a very preliminary state when compared with the extensive study that has been caried out on gyro-tron oscillators.

Figure 3 depicts in very simple schematics, two possible gyrotron am-plifier configurations in addition to the more usual gyromonotron oscilla-tor configuration. The oscillator employs a single cavity where resonant frequency is close to the electron cyclotron frequency. The gyroklystron amplifier employs two resonant cavities separated by a drift space; it has been explored in some detail in the past (Andronov *et al.*, 1978; Jory *et al.*, 1977; Symons and Jory, 1981); however, recent research activity has centered on the gyrotron traveling-wave tube (gyro-TWT) amplifier (Granatstein *et al.*, 1980).

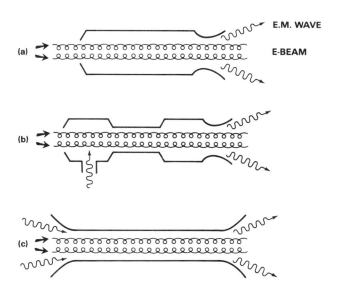

FIG. 3 Types of gyrotrons: (a) gyromontron oscillator, (b) Gyroklystron amplifier, (c) gyro-TWT amplifier. (From Chu, Drobot, Granatstein, and Seftor, 1979. *IEEE Trans. Mi-crowave Theory Tech.* **MTT-27**, 178–187. © 1979 IEEE.)

The circuit employed in a gyro-TWT consists simply of an unloaded waveguide. Since no resonant structures are present, the gyro-TWT is potentially capable of much larger bandwidth than the gyroklystron. A typical specialized radar tube requirement at which gyro-TWT development might aim is for a millimeter-wave amplifier with peak power on the order of a hundred kilowatts, a 10% duty factor, a bandwidth of several percent, and gain in excess of 30 dB; this requirement is much beyond the capability of conventional amplifiers at millimeter wavelengths. Moreover, other performance characteristics such as noise figure and phase linearity should be of sufficiently high quality to be compatible with modern signal processing techniques.

In Section II, the state of art in gyro-TWT development is described together with an assessment of prospects for near-term improvements. Rapid improvement in performance is expected since gyro-TWT development is in a considerably more preliminary state than development of gyromonotron oscillators which are desribed in Section III. It will be noted that performance of both the amplifiers and oscillators can be improved by knowledgeable shaping of magnet field and wall radius along the tube axis; this axial contouring is a topic of intensive current study. Finally, the Appendix describes some of the measurement techniques that have been used.

II.   Gyrotron Traveling-Wave Tube Amplifiers

The gyrotron traveling-wave tube amplifier (gyro-TWT) utilizes a waveguide circuit with a fast-wave propagating mode to interact with the fast cyclotron wave of the electron beam. This electron beam consists of mildly relativistic electrons spiraling in cyclotron orbits and drifting axially down the waveguide. The beam is usually annular as produced by a magnetron injection-type electron gun, with annular radius much larger than the Larmor radius.

Since the gyro-TWT is an amplifier and has substantial gain and bandwidth, it has potential uses as power output amplifiers in millimeter-wave radar and communications systems. To date, there have been only two development programs which have actually tested experimental gyro-TWT's. These programs are at the Naval Research Laboratory (Granatstein et al., 1975; Seftor et al., 1979b; Barnett et al., 1979, 1980a) and Varian Associates (Symons et al., 1979; Ferguson and Symons, 1980; Ferguson et al., 1981). The NRL program has thus far concentrated on experiments in the TE_{01} circular electric mode near 35 GHz, while the Varian program has produced good results near 5 GHz using the TE_{11}^0 circularly polarized mode with a view to scaling to 94 GHz. Hughes Aircraft Co. is also currently developing a 94-GHz model using TE_{01}^0, but as of this time no operation has taken place. Although Soviet publications of gyro-monotron work is extensive, little has been mentioned of gyro-TWT's in the Soviet literature.

A. GYRO-TWT THEORY

The gyro-TWT interaction stems from a waveguide mode whose characteristic equation is

$$\omega^2 - k_z^2 c^2 - \omega_{co}^2 = 0, \tag{1}$$

interacting with a beam cyclotron mode with a characteristic equation.

$$\omega - k_z v_\parallel - s\Omega_c = 0, \tag{2}$$

where ω is the signal frequency, ω_{co} the cutoff frequency of the mode of interest, k_z the axial wave number, and c the speed of light in vacuum, $v_\parallel$ is the axial velocity of the electron beam, Ω_c the relativistic cyclotron frequency, and s the cyclotron harmonic number. For high gain and efficiency the two characteristic curves are adjusted near grazing intersection such that the phase velocities of the two modes are nearly matched and the group velocity of the waveguide mode is nearly equal to $v_\parallel$. Derivations of the complete dispersion relation describing the coupling of the beam and the waveguide modes appear in the literature (Chu *et al.*, 1980a; Sprangle and Drobot, 1977; Lau *et al.*, 1981a,b; Ahn and Choe, 1980). A simplified dispersion relation in normalized form for circular TE_{mn} modes for a cold electron beam and for electron energy $\gg$ 1 keV is given by (Lau *et al.*, 1981a)

$$(\bar{\omega}^2 - \bar{k}^2 - 1)\,(\bar{\omega}^2 - \bar{k}\beta_\parallel - b)^2 = -\epsilon, \tag{3}$$

where $\bar{\omega} = \omega/\omega_{co}$, $\bar{k} = k_z/k_{mn}$, $k_{mn} = \omega_{co}c$, $\beta_\parallel = v_\parallel/c$, $b = s\Omega_c/\omega_{co}$, and the coupling factor

$$\epsilon = 4v\beta_\perp^2\,H_{sm}/\gamma x_{mn}^2 K_{mn}, \tag{4}$$

where $v = Ir_e/ev_\parallel$ (I is the electron beam current, $r_e = 2.8 \times 10^{-13}$ cm the classical electron radius, and e the charge of an electron), $\beta_\perp = v_\perp/c$ ($v_\perp$ is the electron perpendicular velocity), x_{mn} is the nth root of $J_m'(x) = 0$, and γ the relativisitic energy factor of the electrons. Note that $k_{mn} = x_{mn}/r_w$ (r_w is the waveguide radius). H_{sm} and K_{mn} are defined as

$$H_{sm} = [J_{s-m}\,(k_{mn}r_0)J_s'(k_{mn}r_L)], \tag{5}$$

where r_0 is the guiding center radius of the annular beam and r_L the

Larmor radius of the electrons:

$$K_{mn} = J_m'^2(x_{mn}) \left[1 - m^2/x_{mn}^2\right].$$ (6)

Note that by Eq. (4) the electrons are required to have a perpendicular velocity or else $\epsilon = 0$ and the waveguide mode and beam cyclotron mode are uncoupled and no interaction takes place.

In a typical experimental example (Barnett *et al.*, 1979), $m = 0$, $s = n = 1$, $\beta_\perp = 0.4$, $\beta_\parallel = 0.266$, $\omega_{c0}/2\pi = 34.3$ GHz, $r_w = 0.533$ cm, and then

$$\epsilon = 3.78 \times 10^{-6} I,$$ (7)

where I is the beam current in amperes (Lau *et al.*, 1981a).

It is sometimes convenient to find an equivalent interaction impedance in order to compare various waveguide modes, circuits, and beam geometries. By rearranging the coupling term in the above dispersion relation, an interaction impedance can be defined as (Symons *et al.*, 1979)

$$Z = 4\pi\eta H_{sm} / x_{mn}^2 K_{mn},$$ (8)

where η is the impedance of free space. A comparison of Eqs. (8) and (3) shows that if Z is maximized by choice of beam and waveguide geometry and mode, the coupling term ϵ is also maximized. The circular waveguide mode with the strongest interaction impedance for fundamental cyclotron operation is the circularly polarized TE_{11}^0 dominant mode (Symons *et al.*, 1979; Ferguson *et al.*, 1981) and hence this mode has the potential for largest gain per unit length. There is preliminary experimental evidence that gyro-TWT amplifiers become degraded in saturated efficiency as the interaction structure becomes long; this may be an effect of beam velocity spread. In that case, for a given total gain and beam velocity spread the TE_{11}^0 mode requires a shorter amplifier length and thus is expected to yield a greater efficiency than the higher order modes such as TE_{01}^0. Also, employing a higher order mode involves greater complication in launching the mode and in maintaining mode purity. However, for very high-frequency, high-power applications, circuit size is very small for the TE_{11}^0 mode and higher order modes such as TE_{01}^0 may be the preferred choice.

Also of significance to the experiments is the wave growth and gain. The temporal growth rate has been found (Chu *et al.*, 1980a) by solving the dispersion equation (3) for the imaginary part of the frequency, yielding

$$\omega_i = \left[\frac{3^{3/2}\nu x_{mn}^2 H_{sm}\beta_\perp^2 c^4}{4\gamma K_{mn}\omega_0 r_w^4}\right]^{1/3} = 0.7\,\omega_{c0}\left(\frac{\epsilon\omega_{c0}}{\omega_0}\right)^{1/3},$$ (9)

where ω_0 is the frequency at which the waveguide mode, Eq. (1), has a grazing intersection with the beam characteristic curve, Eq. (2). Near the point of grazing intersection the spatial growth rate Γ is given by

$$\Gamma \cong \omega_i/v_\parallel \tag{10}$$

and the total power gain in decibels is

$$G = -10 \log_{10} 9 + 8.686 \, \Gamma L. \tag{11}$$

Equation (11) is valid for $\Gamma L > 1$ where L is the length of the amplifying region.

The first term on the right-hand side of Eq. (11), $-10 \log_{10} 9$, is a loss which occurs because not all of the input rf power couples to the growing wave (Seftor *et al.*, 1979b).

It is clear from the form of Eqs. (9) and (10) that

$$\Gamma \sim I^{1/3}, \tag{12}$$

i.e., the growth rate is proportional to the cube root of the beam current. This relationship has been confirmed in experiments (Seftor *et al.*, 1979; Barnett *et al.*, 1979).

B. GYRO-TWT EXPERIMENTS

The NRL experiments using the TE_{01} circular-electric mode (Seftor *et al.*, 1979b; Barnett *et al.*, 1979, 1980a) use a 5.33-mm radius interaction waveguide corresponding to a cutoff frequency of 34.30 GHz. The electron gun was a magnetron-injection gun (Seftor *et al.*, 1979a) designed to produce an annular electron beam of guiding center radius 2.5 mm with a perpendicular velocity $v_\perp = 0.40c$, and a parallel velocity $v_\parallel = 0.27c$, corresponding to a beam energy of 70 keV. As is usual with gyrotrons, the electron gun was operated temperature limited to minimize velocity spread. The design velocity spread $\sigma(v_\parallel)/\bar{v}_\parallel \approx 8\%$, where $\sigma(v_\parallel)$ and $\bar{v}_\parallel$ are, respectively, the standard deviation and the mean of the electron distribution in $v_\parallel$. Recent NRL gyro-TWT experiments (Barnett *et al.*, 1979, 1980a) use an input coupler which is a TE_{01}-coaxial-mode to TE_{01}-circular-mode junction developed for that purpose (Barnett *et al.*, 1980b). The TE_{01} coxial mode is formed (Fliflet *et al.*, 1980) either by a single 360° coaxial sector taper from TE_{10}-rectangular-mode waveguide (Barnett *et al.*, 1980a) or by a double 180° coaxial sector taper (Barnett *et al.*, 1979). The double coaxial-sector-taper design is shown schematically in Fig. 4 and in the photograph of Fig. 5. The double input is obtained from a rectangular waveguide power divider producing equal and oppositely phased signals in the two rectangular feed arms. The overall length of the interaction waveguide (the inside of the coaxial inner wall) is 21 cm. The

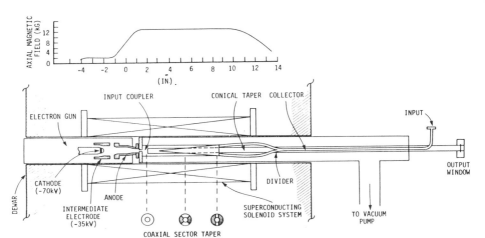

FIG. 4   Gyro-TWT for 35-GHz operation in the TE_{01}^0 mode. (From Barnett, Baird, Fliflet, and Granatstein, 1980b. *IEEE Trans. Microwave Theory Tech.* **MTT-28,** 1477–1481. © 1980 IEEE.)

magnet is a superconducting solenoid system with an axial magnetic field profile as illustrated. Gain curves that were obtained by plotting the output rf power as a function of input drive power at several beam currents are shown in Fig. 6. The frequency was 35.14 GHz. For beam currents of 9, 3, and 1 A, gains obtained were 32, 24, and 12.5 dB, corresponding to linear growth rates of 0.23, 0.18, and 0.12 cm^{-1}, respectively. The growth rate scales as $\Gamma \sim I^{1/3}$ as predicted by theory. Saturation begins at approximately 10 kW for the 9-A case with 30-dB gain. The highest efficiency obtained was 7.8% at 16.6-kW output with 20-dB gain for a 3-A beam current. Figure 7 shows the small signal bandwidth measurement at a beam current of 3 A. The 3-dB bandwidth (FWHM) is approximately 1.4% with useful gain having a much wider extent. The solid curve in Fig. 7 is a

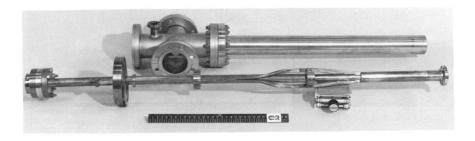

FIG. 5   Photograph of the TE_{01}^0 amplifier input coupler.

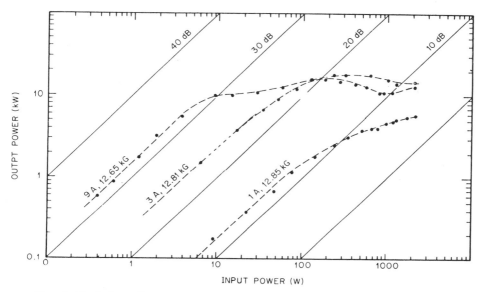

FIG. 6 Typical amplifier gain curves for several values of beam current: 35.14 GHz, 70 kV. The magnetic field was adjusted for optimum that linear gain. These curves are in update of a figure that appeared in Barnett *et al.*, (1979).

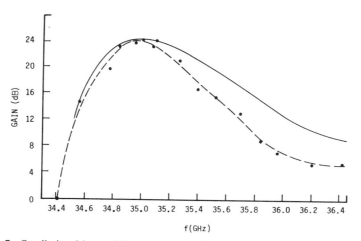

FIG. 7 Small signal bandwidth measurement for the TE_{01}^0 3A, 70 kV, 12.81 kG, gyro-TWT. Solid line, linear theory; dashed line, experiment. (From Barnett, Chu, Baird, Granatstein, and Drobot, 1979. *Technical Digest, IEEE International Electron Devices Meeting, Washington, D.C., December* pp. 164–167. © 1979 IEEE.)

theoretical prediction made by Chu on the basis of growth rate calculations for a Maxwellian distribution of electron velocities (Barnett *et al.*, 1979); the absolute value of the peak gain is in excellent agreement with the experimental data while the predicted bandwidth is somewhat larger than measured.

Problems with oscillations near cutoff occurred in the above experiment. To stabilize the tube a continuous lossy wall was utilized which had a cold tube loss of 0.37 dB/cm at 35.0 GHz (Barnett *et al.*, 1980a). For higher overall gain the interaction length was increased to 43 cm. This longer experiment is illustrated schematically in Fig. 8 without the electron gun, magnet, etc.; those components are the same as Fig. 4 except that the region of uniform high-magnetic field was made longer. The tube was much more stable and produced a best observed small signal gain of 56 dB at 35.12 GHz and a net system gain of 52 dB (after subtracting all circuit losses in the input and output waveguide components). The maximum saturated power was lower than in the lossless wall tube of Fig. 1, with 3.2 kW achieved at a gain of 42 dB. When the lossy wall was removed from the final 7 cm of the interaction region, the output power increased to 6.5 kW. The small signal 3-dB bandwidth at 1-A beam current was 3.4% with a peak gain of 26 dB. Of significance, the resistive wall loading of the gyro-TWT is found to suppress oscillations of both reflective and absolute instability types, the gain–bandwidth product is increased, and the reduction in gain per unit length is only approximately $\frac{1}{3}$ of the cold tube loss (Lau *et al.*, 1981a,b; Barnett *et al.*, 1980a).

Studies of modified gyro-TWT geometry which is predicted to increase the bandwidth to 12%–50% are underway; the wideband modification involves tapering both the wall radius and the magnetic field strength along the axis of the interaction region (Lau and Chu, 1981). Initial experimental results include a small signal 3-dB bandwidth of 12% with a 19-dB peak gain at midband (35 GHz), as shown in Fig. 9 (Barnett *et al.*, 1981).

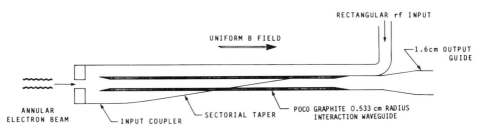

FIG. 8   Schematic of the 35-GHz gyro-TWT using the resistive wall circuit. (From Barnett, Baird, Lau, Chu, and Granatstein, 1980a. *Technical Digest, IEEE International Electron Devices Meeting, Washington, D.C., December* pp. 314–317. © 1980 IEEE.)

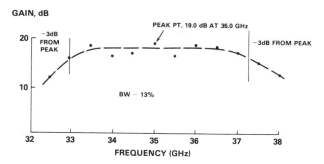

GAIN, dB

PEAK PT, 19.0 dB AT 35.0 GHz

20

−3dB FROM PEAK

−3dB FROM PEAK

10

BW ~ 13%

32   33   34   35   36   37   38

FREQUENCY (GHz)

Fig. 9   Initial small-signal bandwidth measurement of tapered gyro-TWT. Interaction waveguide was 33.3 cm long and had a linear taper that varied its radius from 4.57 to 6.22 mm. Axial magnetic field was also varied so that electron cyclotron frequency was changed in proportion to changes in the cutoff frequency. TE_{01} mode, 1 A, 70 kV (adapted from Barnett *et al.*, 1981.)

Experiments by Varian Associates near 5 GHz using the TE_{11}^0 circularly polarized mode (Symons *et al.*, 1979; Ferguson and Symons, 1980; Ferguson *et al.*, 1981) have produced impressive results. These experiments use a magnetron injection gun operating from 40 to 65 kV. The input coupling uses a sidewall junction to rectangular waveguide. The interaction circuit is 42.5 cm in length and has distributed wall loss over the first two thirds of this length for improved stability.

Several versions of the TE_{11}^0 gyro-TWT were tested. A early model employed a flat magnetic field in the interaction region; the measured small signal gain was 24 dB while saturated peak power was 50 kW at 16.6% efficiency with corresponding saturated gain of 18 dB and saturated bandwidth of 6% (FWHM). Substantial improvement in performance was obtained in a latter model which employed a variable magnetic field profile in the interaction region. Optimized performance was obtained by increasing the magnetic field strength 4.4% from the input to the output end of the tube. With this ramped field, and with beam voltage and current at 65 kV and 7 A, respectively, small signal gain was 26 dB at the center frequency of 5.18 GHz; saturated peak power was 120 kW at 26% efficiency with corresponding saturated gain of 20 dB and saturated bandwidth of 7.25% (FWHM).

The 5 GHz TE_{11}^0 gyro-TWT's have also been extensively characterized as to other aspects of performance important in communications or radar systems. Measurements have been reported of AM and PM modulation coefficient, spectral purity, phase linearity, and noise figure (Ferguson *et al.*, 1981). Almost all of these parameters of the gyro-TWT compared favorably with performance of present-day coupled-cavity TWTs. The one parameter with a somewhat inferior value for gyro-TWTs was noise figure was measured in the range 44–52 dB above thermal; improvement in this parameter may require the development of space charge limited electron guns which are compatible with good gyrotron operation.

C. Summary of Performance and Prospects

In contrast to the more developed gyrotron oscillators which are described in Section III, experience with gyro-TWT amplifiers is in a very preliminary state. Only two sets of experiments have been performed to date, one at a frequency of 5 GHz employing a TE_{11}^0 mode, and the second at 35 GHz with a TE_{10}^0 mode. Because of the mode change as well as the large frequency change between the two sets of experiments, any attempt to derive scaling information must necessarily be very tentative.

However, there is good reason to be encouraged as to the eventual prospects for the practical use of gyro-TWTs in high-power, millimeter-wave communications and radar systems. Very high gain (56 dB) has been achieved in the 35-GHz amplifier, and bandwidths have been in the useful range of 1%–12%. Power and efficiency have been lower than with the gyrotron oscillators but compare very favorably with conventional TWT's (viz., 120 kW with 26% efficiency at 5 GHz and 17 kW with 8% efficiency at 35 GHz), especially when one considers the preliminary state of the gyrotron amplifier work. Other characteristics of the gyro-TWTs which are important in signal processing systems appear to be of acceptable quality even in the preliminary systems and will surely be susceptible to improvement.

The fact that the efficiency of the gyro-TWT amplifiers is lower than the efficiency of gyrotron oscillators has to do with the length of the interaction circuit and the velocity spread in the electron beam. The longer interaction time in the amplifier allows the phase bunching of the electrons to be more degraded by velocity spread. In order to obtain high efficiency in the amplifier, it will be necessary to develop suitable electron guns which have substantially reduced velocity spread. Recent numerical studies by Baird and his colleagues (1981) indicate that magnetron injection guns can be designed for gyrotrons with a two to three times smaller spread than in the magnetron injection guns which are currently being used. Studies on other types of gyrotron electron guns which are not of the magnetron injection variety (e.g., Pierce guns with a transverse kicker coil) and which may prove to have improved performance are underway at a number of laboratories. We note expecially the design of a space charge limited gun at the Raytheon Company with calculated velocity spread $\sigma(v_{\parallel}/\bar{v}_{\parallel}) \cong 4\%$ at $v_{\perp}/v_{\parallel} = 1.5$, V = 25 keV and $I = 3$ A (Dionne, 1981).

Efficiency may also be improved by keeping the gain per unit length high, and thus for a given overall gain, making the interaction region as short as possible. In circular waveguide, the highest gain per unit length can be obtained with the TE_{11}^0 mode. The interaction circuit used for this mode in the 5 GHz gyro-TWT experiments was much shorter in terms of wavelengths than the circuit used in the TE_{01}^0 35-GHz experiments, and efficiency and output power were observed to be considerably higher in the TE_{11}^0 5-GHz experiments. While for the reason stated above, any conclusion must at this point in time be tentative, there is reason to believe that one should employ the TE_{11}^0 mode whenever possible. When power and frequency requirements become very large, one may be forced to use overmoded structures; in that case, the TE_{01}^0 mode would tend to be the logical choice.

Last, we note that dramatic improvements in gyro-TWT performance appear to be achievable by varying the magnetic field strength and/or the wall radius along the axis of the interaction region. In the studies at Varian Associates at 5 GHz, a twofold increase in power and efficiency was achieved by magnetic field shaping. In work in progress at NRL, Lau and Chu (1981) have predicted dramatic increases in bandwidth when both magnetic field and wall radius are varied so that the cutoff frequency and the electron cyclotron frequency track each other over an extensive range; initial experimental results appear to bear out the theoretical predictions (Barnett et al., 1981).

III. Gyromonotron Oscillators

Gyromonotron oscillators appear useful for both energetic applications such as plasma heating (Gilgenbach et al., 1980; Alikaev et al., 1976; Temkin et al., 1979) and material response studies, and for specialized radars. The output characteristics of these devices of interest for energetic applications are the frequency, power, efficiency, pulse duration, and to a lesser extent mode purity. For radar-type applications, the power and efficiency requirements are most modest, but mode purity and linewidth are critical. We therefore examine in this section these pertinent characteristics of gyromonotrons. In doing so, we discuss the nature of their dependence on the design of the various components and on the externally controllable input parameters.

A typical gyromonotron is shown schematically in Fig. 10a (Read et al., 1980). A photograph of the same device mounted in its superconducting solenoid magnet is shown in Fig. 10b, and out of the magnet in Fig. 10c. The design of the labeled components, plus that of the solenoid all influence

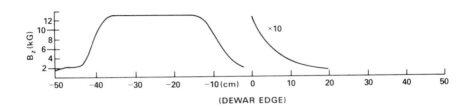

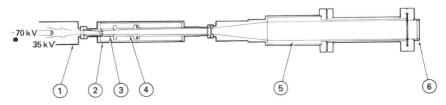

FIG. 10a   Schematic of a 35-GHz gyrotron oscillator: (1) electron gun, (2) vacuum envelope, (3) cavity, (4) output guide, (5) collector, and (6) output window. (From Read, Gilgenbach, Lucey, Chu, Drobot, and Granatstein, 1980. *IEEE Trans. Microwave Theory Tech.* **MTT-28,** 875–877. © 1980 IEEE.)

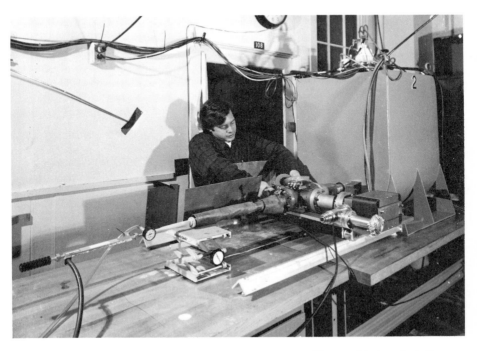

FIG. 10b   Photo of an experimental 35-GHz gyrotron oscillator: with the tube in its superconducting magnet (courtesy Naval Research Laboratory).

282

FIG. 10c Photo of an experimental 35-GHz gyrotron oscillator: tube alone (courtesy Naval Research Laboratory).

the behavior of the gyrotron. Of particular importance are the electron gun and the cavity, since the beam-wave interaction is central to the gyrotron. Therefore in this section emphasis will be placed on measurements of the behavior of gyromonotrons attributable to these components.

A. REVIEW OF THE THEORY

Before discussing the measurements, it is of use to review the basic theory of the single-cavity gyrotron oscillator (gyromonotron). As previously discussed, the gyrotron mechanism involves the interaction of a fast waveguide electromagnetic mode and the fast cyclotron wave on an electron beam. For a single cavity oscillator, these two modes are respectively governed by two dispersion relationships similar to Eqs. (1) and (2) but with k_z set equal to $\pi l/L$, where L is the cavity length and l is the axial eigennumber; viz.,

$$\omega^2 = l^2\pi^2c^2/L^2 + \omega_{c0}^2 \tag{13}$$

and

$$\omega = l\pi v_{\parallel}/L + s\Omega_c. \tag{14}$$

The frequency of the interaction is to first order determined by Eq. (13), which yields the resonance frequency of the cavity. Equations (13) and (14) together give the phase difference between the two waves. This phase difference is then determined by the cavity length and radius, the beam velocity distribution and energy, and the magnetic field magnitude.

The coupling between the two modes given by Eqs. (13) and (14) has been calculated by several authors (Flyagin et al., 1977; Chu, 1978; Sprangle and Drobot, 1977; Kreischer and Temkin, 1980; Bratman and Moiseev, 1975; Antakov et al., 1977). The linear coupling yields the starting current for the oscillator. Chu (1978) has calculated the starting current for a cavity in which the axial profile of the rf electric field is a half-sinusoid. Kreischer and Temkin (1980) have used both sinusoidal and Gaussian profiles. The results differ significantly, with the Gaussian profile yielding substantially lower starting currents.

The calculation of nonlinear coupling leads to prediction of the efficiency of the oscillator. Various calculations have been made of the efficiency (Chu et al., 1980b; Caplan, 1980; Gaponov et al., 1975; Kolosov and Kurayev, 1974; Bratman et al., 1973), the most direct of which employ particle orbit integrating codes. Rf field profiles of sinusoidal, Gaussian, and experimentally measured forms have been used in these calculations. The efficiencies predicted with the nonsinusoidal profiles are substantially above those given for the sinusoidal ones. A pure half-sinusoid form exists only in an empty axially uniform cavity which is closed at both ends and is therefore a limiting case for very high Q cavities. High-power gyrotrons generally use low-Q cavities. Highly output coupled cavities have nonsinusoidal waveforms that can either be approximated by Gaussian profiles or, preferably, found by measurement of the fields excited in a cavity in the absence of the beam. Efficiency enhancement has been accomplished by profiling the cavity walls to obtain a more optimum field shape (Kolosov and Kurayev, 1974; Bratman and Petelin, 1975; Kurayev et al., 1974; Moiseyev et al., 1968).

Another method predicted to increase the efficiency is the profiling of the DC magnetic field (Chu et al., 1980b; Sprangle and Smith 1979; Kurayev and Shestakovich, 1977a,b). Calculations with a linearly rising field (Chu et al., 1980b) appear most useful and predict efficiencies to 78%. Efficiencies to approximately 65% are predicted with the cavity wall profiling (Gaponov et al., 1975).

The quantities that must be measured in order to relate the theory to experiment in addition to those mentioned in conjunction with Eqs. (13)

and (14) are the rf field magnitude and radial and azimuthal waveform, and as discussed above, the axial profile of the rf and dc fields. The rf field magnitudes can be inferred from the relation

$$QP_{\text{beam}} = (\omega\epsilon_0 \int E_0^2 \, dv)/2, \tag{15}$$

which follows simply from the usual definition of the cavity quality factor Q. E_0 is the azimuthal rf electric field, ϵ_0 the vacuum permitivity, dv the differential volume element, and P_{beam} the beam power. Clearly, the form of E_0 must be measured before the magnitude of the integral over E_0^2 can be accurately determined. Figure 11 shows a typical cavity and field profile. Techniques used in the measurement of the field profile, the cavity Q, and other pertinent parameters are discussed in the Appendix.

The efficiency of the beam-wave interaction is of course defined by

$$\eta = P_{\text{out}}/P_{\text{beam}}, \tag{16}$$

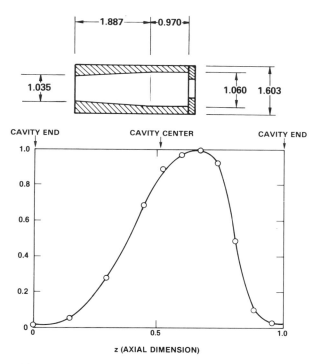

FIG. 11   (a) Diagram of a 35-GHz TE_{011}^0 oscillator cavity with Q of 800 (dimensions are in centimeters) and (b) Electric field profile for the cavity of (a). (From Read, Gilgenbach, Lucey, Chu, Drobot, and Granatstein, 1980. *IEEE Trans. Microwave Theory Tech.* MTT-28, 875–877. © 1980 IEEE.)

where P_{out} is the output power. Equations (15) and (16) must be considered together with the calculation of the nonlinear coupling to obtain an optimized value of efficiency for a realizable Q and for the desired value of P_{out}. Results of some of the calculations are presented below in conjunction with descriptions of the experimental observations.

B.   QUALITATIVE FEATURES OF OSCILLATOR PERFORMANCE

Gyrotron oscillators have been tested in the USSR, the United States, Japan, and the People's Republic of China. Published results are summarized in Table I. Some of these have been discussed in reviews (Flyagin et al., 1977; Jory et al., 1979).

Of interest, as noted above, is the behavior of the starting current, power, mode, and frequency as a function of the controllable input parameters, such as gun voltages, current, and magnetic field magnitude and profile. In addition the effect of cavity length, cavity wall profile, Q, and the spatial and velocity distribution of the electron beam are important.

A given gyromonotron, once constructed, has the variable inputs of voltages and current and magnetic field. (Some devices have been designed such that the shape of the magnetic field at the cavity could be varied. These we shall discuss later.) The range of voltages applied is usually limited by the strong sensitivity of the electron gun to voltage variation, and voltages are generally fixed after initial optimization. The actual gyrotron mechanism itself is predicted to be weakly influenced by changes in the voltage.

The quantitative behavior of a given gyrotron with the remaining inputs, beam current and magnetic field magnitude, is fairly universal. To illustrate it, we take a particular example (Read et al., 1979) with parameters typical of many gyrotrons. This device is an oscillator operating in the TE_{011}^0 mode at 35 GHz. The cavity is given in Fig. 11a, with a length of 3.3 free-space wavelengths. An electron beam with an energy of 70 keV and a ratio of perpendicular to parallel velocities of 1.5 were used.

The behavior of this device is shown in Fig. 12, which is a plot of the experimentally observed output efficiency as a function of the beam current for various magnetic field strengths. Also given are plots of the theoretically predicted efficiency with the magnetic field optimized for given values of the current. Calculations for both half-sinusoid and Gaussian axial profiles of the rf field are given. The observed behavior is qualitatively in agreement with the calculations. (We discuss the quantitative discrepancies below.) The details which are of interest are as follows:

(1)   Oscillation occurs for currents above 0.25 A for a magnetic field of 13.8 kG, but threshold currents are higher for lower fields. This is in

TABLE I

SUMMARY OF PUBLISHED RESULTS ON GYROTRON OSCILLATORS

Reference[a]	Harmonic	Wavelength (mm)	Cavity mode ($TE_{m,n1}$)	Voltage (kV)	Output[b] Current (A)	Output[b] power (kW)	Efficiency[b] (%)	Cavity profiling	Magnetic profiling	Pulse length
1	2	0.9	2,3	27	0.9	1.5	6[c]	No	No	cw
2	2	1.2	0,2	20	1.5	4.3	18[c]	Yes	No	cw
3	2	1.9	2,3	26	1.8	7.0	15[c]	No	No	pulsed
4	2	1.9	0,3	18	1.4	2.4	9.5[c]	No	No	cw
5	1	2.8	0,2	27	1.4	12	31[c]	No	No	cw
6	1	3.0	~15,1	68	48	1100	34	Yes	No	pulsed
7	1	6.7	—	65	55	1250	35	Yes	No	pulsed
8	1	8.6	0,1	70	5–10	150	31	Yes	No	20 msec
9	1	8.6	0,1	70	4	≤100	65[d]	No	Yes	1.5 μsec
10	2	8.6	0,4	70	12	300	40	No	Yes	1.5 μsec
11	1	9.0	0,2	19	1.5	10	40	Yes	No	cw, pulse
12	1	10.7	0,2	80	—	250	45	Yes	No	40 msec
13	1	10.7	0,2	80	—	212	40–50	Yes	No	cw
14	1	20.0	0,1	20	0.5	4	50	Yes	No	cw
15	1	20.0	5,2	40	40	380	45	Yes	No	pulsed

[a] 1, 3, 4, and 5 (Zaytsev et al., 1974); 2, 14 (Gaponov et al., 1975); 6, 7 (Andronov et al., 1978); 8 (Read et al., 1980); 9 (Read et al., 1981b); 10 (Arfin and Read, 1980); 11 (Kisel et al., 1974); 12, 13 (Jory et al., 1980); 15 (Bykov et al., 1975).

[b] Efficiency given may not have been at maximum power.

[c] Substantial thermal losses were expected; the interaction efficiency may have been substantially higher.

[d] Interaction efficiency; the output efficiency was degraded by ohmic losses, and was substantially lower.

287

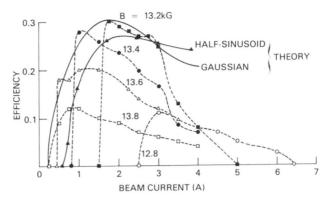

FIG. 12   Efficiency as a function of beam current for various values of the applied magnetic field for a TE_{011}^0 cavity oscillator. The beam voltage was 70 kV (from Read *et al.*, 1979).

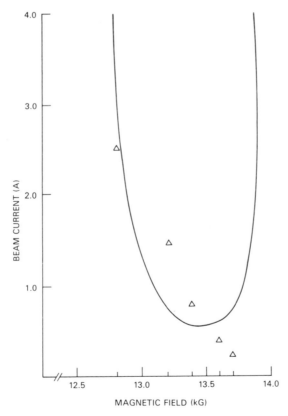

FIG. 13   Starting current as a function of magnetic field for a TE_{011}^0 cavity with a length equal to 5 free-space wavelengths for a beam energy of 70 keV. Solid line, theory (courtesy K. R. Chu and K. J. Kim).

288

qualitative agreement with the linear theory (Fig. 13). The magnetic field at which the starting current is lowest corresponds to that which produces maximum gain for the interaction. This occurs when the phase velocity of the wave is almost exactly equal to the axial beam velocity. Decreasing the magnetic field produces a larger phase shift between the waves; hence lowering the gain.

(2)   The efficiency is maximum for a field value lower than that for a minimum starting current. This is again in qualitative agreement with theory, as shown in Fig. 14. Calculations show that the efficiency is lower when there is higher gain because trapping of the electrons in the wave occurs more readily. This phase trapping results in the reabsorption of some of the wave energy by the electrons near the output end of the cavity.

(3)   The efficiency peaks; then drops with increasing current. Some of this decrease may be again generally attributed to trapping. (However, the rapid drop of the efficiency in Fig. 12 is atypical, and may have been due to disruption of the electron beam before it entered the cavity. Other experiments (Read *et al.*, 1980; Jory *et al.*, 1980) have shown an approximate 25% drop in efficiency for currents double that for peak efficiency.) For further increasing currents, the efficiency is predicted to increase again. This is owing to the cyclical nature of the trapping process. This behavior has been observed, as is shown in Fig. 15. This is then the general behavior pattern of a gyrotron oscillator.

We now turn to the quantitative relationships which govern the design of gyrotrons.

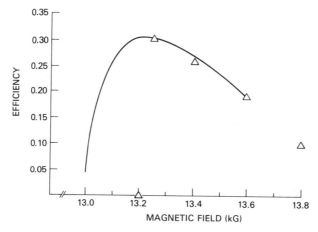

FIG. 14   Plot of the theoretically predicted (solid line) and experimentally observed (points) efficiency as a function of magnetic field for a current of 1.5 A.

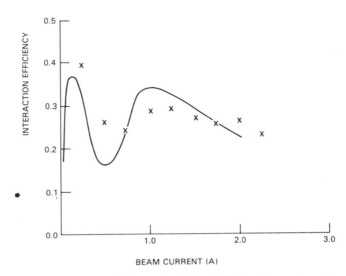

FIG. 15 Plot of the theoretically predicted (solid line) and experimentally observed (points) interaction efficiency as a function of beam current. The cavity length was 9.3 free-space wavelengths, the beam energy equal to 70 keV, and the gum velocity ratio (v/v) equal to 1.8. This is an illustration of a severely overdriven cavity. (From Read, Chu, and Kim, 1981a. *IEEE Trans. Microwave Theory Tech.*, to be published. © 1981 IEEE.)

C. PARAMETRIC DEPENDENCE OF OSCILLATOR PERFORMANCE

1. *Beam Current and Cavity Q*

Theory (Gaponov *et al.*, 1975; Chu *et al.*, 1980b) predicts that the efficiency is a function of the product of the cavity Q and the beam current. Thus, a gyrotron with a cavity Q of 500 and a beam current of 5 A will have (with all other parameters being equal and wall losses being negligible) the same efficiency as a device with a Q of 1000 and a current of 2.5 A. The starting current of the latter device will be one-half that of the former. These features have been observed by one of the authors (Read *et al.*, 1979).

2. *Cavity Length*

The efficiency and output power for optimum efficiency for a given cavity mode are functions of the cavity length. This can be seen from Figs. 16 and 17 (Read *et al.*, 1981a), which show the efficiency and the product of the output power and cavity Q, respectively, as functions of cavity length.

Experimental results are given for the TE_{011}^0 mode. Good agreement with the theory is seen. For higher order modes, higher values of QP are predicted, (Chu *et al.*, 1980b; Read *et al.*, 1981) with no reduction in effi-

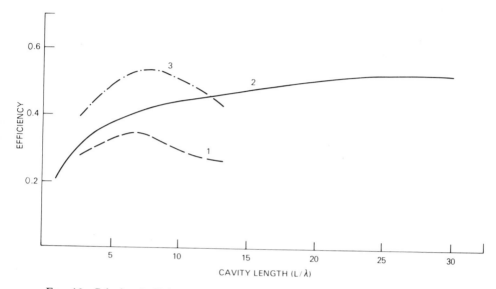

FIG. 16   Calculated efficiency of a gyromonotron as a function of cavity length, for a beam energy of 70 keV, and a ratio of perpendicular to parallel beam velocities equal to 1.5. Results from (1) an rf field with a half-sinusoid axial variation, (2) an rf field with a Gaussian distribution, and (3) an rf field with a half-sinusoid variation and with a profiled dc magnetic field. Results (1) and (3) are from Chu *et al.* (1980b). The data of (2) are from Gaponov *et al.*, 1975.

ciency. An increase of 3.1 in QP over TE_{011}^0 results was observed with a TE_{041}^0 mode (Arfin and Read, 1980); an increase of approximately 3.7 was theoretically predicted (Read *et al.*, 1981a).

3.   Cavity Mode

Modes other than those of the $TE_{0,n,1}^0$ family have been used principally to produce high power with reduced mode competition. An efficiency of 45% was obtained by Bykov *et al.* (1975) with a $TE_{5,2,1}^0$ mode. This efficiency was obtained with a power of 180 kW at a wavelength of 20 mm. Also, 380 kW with an efficiency of 30% was obtained with this device, thus illustrating the utility of this type of mode of strong azimuthal variation (a whispering-gallery mode) for the production of high power. It should be noted that the relatively close proximity of the beam to the cavity wall that was used with this mode allowed the propagation of high perveance beam. (At 380 kW a beam of 30 A, corresponding to 3.8 micropervs, was used.) Even higher power (1100 kW at $\lambda = 3$ mm) has been generated using a whispering gallery mode of higher azimuthal eigennumber: approximately $TE_{15,1,1}$ (Flyagin, 1979).

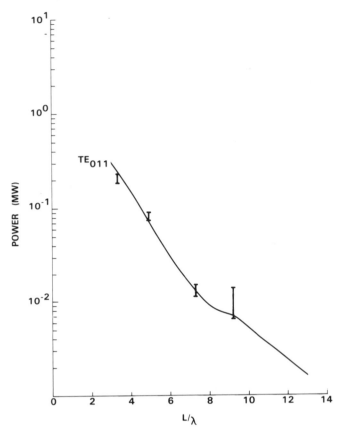

FIG. 17   Calculated and observed output power for TE_{011}^0 gyromonotrons as a function of cavity length. The cavity lengths are given in terms of free-space wavelengths. The observed powers have been normalized by the factor, Q (experiment)$/2 \times Q$ (diffraction). $\alpha = 1.5$ without tapering (from Read *et al.*, 1981a).

4.   *Beam Energy*

The dependence of the efficiency on the beam energy for energies less than 100 keV is predicted to be weak, with a slight increase of efficiency with decreasing energy (Lee *et al.*, 1981). Although comparison between experiments on this single parameter is difficult, owing to the influence of the many other parameters, the fact that efficiencies of over 40% have been achieved at both 20 (Gaponov *et al.*, 1975) and 80 keV (Bykov *et al.*, 1975) tend to support the calculations.

The output power is, however, predicted to be strongly dependent on the beam energy. Two factors are important here: (1) the rf field strength for optimum efficiency rises with voltage, approximately as $V^{(1.5)}$ (Lee *et al.*, 1981); and (2) the current which can be generated and propagated also rises with voltage, roughly as $V^{3/2}$ (Drobot and Kim, 1981). The dependence of the power on voltage has been shown to some degree by Kisel *et al.* (1974), who observed roughly a doubling of output power with a 1.5 increase in beam energy. Similarly, much higher powers have been observed for a given mode (e.g., $TE_{0,2,1}$) at 80 kV (140 kW at peak efficiency) (Bykov *et al.*, 1975) than at 27 kV (12 kW at peak efficiency) (Zaytsev *et al.*, 1974). The observation must, however, be tempered by the possible wide variation of other parameters, particularly the cavity Q and length and the electron gun performance, all of which should have a substantial effect on the output power. These details, particularly in the case of the lower voltage experiment, are not available.

D. EFFICIENCY ENHANCEMENT

Enhancement in the efficiency by the profiling of either the cavity walls or the impressed magnetic field has been reported. Profiling of the cavity walls has been extensively used in the USSR (Gaponov *et al.*, 1975; Zaytsev *et al.*, 1974; Andronov *et al.*, 1978; Kisel *et al.*, 1974) and more recently in the U.S. (Read *et al.*, 1980; Jory *et al.*, 1980). The principal effect of the profiling appears to be that of shifting the peak of the rf electric field toward the output end of the cavity. This shaping of the rf field is also accomplished to a lesser degree by a large output coupling, since in this case the field is large at the output end of the cavity. As indicated in the review of theory (Gaponov *et al.*, 1975), with profiling, perpendicular efficiencies as high as 79% are predicted. (With a gun velocity ratio of $v_\perp/v_\parallel = 2.0$, a maximum overall efficiency of 63% is thus predicted.) However, the maximum efficiency can only be reached for rather long cavities, which will produce relatively low powers owing to their higher diffraction limited Q. For cavities appropriate for high power devices, efficiencies up to approximately 50% are predicted.

Several experiments have been performed on the effect of cavity profiling (Read *et al.*, 1979, 1980; Jory *et al.*, 1980; Gaponov *et al.*, 1975; Zaytsev *et al.*, 1974; Andronov *et al.*, 1978; Kisel *et al.*, 1974). The parameters of some of these are listed in Table II. It is apparent from the table that enhancement over the efficiency expected with a half-sinusoidal field occurs with either profiled or low Q cavities. A Gaussian modeling of the rf fields appears appropriate for either case. From the table is also evident that an enhancement in the efficiency of ≈ 1.2 has been achieved with cavity profiling.

TABLE II

EFFICIENCY ENHANCEMENT BY NONSINUSOIDAL ELECTRIC FIELDS

| | | | Cavity Q | | Calculated Efficiency | |
Reference[a]	Cavity type	Cavity length	Diffraction Q	Observed efficiency[b]	sinusoidal field	Gaussian field
1	Straight	5	High (2000)	0.35	0.35	0.40
2	Profiled	5	Low (500)	0.48	0.40	0.48
3	Profiled	3.3	High (800)	0.35	0.29	0.35
4	Straight	5	Low (500)	0.40	0.35	0.40
5	Profiled	?	?	0.50	0.42	0.44

[a] 1, (Read et al., 1979); 2, (Jory et al., 1980); 3, (Read et al., 1980); 4, (Arfin and Read, 1980); 5, (Gaponov et al., 1975).

[b] Interaction efficiency, calculated from output efficiency by including radiation lost to the cavity walls.

Profiling is perhaps more important in reducing the Q of a cavity, thereby increasing the power that can be generated at peak efficiency. Reductions in the Q of the order of $\frac{1}{2}$ have been observed (Dudas, 1980; Stone, 1980a).

The effect of profiling the dc magnetic field has also been investigated (Read et al., 1981b). Overall efficiencies to 67% (with $v_\perp/v_\parallel = 2.0$) have been predicted (Chu et al., 1980b). Enhancements of 1.2–1.9 in the efficiency have been observed (Read et al., 1981b) with moderate power ($P \leq 100$ kW) devices using fields which rise linearly toward the output end of the cavity. The results are summarized in Table III. Interaction efficiencies (i.e., the efficiency of energy extraction from the beam into both the output wave and ohmic losses) of up to 65% were found in these experiments, in good agreement with theory. As in the case of cavity profiling, fairly long cavities were required for the highest efficiencies and high-power devices may be again limited to efficiencies of 50% to 60%.

TABLE III

EFFICIENCY ENHANCEMENT BY MAGNETIC FIELD TAPERING (EXPERIMENTAL RESULTS) TE$_{011}$ CAVITY @ 35 GHz

Cavity length (free-space wavelengths)	Optimum magnetic field perturbation ($\Delta B/B$)	Efficiency enhancement
3.3	0.10	1.2
4.9	0.07	1.5
7.3	0.06	1.9

E. STABILITY AND COHERENCE

Since gyrotron cavities are overmoded, the degree to which they can be operated in a single, desired mode is of concern. This concern is of particular importance for cw devices with very high ($\gtrsim 1$ MW) power at short wavelengths, where severely overmoded cavities are required (Read, 1980; Kim *et al.*, 1980).

For a TE^0_{mn} mode, mode competition is predicted to some degree by the linear theory (Kreischer and Temkin, 1980) for nearby modes with different radial, azimuthal, and axial eigennumbers. However, no competition owing to modes of different axial mode numbers has been observed. (All reported devices operated in a $TE^0_{m,n,1}$ mode). Recent nonlinear calculations support this observation (Dialetis and Chu, 1980).

Competition owing to modes with different radial eigennumbers is predicted (Read, 1980; Kim *et al.*, 1980) and has been observed (Arfin and Read, 1980; Dialetis and Chu, 1980). An example is given in Fig. 18 (Arfin

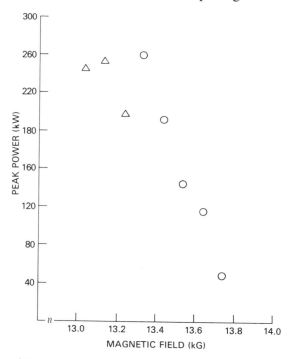

FIG. 18 Plot of the output power of a TE^0_{04} gyromonotron as a function of magnetic field, showing mode switching. (From Arfin and Read, 1980. *Tech. Digest, IEEE International Electron Devices Meeting, Washington, D.C., December* pp. 308–309. © 1980 IEEE.)

and Read, 1980). Here a $TE_{0,4,1}^0$ mode was desired and was achieved for most values of magnetic field. For lower fields, the start current of the $TE_{2,4,1}^0$ mode was less than that of the desired mode, and the $TE_{2,4,1}^0$ mode was observed.

The linewidth of a gyromonotron at 35 GHz was measured to be less than 1 MHz, with the measurement apparently being limited by modulator voltage fluctuations (Read *et al.*, 1980). More recent measurements indicate that the intrinsic linewidth of the gyrotron may be as small as 10 kHz (Epthimion, 1980).

The modal purity of the radiation at the cavity output appears to be quite high. With the TE_{01} mode a measurement of $>90\%$ purity was reported (Read *et al.*, 1980). For high-power cw devices, where a large radius collector (which also serves as the output guide) is required, mode conversion in the transitions between the cavity, collector, and output window can be serious. With a 28-GHz, 212 cw oscillator, (Jory *et al.*, 1980) approximately 37% of the output power was in the TE_{03} and TE_{01} modes. For an oscillator with a nonsymmetric output, extremely high-mode conversion was observed, with less than 50% occurring in the desired code (Loring, 1980).

F.   High-Average Power Performance

One of the recent significant advancements in gyrotron development has been the realization of a 28-GHZ gyrotron which yielded 212 kW on a continuous wave basis (Jory *et al.*, 1980). This device is pictured in Fig. 19. A carefully designed collector/output waveguide and a double-disk window with fluorocarbon cooling was required for the cw operation. Devices with similar cw powers at 60 GHz are being designed.

G.   Summary of Performance and Prospects

From the foregoing, it is clear that the gyrotron oscillator is already a highly developed device. Almost all of the devices discussed produced power at record levels, and have done so with parameters (e.g., efficiency, input voltage, and pulse length) within the range of conventional microwave tubes. Thus the gyrotron oscillator seems destined to become a widely used device.

Agreement of theory and experiment is in general excellent. From the data base now existing it is possible to design devices with parameters over wide ranges. However, the gyrotron oscillator is hardly at the state where its potential has been completely realized. To close this section, we shall make a few comments as to the probable limits of operation of gyrotrons, and the areas of research which will be required for those limits to be reached.

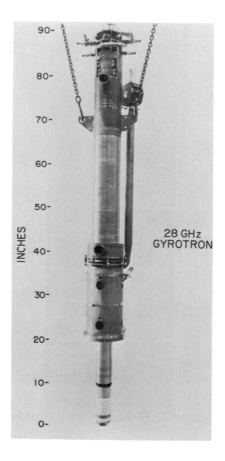

FIG. 19 Photograph of a 28-GHz, 212-kW (cw) gyrotron oscillator. (Courtesy Varian Associates. Contract supported by Oak Ridge National Laboratory.)

For cw applications it appears that powers of 1–3 MW may be possible at frequencies on the order of 100 GHz. Results from the USSR. (Andronov *et al.*, 1978) have already demonstrated that this power and frequency goal is possible with a 100 μsec pulse duration. For a cw or long pulse device, it is likely that a TE_{0n}^0 mode will have to be used, since ohmic losses with any other type of mode will be extremely high. (The megawatt tube developed in the USSR used a whispering gallery mode. It may have been this aspect which limited it to short pulse operation.) Results of on-going studies at the Naval Research Laboratory indicate that a

1-MW (cw), 100-GHz device with a TE_{06}^0 mode may be feasible. The principal limitations are ohmic heating of the cavity walls and beam generation.

For "short" pulse devices ($\tau \leq 100\mu s$), the whispering gallery modes may be suitable, and output powers greatly exceeding the megawatt level should be possible.

For higher power cw operation or for higher frequencies, it may be desirable to utilize an alternate configuration termed a "quasioptical gyrotron" (Sprangle *et al.*, 1980; Rapoport *et al*, 1967). This device is a cyclotron maser with a Fabry–Perot type cavity; the electron beam propagation is perpendicular to the axis of the cavity. Calculations indicate that efficiencies similar to those achieved with the conventional "microwave cavity" gyrotron discussed above are possible, and the large volume quasioptical cavity should have very low ohmic losses, allowing high average power. Operation at the second or third cyclotron harmonic is also predicted (Rapoport *et al.*, 1967), which, with existing superconducting magnet technology, should allow generation of substantial power at over 400 GHz. A proof of principle experiment for the quasioptical gyrotron is currently being performed at NRL.

For any of these advanced oscillator devices to be realized, extensive research on cavities and electron guns will be necessary. For cw devices, the collector, output guide, and output window are expected to prove challenging. Success in this research, however, will produce mm wave sources with output powers adequate for such demanding energetic applications as the heating of plasma in controlled fusion reactors.

Appendix.   Measurement Methods

A.   Cold Tests

The resonant frequency (f_0), Q, and the rf field distribution of the gyrotron cavity are critical in predicting device performance. These quantities can most easily be measured in the absence of the electron beam, i.e., in "cold tests." Measurement of these quantities have been made by adapting methods developed for other microwave devices. In general the methods for determining the resonant frequency f_0 and Q involve driving the cavity with an external source over a range of frequency and measuring the resulting response. Two methods for these measurements are of interest: by transmission and by reflection. In the former, power is injected into the side of the cavity by a waveguide oriented to couple to the desired mode, and the cavity response measured by a waveguide on the opposite side of the cavity (Fig. 20) (Read and Lucey, 1978). Alternatively, the response can be measured through the output of the cavity. Measurement by reflection has the advantage that the signal can be in-

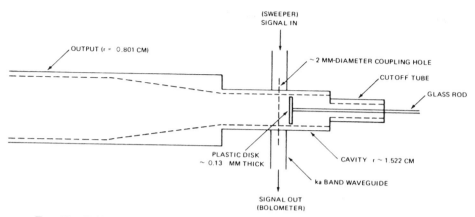

FIG. 20   Cold test arrangement for a gyrotron cavity. △, 241; ○, 041 (from Read and Lucey, 1978).

jected and extracted through the cavity output, thereby avoiding perturbing the cavity walls. However, the low-Q values usually needed in high power gyrotrons make this measurement difficult.

The field distribution has been measured using well known perturbation methods. For gyrotron cavities, a thin disk of dielectric material has been found to work well (Read *et al.*, 1980; Read and Lucey, 1978). An example of a test arrangement for a gyrotron cavity is shown in Fig. 20.

Measurements of the Q and field profiles have been reported for cavities supporting the TE_{01} and TE_{02} modes (Read *et al.*, 1980). An example of the field plot of a profiled cavity (Fig. 11a) is given in Fig. 11b (Read *et al.*, 1980). This measurement was made using the apparatus of Fig. 20 with coupling into and out of the cavity mode via holes in the side wall. This cavity had the relatively high Q of 800, equal to approximately 6 times the diffraction Q. Measurement of cavities with Q near the diffraction Q have been made (Dudas, 1980; Stone, 1980a). The lower limit to Q appears to be, as expected, strongly dependent on the output structure.

The mode structure of the output of gyrotron has also been examined with cold tests. These experiments (Stone, 1980b) designed to determine the degree of conversion of a TE_{02} mode into other TE_{0n} modes by the output structure (i.e., collector, etc.), utilized voltage traveling wave ratios for the measurement.

B.   ELECTRON BEAM PERFORMANCE

Direct measurement of electron-beam characteristic is difficult. Electron-gun performance is usually predicted by computer simulation. To date the only successful reported measurement of electron gun performance has been carried out in the USSR. (Antakov *et al.*, 1975; Avdoshin *et al.*, 1973; Avdoshin and Gol'denberg, 1973). These measurements, which yielded the electron momentum distribution, were made with a gridded energy analyzer with provision for intercepting reflected electrons before they returned to the cathode. This method, however, was used with single anode guns at low energy (1000–3000 eV), and has not been used successfully on high-power beams. Spreads in the perpendicular velocities of 15%–23% were reported by Antakov *et al.* (1975). No comparison with theory was made in that study. An examination of the effects of cathode roughness and initial velocity spread at the cathode was made by Avdoshin and Gol'denberg (1973). Spreads of 9%–17% were found in perpendicular velocities; while 6%–11% (respectively) of the spread was predicted for the cathode effects. No calculation was given for the spread expected owing to space charge.

Unfortunately, theoretical investigations (Seftor *et al.*, 1979a) indicate that beam parameters are very sensitive to gun geometry and voltages, and no accurate extrapolation can be made to the types of geometries and voltages which are commonly used with guns at present.

C. TUBE PERFORMANCE (POWER, FREQUENCY, MODE)

The measurement of the output power of gyrotrons is, in general, complicated by the fact that high-order modes are used in the cavity. Thus, structures in the output or measurement system can cause partial conversion to other modes, making an accurate measurement of the total power generated by the cavity somewhat difficult.

For cw or repetitively pulsed devices calorimetry has been used successfully (Arfin and Read, 1980; Jory *et al.*, 1980). For temporal resolution, the conversion to a fundamental rectangular waveguide mode where calibrated detectors can be deployed unambiguously has been used. (The presence of higher frequency out of band signal components will endanger this lack of ambiguity.) Sidewall directional couplers have also been used, and are successful if adequate discrimination between modes can be achieved.

Standard methods are available for conversion to the $TE_{10}^{\square}$ rectangular mode from the TE_{0n}^0 or TE_{1n}^0 families of modes. Conversion from the TE_{mn}^0 mode to the TE_{ml}^0 mode has been proposed by Kovalev *et al.* (1974) and demonstrated by them for the TE_{13}^0 mode. Conversion to the $TE_{10}^{\square}$ from TE_{11}^0 is accomplished by simple adiabatic deformation of the guide. For conversion from the TE_{01}^0 mode to the $TE_{10}^{\square}$ there are several types of converters (Harvey, 1963). One of these types, manufactured by Hitachi, has been successfully used at 35 GHz with 1 μsec pulses of 200 kW, and 100 kW pulses of 5 msec (Read *et al.*, 1980).

The measurement of frequency is straightforward, assuming signals of more than one frequency are not simultaneously present. Simple calibrated absorption or transmission meters are available for frequencies up to 300 GHz, and optical cavities (e.g., a Fabry-Perot) can be used at higher frequencies.

If there is sufficient frequency separation, identification of the cavity mode can be made by measurement of the frequency. For determination of the output mode, a measurement of the radiation pattern (e.g., Kisel *et al.*, 1974) or the guide wavelength (Stone, 1980b) has been used. For differentiation between classes of modes, such as the TE_{0n}^0 and TE_{mn}^0 ($m \neq 0$) modes, filters in the output can be employed. For the specific example of the TE_{0n}^0/TE_{mn}^0 modes, a filter of either a tightly wound helix waveguide (Read *et al.*, 1980) or a guide fabricated of alternating conductive and insulating disks (Jory *et al.*, 1980) has been used.

ACKNOWLEDGMENTS

The authors are indebted to many people and organizations for the material used in this chapter. In particular we would like to thank Dr. Y. Y. Lau of Science Applications, Inc., Dr. K. R. Chu of the Naval Research Laboratory, Dr. J. M. Baird of B. K. Dynamics, Dr. H. Jory of Varian Associates, Dr. B. Arfin of the Lawrence Livermore Laboratory, and Dr. K. J. Kim of JAYCOR, Inc.

REFERENCES

Ahn, S., and Choe, J. (1980). *IEEE Electron Device Lett.* **1**, No. 1, 8–9.
Alikaev, V. V. *et al.* (1976). *Fiz. Plasmy (Moscow)* **2**, 390 [*English transl.: Sov. J. Plasma Phys.* **2**, 212].
Andronov, A. A. *et al.* (1978). *Infrared Phys.* **18**, 385–393.
Antakov, I. I., Gintsburg, V. A., Zasypkin, E. V., and Sokolov, E. V. (1975). *Radiophys. Quantum Electron.* **18** (8), 884–887.
Antakov, I. I., Ergakov, V. S., Zasypkin, E. V., and Sokolov, E. V. (1977). *Radiophys. Quantum Electron.* **20** (4), 413–418.
Arfin, B., and Read, M. E. (1980). *Tech. Digest, IEEE Int. Electron. Devices Meeting, Washington, D.C., December* pp. 308–309.
Avdoshin, E. G., and Gol'denberg, A. L. (1973). *Radiophys. Quantum Electron.* **16**(10), 1241–1246.
Avdoshin, E. G., Nikolaev, L. V., Platonov, I. N., and Tsimring, Sh. E. (1973). *Radiophys. Quantum Electron.* **16**(4), 461–466.
Baird, J. M., and Attard, A. C. (1981). Gyrotron Gun Study Program Final Report. B. K. Dynamics Inc., Rockville, Maryland (reference contract No. N00173-79-C-0132).
Barnett, L. R., Chu, K. R., Baird, J. M., Granatstein, V. L., and Drobot, A. J. (1979). *Tech. Digest IEEE Int. Electron Devices Meeting, Washington D.C., December* pp. 164–167.
Barnett, L. R., Baird, J. M., Lau, Y. Y., Chu, K. R., and Granatstein, V. L. (1980a). *Tech. Digest, IEEE Int. Electron Devices Meeting, Washington, D.C., December* pp. 314–317.
Barnett, L. R., Baird, J. M., Fliflet, A. W., and Granatstein, V. L. (1980b). *IEEE Trans. Microwave Theory Tech.* **MTT-28**, 1477–1481.
Barnett, L. R., Lau, Y. Y., Chu, K. R., and Granatstein, V. L. (1981). *IEEE Trans. Electron Devices ED-28*, 872–875.

Bratman, V. L., and Moiseev, M. A. (1975). *Radiophys. Quantum Electron.* **18**(7), 772–229.
Bratman, V. L., and Petelin, M. I. (1975). *Radiophys. Quantum Electron.* **18**(10), 1136–1140.
Bratman, V. L., Moiseev, M. A., Petelin, M. I., and Erm, R. E. (1973). *Radiophys. Quantum Electron.* **16**, 474–480.
Bykov, Yu. V., Gol'denberg, A. L., Nikolaev, L. V., Ofitserov, M. M., and Petelin, M. I. (1975). *Radiophys. Quantum Electron.* **18**(10), 1141–1143.
Caplan, M. (1980). Design and Analysis of Gyrotron Oscillator Cavities. Hughes Aircraft Co., Tech. Rep. No. 58.
Chu, K. R. (1978). *Phys. Fluids* **21**, 2354.
Chu, K. R., Drobot, A. T., Granatstein, V. L., and Seftor, J. L. (1979). *IEEE Trans. Microwave Theory Tech.* **MTT-27**, 178–187.
Chu, K. R., Drobot, A. T., Szu, H. H., and Sprangle, P. (1980a). *IEEE Trans. Microwave Theory Tech.* **MTT-28**, 313–317.
Chu, K. R., Read, M. E., and Ganguly, A. K. (1980b). *IEEE Trans. Microwave Theory Tech.* **MTT-28**, 318–325.
Dialetis, D., and Chu, K. R. (1980). NRL Memorandum Rep. 4620.
Dionne, N. (1981). Personal communication.
Drobot, A. T., and Kim, K. J. (1981). *Intern. J. Electronics* (to be published).
Dudas, A. J. (1980). Personal communication.
Epthimion, P. (1980) Personal communication.
Ferguson, P., and Symons R. (1980). *Tech. Digest, IEEE Int. Electron Devices Meeting, Washington, D.C., December* pp. 310–313.
Ferguson, P., Valier, G., and Symons, R. (1981). *IEEE Trans. Microwave Theory Tech.* **MTT-29**, 794–799.
Fliflet, A. W., Barnett, L. R., and Baird, J. M. (1980). *IEEE Trans. Microwave Theory Tech.* **MTT-28**, 1482–1485.
Flyagin, V. A. (1979). Personal communication.
Flyagin, V. A., Gaponov, A. V., Petelin, M. I., and Yulpatov, V. R. (1977). *IEEE Trans. Microwave Theory Tech.* **MTT-25**, 514–521.
Gaponov, A. V., Gol'denberg, A. L., Grigor'ev, K. P., Pankratova, T. B., Petelin, M. I., and Flyagin, V. A. (1975). *Radiophys. Quantum Electron.* **18**(2), pp. 204en.210.
Gaponov, A. V. *et al.* (1980). *Int. J. Infrared Millimeter Waves* **1**, 351–372.
Gilgenbach *et al.* (1980). *Phys. Rev. Lett.* **44**, 647–650.
Gol'denberg, A. L., Petelin, M. I. (1973). *Radio Phys. Quantum Electron,* **16**(1), 106–111.
Granatstein, V. L., Sprangle, P., Herndon, M., Parker, R. K. and Schlesinger, S. P. (1975). *J. Appl. Phys.* **46**, 3800–3805.
Granatstein, V. L., Sprangle, P., Drobot, A. T., Chu, K. R., and Seftor, L. (1980). Gyrotron Travelling-Wave Amplifier. U.S. Patent 4, 224, 576.
Harvey, A. F. (1963). "Microwave Engineering," pp. 101–105. Academic Press, New York.
Hirshfield, J. L. (1979). Gyrotrons, In "Infrared and Millimeter Waves" (K. Button, ed.), Vol. 1, pp. 1–54. Academic Press, New York.
Hirshfield, J. L., and Granatstein, V. L. (1977). *IEEE Trans. Microwave Theory Tech.* **MTT-25**, 522–527.
Jory, H. R., Friedlander, F. Hegji, S. J., Shivley, J. F., and Symons, R. S. (1977). *Digest IEEE Int. Electron Devices Meeting, Washington, D.C., December* pp. 234–237.
Jory, H., Evans, S., Shively, J., Symons, R., and Taylor, N. (1979). Development Program for a 200 kW, cw, 28 GHz Gyroklystron, Quarterly Rep. #13. Varian Associates, Palo Alto, California.

Jory, H., Evans, S., Moran, J., Shirely, J., Stone, D., and Thomas, G. (1980). *Digest IEEE Int. Electron. Devices Meeting, Washington, D.C., December* pp. 304–307.

Kim, K. J., Read, M. E., and Chu, K. R. (1980). NRL Memorandum Rep. 4676.

Kisel, D. V., Korablev, G. S., Navel'yev, V. G., Petelin, M. I., and Tsimring, Sh. Ye. (1974). *Radio Eng. Electron. Phys.* **19** (4), 95–100.

Kolosov, S. V., and Kurayev, A. A. (1974). *Radio Eng. Electron. Phys.* **19**(10, 65–73.

Kovalev, N. F., Orlova, I. M., and Petelin, M. I. (1968). *Izv. Vzz. Radiofiz.* **11**, 783 [*English transl.: Radiophys. Quantum Electron.* **11**, 449].

Kovalev, N. F., Pankratova, T. B., and Shestakov, K. I. (1974). *Radio Eng. Electron. Phys.* **19**(10), 144–145.

Kreischer, K. E., and Temkin, R. J. (1980). *Int. J. Infrared Millimeter Waves* **1**, 195.

Kurayev, A. A., and Shestakovich, V. P. (1977a). *Radio Tekh. Electron.* **22**, 418 [*English transl.: Radio Eng. Electron. Phys.* **22**, 152].

Kurayev, A. A., and Shestakovich, V. P. (1977b). *Radiotekh. Electron.* **22**, 415 [*English transl.: Radio Eng. Electron. Phys.* **22**, 150].

Kurayev, A. A., Shevchenko, F. G., and Shestakovich, V. P. (1974). *Radio Eng. Electron. Phys.* **19**(6), 96–103.

Lau, Y. Y., and Chu, K. R. (1981). Gyrotron traveling wave amplifier: III. A proposed wide band fast wave amplifier. *Int. J. Infrared Millimeter Waves* **2**, 415–425.

Lau, Y. Y., Chu, K. R., Barnett, L. R., and Granatstein, V. L. (1981a). Gyrotron traveling-wave amplifier: I. Analysis of oscillations, *Int. J. Infrared Millimeter Waves* **2**, 373–393.

Lau, Y. Y., Chu, K. R., Barnett, L. R., and Granatstein, V. L. (1981b). Gyrotron traveling-wave amplifier: II. Effects of velocity spread and wall resistivity, *Int. J. Infrared Millimeter Waves* **2**, 395–413.

Lee, R. C., Hirshfield, J. L., Chu, K. R., and Park, S. (1981). NRL Memorandum Rep. (to be published).

Loring, M. (1980). Personal communication.

Moiseyev, M. A., Rogacheva, G. G., and Yuplatov, V. K. (1968). All-Union Scientific Session, in honor of Radio Day. Annotations and Summaries of Papers, Izd. NTORES im A. S. Popova, Moscow, 6.

Rapoport, G. N., Nemak, A., and Zhurakhovskiy, V. A. (1967). *Radio Eng. Electron. Phys.* **12**(4), 587–595.

Read, M. E. (1980). *Proc. Joint Workshop ECE and ECRH, Oxford.*

Read, M. E., and Lucey, R. (1978). Study on the Initial Development of High Power Millimeter Wave Sources, Rep. 2061. Jaycor, Alexandria, Virgina.

Read, M. E., Lucey, R., and Gilgenbach, R. (1979). Experimental and Numerical Studies on Gyrotron Design and Applications, Rep. 6085. Jaycor, Alexandria, Virginia.

Read, M. E., Gilgenbach, R. M., Lucey, R. F. Jr., Chu, K. R., Drobot, A. T., and Granatstein, V. L. (1980). *IEEE Trans. Microwave Theory Tech.* **MTT-28**, 875–877.

Read, M. E., Chu, K. R., and Kim, K. J. (1981a). *Int. J. Infrared Millimeter Waves* **2**, 159–174.

Read, M. E., Chu, K. R., and Dudas, A. J. (1981b). *IEEE Trans. Microwave Theory Tech.* (to be published).

Seftor, J. L., Drobot, A. T., and Chu, K. R. (1979a). *IEEE Trans. Electron Devices* **ED-26**, 1609–1616.

Seftor, J. L., Granatstein, V. L., Chu, K. R., Sprangle, P., and Read, M. E. (1979b). *IEEE. J. Quantum Electron.* **QE-15**, 848–853.

Sprangle, P., and Drobot, A. T. (1977). *IEEE Trans. Microwave Theory Tech.* **MTT-25**, 528–544.

Sprangle, P., and Smith, R. A. (1979). NRL Memorandum Rep. 3983.

Sprangle, P., Manheimer, W. M., and Vomvoridis, J. L. (1980). NRL Memorandum Rep. No. 4366.

Stone, D. S. (1980a). Personal communication.

Stone, D. S. (1980b). *Bull. Am. Phys. Soc.* **25,** 886.

Symons, R. S., and Jory, H. R. (1981). Cyclotron resonance devices, *In* "Advances in Electronics and Electron Physics" (C. Marton, ed.). Academic Press, New York.

Symons, R. S., Jory, H. R., and Hegji, S. J. (1979). *Tech. Digest, IEEE Int. Electron Devices Meeting, Washington, D.C., December* p. 676.

Temkin, R. J., Kreischer, K. E., Shultz, J., and Cohn, D. R. (1979). Design Study of a Tokamak Power Reactor with an Electron Cyclotron Resonance Heating System. MIT Plasma Fusion Center Rep. PFC/RR-79-20.

Tsimring, Sh.E. (1972). *Radio Phys. Quantum Electron.* **16**(8), 952–961.

Zaytsev, N. I., Pankratova, T. B., Petelin, M. I., and Flyagin, V. A. (1974). *Radio Eng. Electron. Phys.* **19**(5), 103–107.

CHAPTER 6

Distributed-Feedback Gas Lasers

F. K. Kneubühl and E. Affolter*

Physics Department
Eidgenössiche Technische Hochschule
Zürich, Switzerland

I.	INTRODUCTION	305
	A. *DFB Dye and Solid-State Lasers*	305
	B. *Waveguiding in Gas Lasers*	306
	C. *History of DFB Gas Lasers*	307
II.	DISPERSION RELATIONS BASED ON HILL'S EQUATION	308
	A. *Wave Equation*	308
	B. *Harmonic Modulation*	309
	C. *Heaviside Modulation*	310
III.	RESONANCE CONDITIONS	311
	A. *Coupled-Wave Theory*	311
	B. *Dielectric Waveguides*	312
	C. *Metallic Waveguides*	313
IV.	DFB IN PERIODIC METAL GUIDES	314
	A. *TM Modes in a Periodic Metallic Waveguide*	314
	B. *Dispersion Relation of the Passive Waveguide*	315
	C. *Dispersion Relation of the Active Waveguide*	317
	D. *Resonance and Threshold Condition*	317
	E. *Numerical Calculations*	318
V.	DFB GAS LASER IN OPERATION	322
	A. *Construction*	322
	B. *The Passive Waveguide*	324
	C. *Variation of the Waveguide Period*	326
	D. *Gas Pressure*	330
	E. *Laser-Beam Characteristics*	331
	F. *Spectral Analysis of the Laser Emission*	332
	G. *Pulse Shape and Energy*	333
VI.	FUTURE DEVELOPMENTS	334
	REFERENCES	335

I. Introduction

A. DFB DYE AND SOLID-STATE LASERS

Distributed feedback (DFB) was first observed in dye lasers in 1971 (Kogelnik and Shank, 1971; Shank *et al.*, 1971; Kaminow *et al.*, 1971).

* Present address: Siemens-Albis AG, CH-8047 Zürich, Switzerland.

DFB structures do not use conventional resonator mirrors, but provide feedback via backward Bragg scattering from periodic variations of the refractive index and/or the gain of the laser medium itself. This is a feedback mechanism distributed over the whole length of the periodic structures. Another important feature of DFB is high spectral selectivity, which originates in the Bragg effect. This allows selection of longitudinal modes even for broad gain profiles of the laser medium.

The DFB dye laser of Shank et al. (1971) was efficient and tunable. DFB was achieved by pumping the dye (Rhodamine 6G in ethanol) with the fringes formed by the interference of two coherent beams. They were obtained by passing the second harmonic of a single-mode ruby laser through a beam splitter. The laser could be tuned between 0.57 and 0.64 μm wavelength by variation of the interference angle of the coherent pumping beams. Single-mode operation was obtained with less than 0.01 Å linewidth.

The construction of the first DFB lasers was based on coupled-wave theory (Kogelnik, 1969; Kogelnik and Shank, 1972). Theory and experimental results of Kogelnik and co-workers initiated a number of theoretical papers discussing various aspects of DFB and coupled-wave theory (Chinn, 1973; De Wames and Hall, 1973a, b; Wang, 1973, 1974a, b, 1975a, b; Wang and Tsang, 1974; Wang et al., 1974; Yariv, 1973; Yariv and Gover, 1975). Furthermore, new periodic waveguide structures suitable for solid-state and semiconductor DFB lasers were proposed by Wang (1973). First solid-state and thin-film DFB lasers were operated by Bjorkholm and Shank (1972), Bjorkholm et al. (1973), Nakamura et al. (1973) and Schinke et al. (1972).

B. WAVEGUIDING IN GAS LASERS

For the present, waveguiding seems to be a prerequisite for the design of DFB gas lasers. Hitherto all published proposals, as well as the experimental realization, incorporate waveguide structures.

Waveguiding in gas lasers was first proposed by Marcatili and Schmeltzer (1964), who determined field configurations and propagation constants of the modes in hollow circular dielectric and metallic tubes. They suggested the use of these waveguides instead of open optical resonators in lasers working at wavelengths between 0.4 and 4 μm. The first experimental evidence of waveguiding in a laser resonator was reported three years later, when Schwaller et al. (1967) observed that resonator mode spectrum and losses of low-Fresnel number far-infrared HCN and ICN gas lasers in the wavelength region between 311 and 774 μm were in essential disagreement with conventional resonator theory. Consequently, they (Steffen and Kneubühl 1968a, b, c; Steffen, 1968) verified waveguide laser action for these far-infrared stimulated emissions. Since they were not aware of the calculations by Marcatili and Schmeltzer, they designated waveguide modes as "tube" modes. An optical waveguide laser constructed with explicit reference to the theory of Marcatili and Schmeltzer was first operated by Smith (1971). Since then a large variety of optical and infrared waveguide gas lasers has been developed. Comprehensive reviews have been published by Degnan (1976), Kneubühl (1977), and Kneubühl and Affolter (1979). Among others, Hodges and Hartwick (1973) gave a strong impetus to waveguiding in gas lasers by introducing a waveguide resonator in an optically pumped far-infrared laser. Subsequently, several hundred new far-infrared emissions have been obtained from optically pumped waveguide lasers (Rosenbluh *et al.*, 1976; Yamanaka, 1976a, 1977).

C. HISTORY OF DFB GAS LASERS

A year after Kogelnik and Shank (1971) demonstrated first operation of a DFB (dye) laser, Marcuse (1972) proposed incorporation of DFB in hollow-core dielectric waveguide gas lasers. He suggested the use of hollow dielectric waveguides with periodic ripples on the inner surface. Although these waveguides only support leaky modes, Marcuse could show that efficient operation of such a DFB laser should be possible. Later, Miles and Grow (1978, 1979) published extensive calculations on the guiding properties of dielectric DFB waveguides for CO_2 lasers. They calculated a loss of 2.75 dB/m at 10.6 μm for a waveguide made of BeO and glass and with a cross section of 0.1×2 mm.

First attempts of an operation of DFB gas lasers were made by Yamanaka (1976b) with an optically pumped 578 μm CH_3OH laser and by Miles and Grow (1976) with a 10.6-μm CO_2 laser.

Yamanaka (1976b) tried the application of a periodic waveguide in a far-infrared optically pumped laser, as well as the construction of a monochromatic far-infrared source using a high-pressure mercury lamp and a periodically corrugated waveguide as a narrow bandwith backward-wave filter, yet without success.

For their experiments with the CO_2 laser Miles and Grow (1976) used a waveguide consisting of two germanium slabs separated by a 100-μm gap. A DFB structure in the form of a grating was milled into the surface of each slab. Periodicity was 5.08 μm and corrugation depth was about 1 μm. Excitation of the laser was produced by externally exciting N_2 in a discharge tube and mixing the excited N_2 with unexcited CO_2 in the waveguide. However, neither this nor subsequent experiments aiming for operation of a DFB CO_2 laser succeeded (Miles and Grow, 1978, 1979, 1981; Miles, 1980).

The goal of the operation of a DFB gas lasers was achieved recently by Affolter and Kneubühl (1979) on the basis of various calculations (Affolter and Kneubühl, 1976a, 1977; Kenubühl, 1977, 1979; Kneubühl and Affolter, 1979, 1980). They demonstrated distributed feedback in an optically pumped 496-μm CH$_3$F laser. For this purpose a rectangular metallic waveguide with mechanically milled corrugations with a period of 248 μm, and a depth of 124 μm was used. It produced sufficient feedback at 496 μm, the emission wavelength of optically pumped CH$_3$F. This gas was chosen because it produces one of the strongest emission lines in the submillimeter region, and since its relatively large wavelength allows mechanical fabrication of the periodic structure. The laser described served for a study of the influence of the waveguide and gas parameters on the pulsed CH$_3$F emission as well as of its spectral and spatial characteristics (Affolter, 1980; Affolter and Kneubühl, 1981).

II. Dispersion Relations Based on Hill's Equation

A. Wave Equation

Periodic perturbations in a waveguide and periodic gain or loss variations can be represented by the wave equation for the amplitude E of the oscillating electric field:

$$\frac{d^2E(z)}{dz^2} + K^2(z)E(z) = \frac{d^2E(z)}{dz^2} + V(z)E(z) = 0 \qquad (1)$$

with the periodic potential

$$V(z) = V(z + L) = V(z + \pi\beta_0^{-1}), \qquad \beta_0 L = \pi. \qquad (2)$$

β_0 indicates the angular Bragg wave number. For active waveguide structures the angular periodic wave number $K(z)$ is complex. In DFB theory it is usually approximated by

$$K(z) = k(z, \omega) + i\alpha(z) = c^{-1}\omega n(z) + i\alpha(z), \qquad (3)$$

where the periodic gain $\alpha(z)$ as well as the periodic effective index $n(z)$ of refraction are assumed to be independent of the angular frequency ω. The angular Bragg frequency ω_0 is usually defined by the average effective index $n = \overline{n(z)}$ of refraction

$$\omega_0/\beta_0 = c/n \qquad \text{or} \qquad \omega_0 = \pi c/nL. \qquad (4)$$

Wave equation (1) represents a Hill differential equation (Magnus and Winkler, 1966; Mc Kean and van Moerbeke, 1975; Mc Kean and Trubowitz, 1976). For a periodic waveguide without loss or gain, the potential $V(z)$ is real. Furthermore, for a real periodic potential $V(z)$, the wave equation corresponds to the Schrödinger equation of an electron in a one-dimensional periodic potential. This equation yields the band structure well known in solid-state physics. (Kohn, 1959; Casey and Weaver, 1973; Reynaud, 1978; Harrell, 1979; Vigneron and Lambin, 1979).

If the loss or gain due to the laser medium in the waveguide (Elachi, 1976) is taken into account, the potential $V(z) = K^2(z)$ is complex. Little is known about the solutions of the wave equation with complex $V(z)$, except for the most recent results obtained by Meiman (1977). Therefore, most theories on DFB lasers are restricted to small periodic gain variations and corrugations of the waveguide (Kogelnik and Shank, 1972; Marcuse, 1972; Wang, 1973, 1974a, b, 1975a, b; Kneubühl, 1977; Miles and Grow, 1978, 1979).

In order to gain insight into dispersion relations of DFB structures we consider two simple wave equations in the following sections, the Mathieu equation with a harmonic potential $V(z)$ and the Hill equation with a Heaviside potential $V(z)$. For active structures with strong modulations of the complex potential $V(z)$ little is known about dispersion relations originating in Mathieu's equation, while the Hill equation with Heaviside potential provides simple explicit dispersion relations.

B. HARMONIC MODULATION

The Mathieu equation is characterized by harmonic variation of the periodic potential $V(z)$. Harmonic variation of gain and index of refraction in DFB structures was studied first with coupled-wave theory (Kogelnik and Shank, 1972) and later by perturbation theory (Kneubühl, 1977; Kneubühl and Affolter, 1979). Both studies are restricted to small variations:

$$n(z) = n + n_1 \cos 2\beta_0 z, \qquad \alpha(z) = \alpha + \alpha_1 \cos 2\beta_0 z. \tag{5}$$

Assuming $n_1 \ll n$ and $\alpha, \alpha_1 \ll \beta_0$, Kogelnik and Shank (1972) derived the approximate periodic potential

$$V(z) = K^2(z) \simeq k^2 + 2\alpha k i + 4\kappa k \cos 2\beta_0 z \tag{6}$$

with

$$k = c^{-1} n\omega, \qquad \kappa = (\pi/2)(n_1/nL) + i(1/2)\alpha_1,$$

where κ indicates the coupling constant. For this potential $V(z)$ they find the following dispersion relation, i.e., the propagation constant β as a function of the normalized frequency k:

$$(\beta - \beta_0)^2 \simeq (k - k_0 + i\alpha)^2 - \kappa^2, \qquad k_0 L = \beta_0 L = \pi. \tag{7}$$

This relation allows a classification of DFB for $\alpha \gtrless 0$:

(a) index modulation without gain or loss: $\alpha_1 = \alpha = 0$;

(b)   index modulation with gain: $\alpha_1 = 0$;
(c)   weak gain modulation: $n_1 = 0$; $\alpha_1 < 2\alpha$; and
(d)   strong gain modulation: $n_1 = 0$; $\alpha_1 > 2\alpha$.

Figure 1 shows real and imaginary parts of the propagation constant β as function of the normalized frequency $k = c^{-1}n\omega$ near the Bragg condition for the four different cases. Index modulation without gain or loss (case a) shows the well known frequency gap. Index modulation (case b), as well as gain modulation, exhibit an effective gain Im β which differs near the Bragg condition from the standard laser gain α.

This difference of the effective gain Im β from the standard laser gain α near the Bragg condition is the essential feature of DFB. It provides low laser thresholds for longitudinal modes near the Bragg frequency. For extreme gain modulation (case d), which includes lossy sections of the periodic structure a gap appears in the real part of the propagation constant β.

C. HEAVISIDE MODULATION

Another approach to study DFB makes use of step modulations (Heaviside functions) of gain and index of refraction (e.g., Wang, 1974b). A relevant example is

$$0 \leqq z < L/2: \qquad n(z) = n + n_1, \qquad \alpha(z) = \alpha + \alpha_1,$$
$$L/2 \leqq z < L: \qquad n(z) = n - n_1, \qquad \alpha(z) = \alpha - \alpha_1, \tag{8}$$

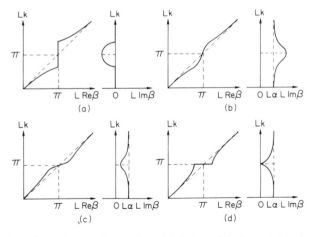

FIG. 1   Dispersion relations for different DFB modulations: (a) index modulation without gain, (b) index modulation with gain, (c) weak gain modulation, and (d) strong gain modulation. The propagation constant β is plotted versus normalized frequency $k = c^{-1}n\omega$ for $\alpha \geq 0$.

and $k = c^{-1}n\omega$, $K = k + i\alpha$, $K_1 = (n_1/n)k + i\alpha_1$. The corresponding dispersion relation $\beta(k)$ can be deduced from Eq. (44.12) in the book by Brillouin (1946) on periodic structures:

$$\cos \beta L = [K^2 \cos KL - K_1^2 \cos K_1 L]/(K^2 - K_1^2) \qquad (9)$$

or by introducing the first-order Bragg condition

$$\cos(\beta - \beta_0)L = [K^2 \cos(K - k_0)L + K_1^2 \cos K_1 L]/(K^2 - K_1^2). \qquad (10)$$

Equation (9) serves for the evaluation for the complete band structure, whereas Eq. (10) is relevant for the first-order Bragg reflection, which is essential for many DFB lasers. Both equations are valid for weak and strong modulations.

For a comparison with the coupled-wave theory by Kogelnik and Shank (1972) we assume $|K_1| \ll |K|$ and find from Eq. (10)

$$(\beta - \beta_0)^2 \simeq (k - k_0 + i\alpha)^2 - \kappa^{*2} \qquad (11)$$

with $\kappa^* = 2(n_1/nL) + i(2/\pi)\alpha_i$. Hence Eqs. (11) and (7) are identical except for the expected difference between κ and κ^*. However, Eq. (9) embraces a considerably larger gamut of gain and refraction-index modulations.

III. Resonance Conditions

A. COUPLED-WAVE THEORY

Recent experiments (Kneubühl and Affolter, 1980) on the first DFB gas laser in operation indicate that coupled-wave theory provides approximate resonance conditions. The basis of coupled-wave theory was laid by Brillouin in his book (Brillouin, 1946) on wave propagation in periodic structures. The coupled-wave theory on DFB lasers near laser threshold was first published by Kogelnik and Shank (1972). Subsequently, this theory has been refined and extended by several authors: e.g., Marcuse (1972), Chinn (1973), De Wames and Hall (1973a, b), Kogelnik et al. (1973), Shubert (1974), Streifer et al. (1975), Wang (1973, 1974a, b, 1975a, b), Yariv (1973), Yariv and Gover (1975).

If distributed feedback is achieved by periodic corrugations in a hollow waveguide, one expects distributed feedback with index modulation in the high-gain approximation (Kogelnik and Shank, 1972). This results in a resonance condition for large gain oscillations:

$$2/\lambda_g = (1/L) + (1/R)(q + \tfrac{1}{2}), \qquad q = \text{integer} \qquad (12)$$

in agreement with the corresponding equations given by Kogelnik and

Shank (1972) and by Marcuse (1972). The index q of the DFB resonant mode is not to be confused with the index of a waveguide mode. R represents the length of the periodic waveguide structure, L is its period, and $\lambda_g = 2\pi/\text{Re }\beta$ the wavelength in the waveguide. The Bragg wavelength λ_B is defined by

$$\lambda_B = 2\pi/\beta_0 = 2L, \tag{13}$$

where β and β_0 are the corresponding propagation constants. According to Eq. (12), there is no resonance at the Bragg wavelength $\lambda_g = \lambda_B$ for index coupling.

Introduction of the number $M = R/L$ of periodic corrugations allows to transform Eq. (12) into the DFB laser resonance condition

$$R = (\lambda_g/2)(M + q + \tfrac{1}{2}), \qquad q = \text{integer}. \tag{14}$$

This can be compared with the resonance condition of a Perot–Fabry resonator formed by a waveguide with ideally reflecting mirrors at both ends:

$$R = (\lambda_g/2)p, \qquad p = \text{integer}. \tag{15}$$

Thus the resonance conditions of a Perot–Fabry and a periodic waveguide structure are similar. The basic difference, however, is not in the resonance conditions, but in different threshold gains. In a laser with a Perot–Fabry resonator, the modes corresponding to different p are equal with respect to threshold gain. On the other hand, in a DFB laser the modes with the lowest absolute values $|q|$ also have the lowest threshold gains (Kogelnik and Shank, 1972).

B.   DIELECTRIC WAVEGUIDES

Hollow dielectric waveguides with periodic corrugations for DFB lasers have been discussed at length by Marcuse (1972), Miles and Grow (1978, 1979). In order to apply (14) to DFB lasers with a periodic hollow dielectric waveguide one has to determine the wavelength λ_g in the waveguide according to the theory by Marcatili and Schmeltzer (1964). They assume two inequalities:

$$a \gg \lambda(\nu_e/\nu_i) \tag{16}$$

and

$$[(\lambda_g\nu_i/\lambda) - 1] \ll 1. \tag{17}$$

In these inequalities $\lambda = c/f$ indicates the free-space vacuum wavelength, ν_i the approximately real index of refraction of the internal gas, and ν_e the complex index of refraction of the external medium. Inequality (16) states that for dielectric hollow waveguides the free-space wavelength λ/ν_i of the gas should be considerably smaller than the diameter $2a$ of the guide. The second inequality (17) restricts the use of waveguides to low-loss modes, which are those whose propagation constants are nearly equal to that of free space.

For a circular dielectric waveguide with diameter $2a$ the first approximation of the mode nm results in

$$\lambda_{g,nm} \simeq (\lambda/\nu_i)[1 - \tfrac{1}{2}(u_{nm}/2\pi)^2(\lambda/\nu_i a)^2]^{-1}, \tag{18}$$

where u_{nm} is the mth root of the equation:

$$J_{n-1}(u_{nm}) = 0. \tag{19}$$

Inequalities (16) and (17) require that the guide wavelengths $\lambda_{g,nm}$ of the modes nm be very close to the free-space wavelength λ/ν_i of the gas in the waveguide. As a consequence, the Bragg wavelengths $\lambda_{B,nm}$ and periods L_{nm} corresponding to the modes nm are also near this free-space wavelength:

$$\lambda_{B,nm} = 2L_{nm} \simeq \lambda/\nu_i. \tag{20}$$

Furthermore, inequality (17) and experience with hollow dielectric waveguides show, that for low transmission losses the condition $2a \simeq 50\lambda/\nu_i$ should be fulfilled. This is disadvantageous for distributed feedback in hollow dielectric waveguides generated by periodic wall corrugations since it was found that the periodic field modulations caused by wall corrugations exhibit (Affolter and Kneubühl, 1976a, b, 1977, 1979; Miles and Grow, 1978, 1979) a penetration depth of about $10\lambda/\nu_i$ in the adjacent medium. On the other hand, a diameter of $2a \simeq 10\lambda/\nu_i$ leads to high transmission losses in a hollow dielectric guide.

C. METALLIC WAVEGUIDES

For a metallic waveguide filled with a gas of refractive index ν_i the guide wavelength λ_g is

$$\lambda_g = \lambda[\nu_i^2 - (\lambda/\lambda_c)^2]^{-1/2}, \tag{21}$$

where f_c and $\lambda_c = c/f_c$ denote cutoff frequency and wavelength. For a rectangular guide of width a and height d the cutoff wavelength $\lambda_{c,nm}$ with the subscripts nm defining the transverse modes are evaluated from the equation

$$\lambda_{c,nm}/2 = [(m/a)^2 + (n/d)^2]^{-1/2}. \tag{22}$$

This results in the coupled-wave DFB laser condition

$$R_{qmn} = (\lambda/2)\{\nu_i^2 - (\lambda/2)^2[(m/a)^2 + (n/d)^2]\}^{-1/2}(M + q + \tfrac{1}{2}). \tag{23}$$

The corresponding Bragg wavelengths $\lambda_{B,nm}$ differ from the free-space wavelength λ/ν_i in the guide:

$$\lambda_{B,nm} = 2L_{nm} = \lambda\{\nu_i^2 - (\lambda/2)^2[(m/a)^2 + (n/d)^2]\}^{-1/2} > \lambda/\nu_i. \quad (24)$$

The optically pumped CH_3F 496-μm DFB laser (Affolter and Kneubühl, 1979) equipped with a parallel-plate or flat rectangular metal waveguide fulfills approximately the conditions $\nu_i \simeq 1$ and $(m/a)^2 \ll (n/d)^2$. Consequently, the coupled-mode DFB laser condition is

$$R_{qn} = (\lambda/2)[1 - (\lambda/2)^2(n/d)^2]^{-1/2}(M + q + \tfrac{1}{2}). \quad (25)$$

The above equation can be tested experimentally by temperature variation ΔT of the periodic waveguide structure. For a brass waveguide we have

$$\Delta R(\Delta T)/R = 1.85 \times 10^{-5} \, \Delta T \quad (°C). \quad (26)$$

This variation ΔR must be related to the difference between resonance lengths of longitudinal DFB laser modes

$$\Delta R(\Delta q = 1, \Delta n = 0) = (\lambda/2)\{1 - (\lambda/2)^2(n/d)^2\}^{-1/2}. \quad (27)$$

However it should be noticed that this difference is not characteristic for a DFB waveguide, but also for a Perot–Fabry resonator described by Eq. (15).

IV. DFB in Periodic Metal Guides

A. TM MODES IN A PERIODIC METALLIC WAVEGUIDE

In their optically pumped CH_3F DFB laser Affolter and Kneubühl (1979) used a periodic metallic waveguide as shown in Fig. 2. It represents essentially a periodic parallel-plate guide. Hence the electromagnetic fields are independent of the transverse coordinate x. According to Floquet's theorem, the fields of TM modes in this type of periodic waveguide

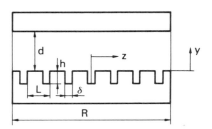

FIG. 2   Longitudinal section of the DFB waveguide.

with propagation in the z direction can be written for $y > 0$ as (Watkins, 1958)

$$E_z(y, z) = \sum_{n=-\infty}^{\infty} E_{n0}(Ae^{i\beta_n z} + Be^{-i\beta_n z}) \sin[\tau_n(d - y)],$$

$$E_y(y, z) = -\sum_{n=-\infty}^{\infty} E_{n0} \frac{i\beta_n}{\tau_n} (Ae^{i\beta_n z} - Be^{-i\beta_n z}) \cos[\tau_n(d - y)], \quad (28)$$

$$H_x(y, z) = -\sum_{n=-\infty}^{\infty} E_{n0} \frac{ik}{\tau_n} (Ae^{i\beta_n z} + Be^{-i\beta_n z}) \cos[\tau_n(d - y)],$$

where $\beta_n = \beta + 2\pi n/L$, $\tau_n = (k^2 - \beta_n^2)^{1/2}$, $k = \omega/c$, and β indicates the propagation constant.

In the slots one assumes a standing wave in the y direction:

$$E_z(y) = \begin{cases} E_0 \dfrac{\sin[k(y + h)]}{\sin(kh)} (Ae^{i\beta_0 mL} + Be^{-i\beta_0 mL}) \\ \qquad \text{for} \quad mL - \tfrac{1}{2}\delta < z < mL + \tfrac{1}{2}\delta \\ 0 \\ \qquad \text{for} \quad mL + \tfrac{1}{2}\delta < z < (m + 1)L - \tfrac{1}{2}\delta, \end{cases}$$

$$E_y(y) = 0, \qquad\qquad\qquad\qquad\qquad\qquad\qquad\qquad (29)$$

$$H_x(y) = \begin{cases} iE_0 \dfrac{\cos[k(y + h)]}{\sin(kh)} (Ae^{i\beta_0 mL} + Be^{-i\beta_0 mL}) \\ \qquad \text{for} \quad mL - \tfrac{1}{2}\delta < z < mL + \tfrac{1}{2}\delta \\ 0 \\ \qquad \text{for} \quad mL + \tfrac{1}{2}\delta < z < (m + 1)L - \tfrac{1}{2}\delta, \end{cases}$$

where m is an integer. The width of the slots is less than half the wavelength. Thus the electric field exhibits no component in the y direction. We also neglect the z dependence of both the electric and the magnetic fields, which is a good approximation for $\delta \ll L$.

Since the electric field in the slots has a component in the z direction, one expects a TM mode to oscillate in the waveguide. For this reason we have chosen the field statement according to (28).

B. DISPERSION RELATION OF THE PASSIVE WAVEGUIDE

In order to determine the dispersion relation $\beta(k)$ of the waveguide in Fig. 2, one first calculates the coefficients E_{n0} occurring in (28). Defining a scalar product

$$(f, g) = \lim_{T \to \infty} \frac{1}{2T} \int_{-T}^{T} \overline{f(z)}g(z) \, dz, \qquad (30)$$

one finds the orthogonality relation

$$(Ae^{i\beta_n z} + Be^{-i\beta_n z}, Ae^{i\beta_m z} + Be^{-i\beta_m z}) = (|A|^2 + |B|^2)\delta_{nm}, \qquad (31)$$

where

$$\delta_{nm} = \begin{cases} 1 & \text{for} \quad m = n \\ 0 & \text{for} \quad m \neq n \end{cases}.$$

With the aid of Eq. (31) we determine the coefficients

$$E_{n0} = \frac{1}{|A|^2 + |B|^2} \frac{1}{\sin[\tau_n(d-y)]} \lim_{T \to \infty} \frac{1}{2T}$$
$$\int_{-T}^{T} E(z)(A^* e^{-i\beta_n z} + B^* e^{i\beta_n z}) \, dz. \qquad (32)$$

Matching the electric field at the boundary $y = 0$ between the waveguide region and the slots results in

$$E_{n0} = \frac{1}{\sin(\tau_n d)} E_0 \frac{\delta}{L} \frac{\sin(\beta_n \delta/2)}{\beta_n \delta/2}. \qquad (33)$$

In a next step one has to match the magnetic field H_x for $y = 0$. Since the z dependence of the field components in the slots is neglected, one can reach the goal only approximatively. We consider two different possibilities:

(1)   Matching the magnetic field in the center of the slots $z = mL$ leads to the dispersion relation

$$\frac{1}{kh \tan(kh)} = -\frac{\delta}{L} \sum_{n=-\infty}^{\infty} \frac{\cot(\tau_n d)}{\tau_n h} \frac{\sin(\beta_n \delta/2)}{\beta_n \delta/2}. \qquad (34)$$

(2)   Another possibility to match the magnetic field is to make the average value of the field in the waveguide equal to the field in the slots:

$$\frac{1}{\delta} \int_{mL-(\delta/2)}^{mL+(\delta/2)} H_{y>0}(z) \, dz = H_{y<0}.$$

This leads to a somewhat different dispersion relation

$$\frac{1}{kh \tan(kh)} = -\frac{\delta}{L} \sum_{n=-\infty}^{\infty} \frac{\cot(\tau_n d)}{\tau_n h} \left(\frac{\sin(\beta_n \delta/2)}{\beta_n \delta/2}\right)^2. \qquad (35)$$

Equation (34) is the dispersion relation given by Watkins (1958). Numerical calculations show that no significant difference exists between Eqs. (34) and (35). Therefore, we shall consider only Eq. (34) in our further investigations.

C. Dispersion Relation of the Active Waveguide

All the preceding formulas are derived under the assumption that no loss or gain exists inside the waveguide. In a distributed feedback laser, however, we have a net gain α and therefore we have to modify our dispersion relation. Fortunately, Butcher (1956) has published a procedure to calculate the dispersion behavior of a lossy periodic waveguide if the dispersion relation for the lossless guide is known. He showed that

$$\omega(\beta) = \omega_l(\beta)[1 - (i/2Q_c)], \tag{36}$$

where $\omega(\beta)$ is the dispersion relation for a lossless guide ($\alpha = 0$), $\omega_l(\beta)$ the dispersion relation for a lossy guide ($\alpha < 0$), β the propagation constant, and Q_c the Q factor of the waveguide. Here one assumes that the walls of the waveguide are perfectly conducting and that all the loss (or gain) is caused by the dielectric filling of the waveguide.

Butcher (1956) also calculated the Q factor of the waveguide, which in our case is

$$Q_c = -\omega_l/2\alpha v_g. \tag{37}$$

Here v_g is the group velocity of the waveguide mode considered. In high-gain media, such as CH_3F, the group velocity nearly equals c even in the Bragg region (Kogelnik and Shank, 1972). The deviation of v_g from c is calculated to be smaller than 1% for waveguide heights greater than 1.5 mm. Therefore we replace v_g by c in Eq. (37). Equation (36) can therefore be written

$$k(\beta) = k_l(\beta) + i\alpha, \tag{38}$$

assuming $k = \omega/c$.

On the basis of relation (38) we can derive the dispersion relation $\beta(k)$ of the active waveguide from Eq. (34):

$$\frac{1}{(k + i\alpha)h \, \tan[(k + i\alpha)h]} = -\frac{\delta}{L} \sum_{n=-\infty}^{\infty} \frac{\cot(\tau_n d)}{\tau_n h} \frac{\sin(\beta_n \delta/2)}{\beta_n \delta/2} \tag{39}$$

with $\tau_n = [(k + i\alpha)^2 - \beta_n^2]^{1/2}$.

D. Resonance and Threshold Condition

In order to evaluate the resonance and threshold condition of a DFB laser, we consider a periodic waveguide of finite length R. Outside the waveguide we assume a plane wave. By applying the condition of continuity at both ends $z = \pm R/2$ of the guide we find

$$\sum_{n=-\infty}^{\infty} E_{n0}[\beta_n \cos(\tfrac{1}{2}\beta_n R) + ik \sin(\tfrac{1}{2}\beta_n R)]$$

$$\times \sum_{n=-\infty}^{\infty} E_{n0}[k \cos(\tfrac{1}{2}\beta_n R) + i\beta_n \sin(\tfrac{1}{2}\beta_n R)] = 0. \qquad (40)$$

Together with Eqs. (33) and (39), we can calculate in principle the resonance frequencies k_q and the corresponding threshold gain α_q of the longitudinal modes in our DFB laser.

E. Numerical Calculations

Figure 3 shows the results of a computer calculation of the dispersion relation (34) for a passive waveguide. As expected the cutoff frequency is the same as for a planar parallel-plate guide with the same height d. Near the Bragg frequency the imaginary part of the propagation constant becomes negative and a stop band occurs. This means that the amplitude of a propagating wave shows an exponential decay and that the wave is reflected back. This behavior is quite similar to that of electrons moving in a periodic potential. There, the waves are known as Bloch waves.

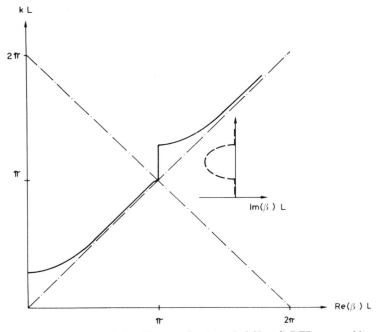

Fig. 3   Dispersion relation for a passive ($\alpha = 0$, $h/L = \tfrac{1}{2}$) DFB waveguide.

A further interesting detail is the shift of the band gap relative to the Bragg frequency. This corresponds to a change in the average phase velocity and it is also observed in dielectric periodic waveguides (Jaggard and Elachi, 1976).

If we introduce a gain α, the dispersion relation alters completely as shown in Fig. 4. The band gap disappears and the imaginary part of the propagation constant exhibits a positive resonance. It is in the region of this resonance where the DFB CH_3F laser is operated (Section V).

In Figs. 3 and 4 we have only plotted the lowest mode. The curves for the higher modes are quite similar but shifted towards higher frequencies.

Distributed feedback was previously used in semiconductor and in dye lasers. Typically these media have a relative gain bandwidth between 1% and 10% (Kuhl and Schmidt, 1974). This value determines the precision needed for the manufacture of the periodic structure and can be achieved without problems. Closer tolerances have to be met only if one requires a well-defined emission wavelength.

If the active medium in a DFB laser is a gas the relative gain bandwidth is narrow. For pulsed CO_2, as well as for optically pumped CH_3F lasers it is only in the order of 10^{-4}. It is therefore of special interest to know the bandwidth Δk of the Bragg region with efficient feedback. Figure 5 shows

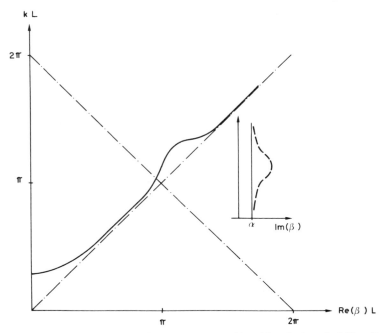

FIG. 4   Dispersion relation for a DFB waveguide with gain ($a > 0$, $h/L = \frac{1}{2}$).

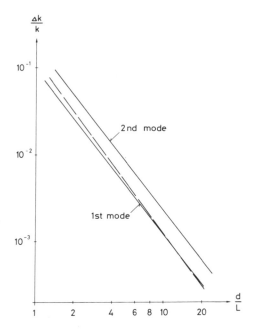

FIG. 5   Relative bandwidth $\Delta k/k$ of the stop band as a function of the waveguide height d. The dotted line is an approximation for the dominant mode according to (41). δ/L is assumed to be 0.2.

the relative bandwidth $\Delta k/k$ of the band gap as a function of the ratio d/L for the two lowest modes. $\Delta k/k$ is approximately proportional to $1/d^2$.

Since the right-hand side of Eq. (28) is a very slowly converging sum, numerical calculations of the dispersion relation are rather time consuming. For this reason we tried to find an approximative solution for the relative bandwidth. Taking only the terms with $n = 0$ and $n = -1$ in Eq. (39) and putting $\beta_0 = \pi/L$, we obtain for $\alpha = 0$,

$$\frac{1}{k \tan(kh)} + \frac{2\delta}{L} \frac{\sin(\pi\delta/2)}{\pi\delta/2} \frac{1}{(k^2 - \pi^2)^{1/2} \tan[(k^2 - \pi^2)^{1/2} d]} = 0. \quad (41)$$

The first zero k_1 of this equation gives us the bandwidth $\Delta k = k_1 - \pi/L$ for the lowest mode. The dotted line in Fig. 5 is a result of this approximation. The agreement with the precise curve is excellent for $d/L < 100$.

A further figure of merit is the maximum of the imaginary part of the propagation constant inside the stop band. This represents a measure for the strength of the feedback. As indicated in Fig. 6 the feedback decreases with the square of the waveguide height.

From the preceding calculations it seems advantageous to take the

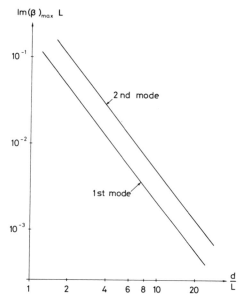

FIG. 6   Maximum of the imaginary part of the propagation constant versus waveguide height d. $Im(\beta)_{max}$ is a measure for the strength of the feedback.

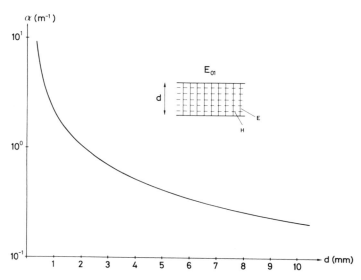

FIG. 7   Wall losses for the TM_{01} (or E_{01}) mode in a metallic parallel plate waveguide at a wavelength of 496 μm ($R = 0.39\ \Omega$).

waveguide height as small as possible. On the other hand, there are two effects which favor a large waveguide height. First, we should have a large gas volume to obtain high output energies. Second, the waveguide losses increase with decreasing height d. The wall losses of the lowest TM mode in a planar parallel plate guide are given by Marcuvitz (1951).

$$\alpha_{TM} = -(2\bar{R}/dZ_0)[1 - (\lambda/2d)^2]^{-1/2}, \tag{42}$$

where $\bar{R}$ is the surface impedance of the metallic walls and $Z_0 = 377 \ \Omega$. In Fig. 7 we plot the wall losses for the TM_{01} mode (also called E_{01} mode) in a brass waveguide at a wavelength of 496 μm.

From the above analysis we expect an optimum waveguide height which depends on the gain of the laser medium and on the waveguide material.

V. DFB Gas Laser in Operation

This Section is devoted to the optically pumped CH_3F 496-μm DFB laser (Affolter and Kneubühl, 1979), which is still the only DFB gas laser in operation. Its operating characteristics have been studied in detail (Affolter and Kneubühl, 1981; Affolter, 1980) in order to understand its mechanisms for future developments.

A. CONSTRUCTION

The design of the optically pumped CH_3F 496-μm DFB laser was based on the theories in Sections II to IV. Figure 8 shows a schematic diagram of the experimental arrangement. The active laser medium is CH_3F, optically pumped by the $9P(20)$ line of a grating tuned single mode CO_2 laser. the maximum available pump energy is 1 J, which allows power densities of several MW/cm². The pump beam is adapted to the waveguide by a condensing telescope with either circular or cylindrical lenses.

The waveguide for the far-infrared (FIR) radiation consists of the periodic structure of brass at the bottom and three plane walls covered with gold by vapor deposition. Although the rectangular metallic waveguide exhibits remarkable losses in the far infrared, we have chosen this configuration. It is easy to manufacture and exhibits good polarization characteristics (Yamanaka *et al.*, 1975). Usually only a fraction of the waveguide cross section was pumped. Consequently, it acted more as a parallel-plate guide than as a rectangular guide.

Figure 9 shows the corrugated wall of the DFB waveguide with a length of 32 cm. According to Eq. (20) the period of the structure should be close to half the wavelength. For the dominant emission line of CH_3F this corresponds to a period of 248.05 μm. The allowed deviation from this value is

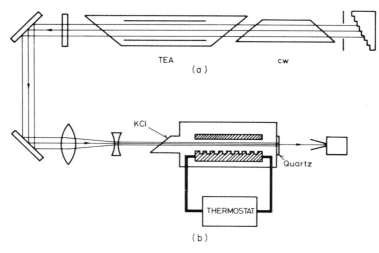

FIG. 8 Experimental arrangement. The DFB waveguide is rotated by 90° for clarity. (a) CO_2 laser, (b) DFB-CH_3F laser.

FIG. 9 Periodic structure used in the DFB laser. The mean period is 247.99 ± 0.02 m corresponding to half the wavelength of the optically pumped CH_3F emission

determined by the width of the band gap in the dispersion relation. According to Fig. 5, this bandwidth is only 8×10^{-4} for a waveguide height of 3 mm and becomes even smaller for greater heights. For this reason a temperature control was incorporated which allowed a temperature variation of the waveguide between -32 and $95°$ C. Thus the period of the structure could be tuned by a factor of about 2×10^{-3}, which corresponds to 0.5 μm.

The period of the structure was measured at room temperature by means of a projecting microscope with a reading precision of 1 μ. In order to increase the accuracy 100 periods were measured and an average value for the period L was determined. The result was $L = 247.99 \pm 0.02$ μm instead of the desired 248.05 μm. This corresponds to a relative deviation of only 2.4×10^{-4}, which lies well within our tuning range. It should be noted that the determination of an average value for the period is reasonable since small statistical deviations from the correct period cause no significant alteration in the DFB characteristics of the waveguide (Shellan *et al.*, 1978).

The depth of the corrugations is 124 μm, a quarter of the emitted wavelength. In general it should be an odd multiple of $\lambda/4$. If the depth is an even multiple of $\lambda/4$, the standing wave (29) has a node at the boundary between the slots and the guiding region, and the structure acts like a homogeneous waveguide. As we have seen, the waveguide properties depend strongly on the height d. In the laser this height was adjustable between 0 and 20 mm.

Table I gives a summary of the characteristics of the DFB laser. For the calculations of the gain bandwidth in CH_3F Raman contributions (Biron *et al.*, 1979) have been neglected.

B. THE PASSIVE WAVEGUIDE

The transmissivity and reflectivity of DFB waveguides in the visible or near infrared are normally measured with a dye laser. Since no tunable source is available in our wavelength region, we have taken a standard

TABLE I

CHARACTERISTICS OF THE DFB CH_3F LASER[a]

Gain bandwidth ($p = 5$ torr)	3×10^{-4} (200 MHz)
Band gap ($d = 3$ mm)	8×10^{-4} (480 MHz)
Longitudinal mode spacing	8×10^{-4} (470 MHz)
Waveguide tuning ($\Delta T = 100$ K)	1.8×10^{-3} (1100 MHz)
Measured difference between	
Waveguide period and $\lambda/2$	2.4×10^{-4}

[a] For the calculation of the gain bandwidth, Raman contributions are neglected.

optically pumped CH_3F laser. Instead of tuning the wavelength, we have varied the temperature, and therefore the period of the waveguide, and monitored the transmitted signal during the heating cycle.

Figure 10 shows the measured transmissivity for two different diameters of the waveguide. The polarization of the laser beam is perpendicular to the grooves. We observe a remarkable decrease in the transmitted signal for temperatures above 40° C. The width ΔT of this resonances can be converted into a corresponding bandwidth $\Delta\lambda$. The measured bandwidth, shown in Fig. 11, depends on the waveguide height d. For an infinitely long waveguide, a lower limit for $\Delta\lambda$ can be taken from Fig. 4. Variations in the grating period and in the guide height d cause an increase of the bandwidth (Schmidt et al., 1974). For a waveguide of finite length R the bandwidth will be further enlarged. For weak coupling we can give a rough estimate for $\Delta\lambda$ from Flanders et al. (1974):

$$\Delta\lambda/\lambda \gtrsim L/R = 1/M, \tag{43}$$

where L is the grating period and M is the number of grooves. For the waveguide with $L = 248\ \mu m$ and $R = 32$ cm, we obtain $\Delta\lambda \gtrsim 0.19\ \mu m$. The condition of weak coupling implies a transmissivity near 1 at the Bragg frequency and is met in our configuration only for heights $d \gtrsim 3$ mm. For

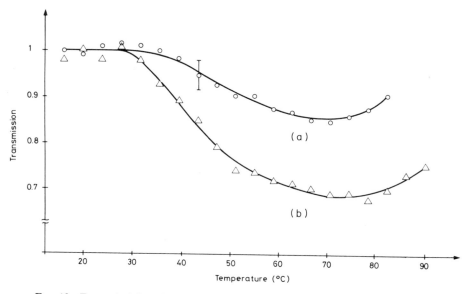

FIG. 10 Transmissivity of the DFB waveguide for various diameters. The transmissivity at room temperature is normalized to 1. $\lambda = 496\ \mu m$, $L = 32$ cm. (a) $d = 0.3$ cm, (b) $d = 0.2$ cm.

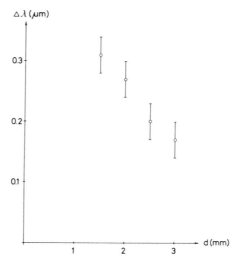

FIG. 11 Measured bandwidth $\Delta\lambda$ of the stop band versus waveguide height d. The waveguide length is 32 cm.

greater waveguide heights the transmission remains close to 100% even at the Bragg resonance. For the determination of the bandwidth under this condition, it is favorable to measure the reflectivity of the structure. However precise measurements were not possible because of pulse to pulse fluctuations of the laser output and the incomplete suppression of reflections from the front end of the guide.

C. VARIATION OF THE WAVEGUIDE PERIOD

The results of Section V.B demonstrate that the waveguide exhibits distributed feedback. Laser action can be expected if one overcomes the threshold for the given length of the structure. In the subsequent experiment the waveguide was pumped with the single mode output of a CO_2 laser at an intensity of 23 MW/cm^2. This is the intensity measured at the input of the glass tube which contains the DFB laser. Owing to the construction of the laser, the pump beam first passes a distance of about 30 cm in the tube before entering the waveguide. This reduces the power by about 35% at typical operating pressures of 5–6 torr. Therefore, the effective pump intensity was 15 MW/cm^2. At this power level the absorption is highly saturated (Temkin and Cohn, 1976).

The pulse energy measured with a Golay detector is shown in Fig. 12 as a function of the waveguide temperature for a height of about 2.5 mm. Below 30° C one observes a constant pulse energy. This signal is caused

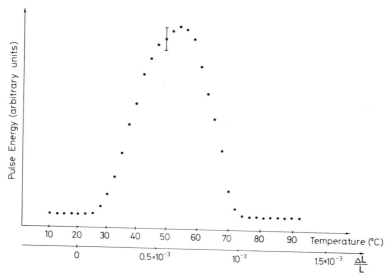

FIG. 12   Output energy as a function of the waveguide temperature T. The active pump intensity is 15 MW/cm² and the waveguide diameter is about 2.5 mm. $P = 6$ torr, $E_{pump} = 550$ mJ. Error bars indicate pulse to pulse fluctuations.

by superradiant emission which does not depend on the waveguide period. When the temperature is increased, we observe a resonance located at 55° C. This resonance clearly shows the resonator effect in our waveguide. The half-width of the peak is 27 K. It is difficult to predict the bandwidth since one has to take into account the possibility of Raman contributions to the gain. At the high pump intensities used on our experiment a significant part of the emission is expected to be of this type (Biron *et al.*, 1979). Furthermore, several K levels can be excited by a high-power pump laser. This also increases the gain bandwidth (Drozdowicz *et al.*, 1979).

Figure 13 shows the laser output as function of the waveguide temperature for a fixed waveguide height $d = 2$ mm. Two peaks separated by about 40 K are observed. They correspond to longitudinal modes. Their spacing is $\Delta R(\Delta q = 1, \Delta n = 0) = 250$ μm according to Eq. (27) for $\lambda = 496$ μm, $d = 2$ mm, and $n = 1$. This value corresponds to a relative length change of 7.8×10^{-4} or a 42 K temperature variation of brass with a thermal expansion coefficient of 1.85×10^{-5} K^{-1}.

DFB theory also predicts different thresholds for the longitudinal modes. This can be also observed in Fig. 13. If the laser is pumped with an energy of 320 mJ, corresponding to 14 MW/cm², one observes only one

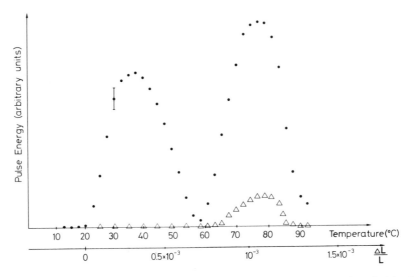

FIG. 13   Output energy as a function of the waveguide temperature T for a height $d =$ 2 mm. Two longitudinal modes with different thresholds are observed. ●, $P = 4.5$ torr; $E_{pump} = 320$ mJ; △, $P = 4$ torr; $E_{pump} = 210$ mJ.

mode compared to the two modes for a pump energy of 550 mJ. Since the threshold increases with increasing deviation from the Bragg frequency, it is obvious that this DFB laser operates on the high-frequency side of the stop band. This was expected since the period L has been chosen to fit half the free-space wavelength λ of the CH_3F emission.

Since the ground-state population of the lasing molecule is temperature sensitive the output power of all optically pumped gas lasers varies with temperature. In CH_3F one expects a decreasing pulse energy when the temperature is raised over room temperature. Walzer and Tacke (1980) observed a decrease of about 50% between 40 and 80° C in cw operation. Thus the effect of different threshold for various longitudinal modes is even stronger than indicated in Fig. 13.

As we see in Figs. 12 and 13, the location of the peaks depends on the waveguide height. This can be explained by the following consideration. An increase of the height d decreases the transverse wave number $n\pi/d$. This leads to an increase of the longitudinal wave number $2\pi/\lambda_g$ since

$$(n\pi/d)^2 + (2\pi/\lambda_g)^2 = (2\pi/\lambda)^2. \tag{44}$$

A greater longitudinal wave number means a shorter wavelength in the guide and therefore a shift of the peak in Fig. 11 to lower temperatures. This is observed when the waveguide diameter is varied between 1.75 and

2.5 mm (Fig. 14). The values of d in Fig. 14 are accurate only to 1/100 mm with respect to the reference height $d = 2.00$ mm. The absolute precision is ± 0.1 mm.

For an experimental prove of DFB operation of the laser care had to be taken to avoid reflections from the ends of the waveguide. The impedance mismatch and therefore the reflection coefficient is expected to be high for higher order transverse modes. DFB operation was confirmed by replacing the corrugated waveguide wall by a plane metal plate. As a consequence of this replacement the laser emission disappeared.

As mentioned in Section IV.A, a TM mode was expected to oscillate in

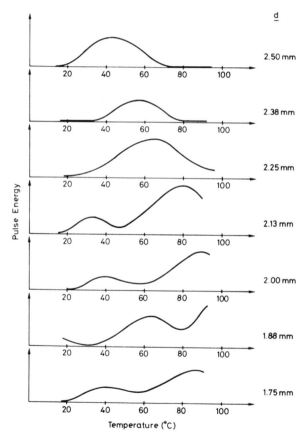

FIG. 14   Output energy versus waveguide temperature for various waveguide heights d. The values of d are accurate to 1/100 mm only with respect to the reference height $d = 2.00$ mm.

the CH_3F DFB laser. For this reason the polarization of the pump beam was chosen parallel to the grooves. The output beam is then polarized perpendicular to the grooves. The degree of polarization was greater than 97%, which was the limiting value of the polarizer. When the polarization of the pump beam was rotated by 90°, no output was observed although the losses for TE modes in a parallel plate guide are smaller than for TM modes. This shows that the field statements (28) in Section IV.A are correct.

The highest output energies were measured at temperatures above 70° C. As mentioned in Section IV.E, one expects a limited range for the waveguide height where lasing is possible. This could be verified. For diameters approximately below 1 mm and above 2.7 mm, laser action ceased completely.

D. GAS PRESSURE

Figure 15 illustrates the pressure dependence of the output signal for three different pump intensities. The optimum pressure lies between 4 and 7 torr. This is several times higher than for conventional optically pumped CH_3F lasers (Cohn *et al.*, 1976; Evans *et al.*, 1977). The reason for these high optimum pressures is the very short cavity length of only 32 cm. Plant *et al.* (1974) have shown that the optimum cell pressure increases when the resonator length is reduced. Operating pressures exceeding 20 torr were found for a 39-cm long, unstable resonator pumped with a CO_2 energy of 1.14 J (Ewanizky *et al.*, 1979).

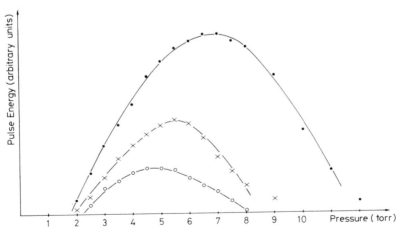

FIG. 15   Pressure dependence of the Golay signal for three different pump intensities. The indicated pump intensities are measured at the input window of the glass tube and must be reduced about 35% to obtain the intensities reaching the DFB waveguide, ●, 23 MW/cm²; ×, 14 MW/cm²; ○, 10 MW/cm².

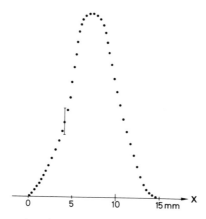

FIG. 16 Transverse mode pattern in horizontal direction measured at a distance of 15 cm from the waveguide end. $d = 2.13$ mm, $T = 74°$ C.

E. LASER-BEAM CHARACTERISTICS

The transverse energy distributions of the output beam were measured by scanning a Golay cell with a 1.5-mm diam diaphragm across the beam profile. The horizontal and vertical field distributions are given in Figs. 16 and 17 for a temperature of 74° C and a waveguide height of 2.13 mm. Measurements were made at a distance of 150 mm from the end of the wave-guide. The single peaks in Figs. 16 and 17 indicate mode numbers $n = m = 1$ in Eq. (23).

The fact that only the lowest order mode exists once again proves that

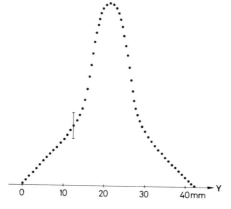

FIG. 17 Transverse mode pattern in vertical direction measured at a distance of 15 cm from the waveguide end. $d = 2.3$ mm, $T = 74°$ C.

reflections from the waveguide ends can be neglected. The power reflection coefficient for the lowest order TM mode is (Klages, 1956)

$$r = [\sqrt{1 - (\lambda/2d)^2} - 1]/[\sqrt{1 - (\lambda/2d)^2} + 1]. \tag{45}$$

A waveguide height d of 2.1 mm leads to a reflection coefficient $r = 1.2 \times 10^{-5}$, which is orders of magnitude too small for laser action.

The beam divergence in both directions was calculated from the field distributions measured at various distances from the waveguide end. The evaluated full-angle divergence in the horizontal plane was 38 mrad, which is near the diffraction limit. The divergence in the vertical direction depends on the waveguide height and was also found to be diffraction limited.

F.   SPECTRAL ANALYSIS OF THE LASER EMISSION

The wavelength and the spectral content of the submillimeter radiation were measured with the aid of a scanning Fabry–Perot interferometer. The plate spacing was set to 30.5 cm, giving a free spectral range of 492 MHz. The grating constant of the meshes was 63.5 μm and the measured finesse was 22. A typical interferogram is shown in Fig. 18. The wavelength is found to be 496 μm as expected. A strongly dominant mode was observed in agreement with the measurements of the transverse field distributions in Figs. 16 and 17. The full width at half maximum of the

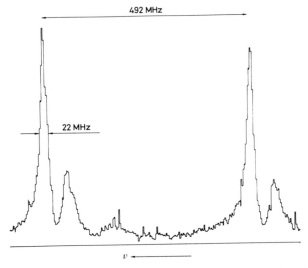

FIG. 18   Fabry–Perot interferogram of the DFB laser emission. The linewidth is limited by the spectral resolution of the interferometer. $d = 2.3$ mm, $T = 91°$ C.

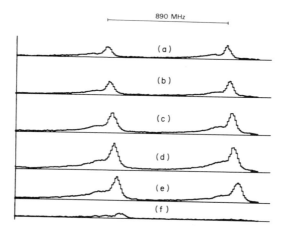

FIG. 19   Fabry–Perot interferograms of the laser emission at various waveguide temperatures. (a) 72, (b) 63, (c) 53, (d) 45, (e) 32, and (f) 22° C. The waveguide height is 2.5 mm; $P = 5$ torr.

dominant mode is about 22 MHz, a value that corresponds to the spectral resolution of the interferometer.

Since the temperature of the waveguide determines the exact period, a temperature change should result in an alteration of the emission frequency. This is shown in Fig. 19 for a waveguide height of 2.5 mm. The reduction in the finesse compared to the interferogram in Fig. 18 is probably caused by a slight misalignment of the meshes.

The strong dependence of the emission frequency on waveguide temperature offers a simple possibility of frequency stabilization of the DFB laser. A temperature stability of 0.1 K, which is relatively simple to achieve, keeps the frequency within 230 kHz or 4×10^{-7}. This is a good value for a passive stabilization. It could be interesting for a cw source used as a local oscillator in a heterodyne experiment.

G.   PULSE SHAPE AND ENERGY

Studies of the temporal behavior of the pulses were performed using a photon drag detector for the CO_2 beam and a Schottky barrier diode for the submillimeter beam. Figure 20 shows the time resolved shapes of the pump and submillimeter pulses for 5.5-torr CH_3F pressure. The pulsewidth of the CH_3F pulse is about 50 ns. This corresponds to a minimum linewidth of 20 MHz, which is slightly less than the value observed in Fig. 18.

The laser pulse energy was measured with a pyroelectric Joulemeter (Molectron J3-05). The uncorrected reading corresponds to an energy of

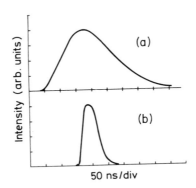

FIG. 20   Temporal shape of the CO_2 (a) and CH_3F (b) pulses. The CH_3F cavity pressure was 5.5 torr.

2 μJ when pumped with a single mode pulse. Pumping with the multimode output of the CO_2 laser leads to an energy about a factor of 5 higher but the shot to shot fluctuations also increase. A pulse with a half-width of 50 ns and an energy of 2 μJ reaches a peak power near 40 W. Yet it was not the aim of this study to achieve high pulse powers in the far infrared.

VI.   Future Developments

An optically pumped CH_3F 496-μm DFB laser as described in Section V allows the experimental study of DFB in laser waveguides with periodic corrugations. Besides, optically pumped DFB far-infrared lasers exhibit two advantages. First, they avoid spectrally selective resonator mirrors adapted to both the pumping and the laser wavelength. Second, their compact and short structures allow an increase of the operating gas pressure as demonstrated in Section V.D. On the other hand, when aiming for high output powers in the far infrared, it may be advantageous to replace the DFB structure by a standard waveguide containing the laser-active gas and distributed Bragg reflectors (DBR). They correspond to the passive waveguide described in Sections IV.B and V.B.

The real goal for research in DFB gas lasers is probably the operation of cw and TEA CO_2 lasers as proposed and attempted by Marcuse (1972) and Miles and Grow (1978, 1979). However, a different approach may be necessary. Miniature high-pressure TEA CO_2 DFB lasers would exhibit high mode selectivity even for very broad gain widths.

Finally, it should be mentioned that gas lasers are suited for the development of DFB with helical structures as proposed by Affolter and Kneubühl (1976b). This should be attempted with connection to the techniques and structures used in microwave tubes and in free-electron lasers.

ACKNOWLEDGMENTS

We are indepted to K. J. Button (Cambridge, Massachusetts), P. D. Coleman (Urbana), E. Danielewicz (Los Angeles), R. Daendliker (Neuenburg), D. T. Hodges (Los Angeles), B. Lax (Cambridge, Massachusetts), W. Lukosz (Zürich), R. O. Miles (Washington), Ch. Sturzenegger (Aarau), M. Yamanaka and H. Yoshinaga (Osaka) for many fruitful discussions on DFB or optically pumped gas lasers and related problems. Parts of the experiments described in Section V were performed with the support by B. Barbisch, S. Gnepf, H. Jaeger, and H. P. Preiswerk. We are also indebted to Henry Hauser AG, Biel, for the manufacture of periodic brass structures and to M. V. Schneider (Holmdel) for supplying the Schottky barrier diode chip we used for the pulse measurements mentioned in Section V,G.

Our research on DFB gas lasers is supported by the ETH Zürich, the Swiss National Science Foundation, the Kommission zur Förderung der wissenschaftlichen Forschung, and the GRD of the EMD.

REFERENCES

Affolter, E. (1980). *PhD Thesis, ETH Zürich.*
Affolter, E., and Kneubühl, F. K. (1976a). *Phys. Lett.* **58A,** 91–92.
Affolter, E., and Kneubühl, F. K. (1976b). *Phys. Lett.* **58A,** 183–184.
Affolter, E., and Kneubühl, F. K. (1977). *J. Appl. Math. Phys.* (*ZAMP*) **28,** 1171–1175.
Affolter, E., and Kneubühl, F. K. (1979). *Phys. Lett.* **74A,** 407–408.
Affolter, E., and Kneubühl, F. K. (1971). *IEEE J. Quantum Electron.* **QE-17,** 1115–1122.
Biron, D. G., Temkin, R. J., Lax, B., and Danly, B. G. (1979). *Opt. Lett.* **4,** 381–383.
Bjorkholm, J. E., and Shank, C. V. (1972). *IEEE J. Quantum Electron.* **QE-8,** 833–838.
Bjorkholm, J. E., Sosnowski, T. P., and Shank, C. V. (1973). *Appl. Phys. Lett.* **22,** 132–134.
Brillouin, L. (1946). "Wave Propagation in Periodic Structures." McGraw Hill, New York and Dover, New York (1953).
Butcher, P. N. (1956). *Proc. IEEE* **103,** 301–306.
Casey, K. F., and Weaver, L. (1973). *J. Math. Phys.* **14,** 647–650.
Chinn, S. R. (1973). *IEEE J. Quantum Electron.* **QE-9,** 574–580.
Cohn, D. R., Button, K. J., Temkin, R. J., and Drozdowicz, Z. (1976). *Infrared Phys.* **16,** 429–434.
Degnan, J. J. (1976). *Appl. Phys.* **11,** 1–33.
De Wames, R. E., and Hall, W. F. (1973a). *Appl. Phys. Lett.* **23,** 28–30.
De Wames, R. E., and Hall, W. F. (1973b). *Appl. Phys. Lett.* **44,** 3638–3640.
Drozdowicz, Z., Temkin, R. J., and Lax, B. (1979). *IEEE J. Quantum Electron.* **QE-16,** 170–178, 865–869.
Elachi, C. (1976). *Proc. IEEE* **64,** 1666–1698.
Evans, D. E., Sharp, L. E., Peebles, W. A., and Taylor, G. (1977). *IEEE J. Quantum Electron.* **QE-13,** 54–58.
Ewanizky, T. F., Bayha, W. T., and Rohde, R. S. (1979). *IEEE J. Quantum Electron.* **QE-15,** 538–540.
Flanders, D. C., Kogelnik, H., Schmidt, R. V., and Shank, C. V. (1974). *Phys. Lett.* **24,** 1994–1996.
Harrel, E. M. (1979). *Ann. Phys.* **119,** 351–369.
Hodges, D. T., and Hartwick, T. S. (1973). *Appl. Phys. Lett.* **23,** 252–253.
Jaggard, D. L., and Elachi, C. (1976). *J. Opt. Soc. Am.* **66,** 647–682.
Kaminow, I. P., Weber, H. P., and Chandross (1971). *Appl. Phys. Lett.* **18,** 497–499.
Klages, G. (1956). "Einführung in die Mikrowellenphysik." Steinkopff Verlag, Darmstadt.

Kneubühl, F. K. (1977). *J. Opt. Soc. Am.* **67**, 959–963.

Kneubühl, F. K. (1979). *Proc. Int. Conf. Lasers, Orlando, Florida* pp. 812–816.

Kneubühl, F. K., and Affolter, E. (1979). *In* "Infrared and Millimeter Waves" (K. J. Button, ed.), Vol. I, pp. 235–278. Academic Press, New York.

Kneubühl, F. K., and Affolter, E. (1980). *J. Opt.* **11**, 449–453.

Kogelnik, H. (1969). *Bell Syst. Tech. J.* **48**, 2909–2947.

Kogelnik, H., and Shank, C. V. (1971). *Appl. Phys. Lett.* **18**, 152–154.

Kogelnik, H., and Shank, C. V. (1972). *J. Appl. Phys.* **43**, 2327–2335.

Kogelnik, H., Schank, C. V., and Bjorkholm, J. E. (1973). *Appl. Phys. Lett.* **22**, 135–137.

Kohn, W. (1959). *Phys. Rev.* **115**, 809–821.

Kuhl, J., and Schmidt, W. (1974). *Appl. Phys.* **3**, 251–270.

Magnus, W., and Winkler, S. (1966). "Hill's Equation." Wiley, New York.

Marcatili, E. A. J., and Schmeltzer, R. A. (1964). *Bell Syst. Tech. J.* **43**, 1783–1809.

Marcuse, D. (1972). *IEEE J. Quantum Electron.* **QE-8**, 661–669.

Marcuvitz, N., (1951). "Waveguide Handbook." McGraw Hill, New York.

Mc Kean, H. P., and Trubowitz, E. (1976). *Commun. Pure Appl. Math.* **29**, 143–226.

Mc Kean, H. P., and van Moerbeke, P. (1975). *Inv. Math.* **30**, 217–274.

Meiman, N. N. (1977). *J. Math. Phys.* **18**, 834–848.

Miles, R. O. (1980). Private communication.

Miles, R. O., and Grow, R. W. (1976). *J. Opt. Soc. Am.* **66**, 292.

Miles, R. O., and Grow, R. W. (1978). *IEEE J. Quantum Electron.* **QE-14**, 275–283.

Miles, R. O., and Grow, R. W. (1979). *IEEE J. Quantum Electron.* **QE-15**, 1396–1401.

Miles, R. O., and Grow, R. W. (1981). *IEEE J. Quantum Electron.* **QE-17**, 1071–1074.

Nakamura, M., Yen, H. W., Yariv, A., Garmire, E., and Garvin, H. L. (1973). *Appl. Phys. Lett.* **22**, 515–516; **23**, 224–225.

Plant, T. K., Newman, L. A., Danielewicz, E. J., De Temple, T. A., and Coleman, P. D. (1974). *IEEE Trans. Microwave Theory Tech.* **MTT-22**, 988–990.

Reynaud, F. (1978). *Rad. Effects* **39**, 181–188.

Rosenbluh, M., Temkin, R. J., and Button, K. J. (1976). *Appl. Opt.* **15**, 2635–2644.

Schinke, D. P., Smith, R. G., Spencer, E. G., and Galvin, M. F. (1972). *Appl. Phys. Lett.* **21**, 494–496.

Schmidt, R. V., Flanders, D. C., Shank, C. V., and Standley, R. D. (1974). *Phys. Lett.* **25**, 651–652.

Schwaller, P., Steffen, H., Moser, J. F., and Kneubühl, F. K. (1967). *Appl. Opt.* **6**, 827–829.

Shank, C. V., Bjorkholm, J. E., and Kogelnik, H. (1971). *Appl. Phys. Lett.* **18**, 395–396.

Shellan, J. B., Agmon, P., Yeh, P., and Yariv, A. (1978). *J. Opt. Soc. Am.* **68**, 18–27.

Shubert, R. (1974). *J. Appl. Phys.* **45**, 209–215.

Smith, P. W. (1971). *Appl. Phys. Lett.* **19**, 132–134.

Steffen, H. (1968a) PhD Thesis, ETH, Zurich.

Steffen, H., and Kneubühl, F. K. (1968b). *Phys. Lett.* **27A**, 612–613.

Steffen, H., and Kneubühl, F. K. (1968c). *IEEE J. Quantum Electron.* **QE-4**, 992–1008.

Streifer, W., Burnham, R. D., and Scifres, D. R. (1975). *J. Appl. Phys.* **46**, 946–948.

Temkin, R. J., and Cohn, D. R. (1976). *Opt. Commun.* **16**, 213–217.

Vigneron, J. P., and Lambin, P. (1979). *J. Phys. A* **12**, 1961–1970.

Walzer, K., and Tacke, M. (1980). *IEEE J. Quantum Electron.* **QE-16**, 225–257.

Wang, S. (1973). *J. Appl. Phys.* **44**, 767–780.

Wang, S. (1974a). *Opt. Commun.* **10**, 149–153.

Wang, S. (1974b). *IEEE J. Quantum Electron.* **QE-10**, 413–427.

Wang, S. (1975a). *Appl. Phys. Lett.* **26**, 89–91.

Wang, S. (1975b). *Wave Electron.* **1**, 31.

Wang, S., and Tsang, W. T. (1974). *J. Appl. Phys.* **45,** 3978–3980.

Wang, S., Cordero, R. F., and Tsang, Ch.-Ch. (1974). *J. Appl. Phys.* **45,** 3975–3977.

Watkins, D. A. (1958). "Topics in Electromagnetic Theory." Wiley, New York.

Yamanaka, M. (1976a). *Rev. Laser Eng. (Jpn.)* **3,** 253–294.

Yamanaka, M. (1976b). Private communication.

Yamanaka, M. (1977). *J. Opt. Soc. Am.* **67,** 952–958.

Yamanaka, M., Tsuda, H., and Mitani, S. (1975). *Opt. Commun.* **15,** 426–428.

Yariv, A. (1973). *IEEE J. Quantum Electron.* **QE-9,** 919–933.

Yariv, A., and Gover, A. (1975). *Appl. Phys. Lett.* **26,** 537–539.

INDEX

A

Absorption lines, measurement of, 44–46
AID effect, effect on TUNNETT diode, 228
Ammonia, Stark effect in, 136–137
Ammonia FIR laser, saturation effect in output of, 172
Avalanche diode, distributed structure of, 261
Avalanche multiplication phenomenon, 215–216
  dispersion effects in, 217–227

B

Ballistic transistor, operation characteristics of, 263
Bipolar transistor (BPT), comparison with other transistors, 263
Brillouin semiconductor lasers, uses of, 264

C

Cadmium compounds, cyclotron resonance studies on, 13
Carbon dioxide lasers, low-pressure type, 32–34
Carcinotron, applications of, 6
CARS (Coherent Antistokes Raman Scattering) technique, 21
Coherent sources
  for resonance spectroscopy, 1–28
  electronic devices, 6–7
  free-electron laser, 8–9
  harmonic generation, 2–5
  optical pumping, 5–6
  scientific applications, 9–25

Cold tests on gyrotrons, 298–299
Coupled-wave theory applied to DFB gas lasers, 311–312
Cryogenic detectors for FIR emission, 43
Cyclotron resonance of microwaves, 9–14
Cyclotron resonance maser (CRM) use to extend coherent sources, 6–7

D

Diamond, cyclotron resonance trace in, 12
Dielectric waveguides, for DFB gas lasers, 312–313
Diode laser spectra, of methyl alcohol, 64–65
Dispersion effects, in avalanche phenomena, 217–227
Distributed-feedback (DFB) gas lasers
  construction of, 322–324
  coupled-wave theory for, 311–312
  dielectric waveguides for, 312–313
  dispersion relations of, 308–311
  future developments in, 334
  gas pressure of, 330
  heaviside modulation studies of, 310–311
  historical aspects of, 306, 307–308
  laser-beam characteristics of, 331–332
  metallic waveguides for, 313–314
  operation of, 322–333
  in periodic metal guides, 314–322
  pulse shape and energy of, 333–334
  resonance and threshold conditions of, 311–314, 317–318
  spectral analysis of emission of, 332–333
  waveguide-period variation of, 326–330
  waveguiding in, 306–307
Double resonance spectrometer, schematic diagram of, 3

Double resonance techniques, for FIR
    emission studies, 56–57

E

Electronic devices, as coherent sources for
    resonance spectroscopy, 6–7
Emission lines, measurement of, 46–49
Esaki diode, distributed structure of, 261

F

Far-infrared (FIR) lasers
    in ac electric fields, 204–209
        effects on output power, 205–206
        high-speed FM modulation by, 206–
            208
    conventional and Stark waveguides of,
        159–166
    in dc electric fields, 176–177
    electrical breakdown in, 166–169
    general performances of, 155–176
    with high efficiency and stability, 155–
        159
    Lamb dip generation by, 172–176
    power and frequency measurements by,
        169–171
    saturation effects in output of, 171–176
    small frequency tuning of, 187–188
    Stark operation of, 129–213
Far-infrared (FIR) laser emission
    detection of, 42–44
    development of, 30–31
    diagnostic techniques by, 44–57
    double resonance techniques for, 56–57
    electric field effects on, 192–196
    instrumentation for, 32–44
    limitations of, 93–94
    methyl alcohol in studies of, 57–94
    molecular spectroscopy by, 29–128
    optoacoustic Stark spectroscopy investi-
        gation of, 196–200
    polarization effects of, 49–52
    pump laser use in, 32–37
    resonators for, 37–42
    spectrometer for, schematic diagram, 32
    Stark effect on, 53–55, 87, 177–186
    TEA laser use in, 34–35
    two-photon pumping in, 36–37

wavelength and frequency measurements
    by, 44–49
Field effect transistor (FET), frequency of
    operation of, 263
FM modulation, high-speed, generation by
    FIR lasers, 206–208
Free-electron laser, as coherent source, 8–
    9
Frequency measurements of, 44–49

G

GaAs TUNNETT diodes, 215–266
Gas lasers, distributed feedback type, *see*
    Distributed-feedback gas lasers
Ge:Ga detector, for FIR emission, 43
Germanium, cyclotron resonance trace in,
    10, 11
Golay cell, as FIR detector, 43
Gunn diode, distributed structure of, 261
GUNNETT diode
    operation principle of, 261
    schematic illustration of, 262
GUNPATT diode, operation principle of,
    261
Gyrotrons
    cold tests of, 298–299
    complications in technology of, 269–270
    electron-beam performance of, 299–300
    tube performance of, 300–301
    types of, 271
Gyrotron amplifiers, 267–304
    experiments on, 275–280
    performance and prospects for, 280–281
    theoretical aspects of, 273–275
Gyrotron oscillators, 267–304
    applications of, 281
    beam current of, 290
    beam energy of, 292–293
    cavity length of, 290–291
    cavity mode of, 291
    efficiency enhancement of, 293–294
    performance of
        high average, 296
        parametric dependence of, 290–293
        qualitative aspects of, 286–289
        summary, 296–298
    schematic of, 282
    stability and coherence of, 295–296
    theoretical aspects of, 283–286

H

Harmonic generation, nonlinear mixing and, 2–5
Harmonic modulation, in DFB structures, 309–310
Helium, cyclotron resonance studies on, 10
Hill differential equation, dispersion relations based on, 308–311
Hollow dielectric waveguide resonators, for FIR laser signals, 41–42
Hybrid resonators, for FIR laser signals, 42

I

IMPATT diode
    equivalent circuit of, 236
    operation principle of, 215–216
Indium antimonide, cyclotron resonance studies on, 12, 16
Isotopic lasers, use in FIR studies, 37

L

Lamb dip
    generation of, 172–176
    spectroscopy, IR-FIR transferred, 188–192

M

Magneto-optical effects, spin-flip laser and, 14–15
Maser, development of, 2
MASER oscillators, optically pumped FIR sources as, 132
Mathieu equation, 309
Metal Insulator Semiconductor (MIS)
    diode structures for tunnel injection as a, 240
Metallic waveguides, for DFB gas lasers, 313–314
Metallic waveguide resonators, for FIR laser signals, 40–41
Methyl alcohol
    equilibrium structure of, 58
    in FIR emission studies, 57–94
        data on, 94–100
        diode laser spectra, 64–65
        internal rotation, 62–64

torsion-rotational model, 65–70
    vibration-rotation spectrum, 61–62
    with small frequency tuning, 187–188
molecular parameters for, 67
Paschen curve for, 168
Stark effect in, 134–136, 178–179, 183–186
    expected line shapes, 148–154
    vibrational modes of, 59
Methyl alcohol laser, application of, 18
Methyl fluoride
    FIR lines of, Stark effect on, 178–180
    Paschen curve for, 168
Methyl fluoride DFB laser, characteristics of, 324
Methyl fluoride laser
    cavity length in, 163–164
    stability of, 158–159
Microwaves
    cyclotron resonance of, 9
    application, 10
Microwave tubes, state of the art, 268
MITATT diode, development of, 216
Molecules, off-resonant pumping of, 176–177
Molecular spectroscopy, by FIR laser emission, 29–128
Motional Stark spectroscopy, applications of, 21–25

N

Nickel-tungsten contacts, for frequency measurements, 47–48
Nonlinear mixing, harmonic generation and, 2–5

O

Open resonators for FIR laser signals, 38–40
Optical pumping, in submillimeter region, 5–6
Optoacoustic technique, for absorption-line measurement, 44–45

P

Paschen law, equation for, 166
Plasmas, ion temperature measurement of, 19–21

Plasma diagnostics, scientific applications of, 18–21

p-n diode, structures for tunnel injection as a, 240

Poisson's equation, 220, 232

Polarization effects, of FIR emission, 49–52

Polar molecule
  energy level diagram of, 131–132
  Stark effect in, 133–134

Pump laser, for FIR studies, 32–37

R

Raman semiconductor lasers, uses of, 264

Raman transitions, in submillimeter lasers, 15–18

Resonance conditions, in DFB gas lasers, 311–314

Resonance spectroscopy, coherent sources for, 1–28

Resonators, for FIR laser signals, 37–38

Rotovibrational transitions, Stark effect on, 137–138

Ruby laser, Thomson scattering by, 19

S

SAW (Surface Acoustic Wave) dispersive delay line, 21

Schottky barrier diode
  fabrication of, 249–250
  structures for tunnel injection as a, 240
  uses of, 2–3, 43

Semiconductors
  avalanche multiplication phenomenon in, 215–216
  cyclotron resonance studies on, 10

Sequence band lasers, use in FIR studies, 37

Solid-state lasers, distributed-feedback gas lasers and, 305–306

Spin-flip laser, magneto-optical effects and, 14–15

Stark effect
  in ammonia, 136–137
  on FIR emission, 53–55, 87

in methyl alcohol, 134–136
  on rotovibrational transitions, 137–138

Stark energy, equations for, 133, 135

Stark frequency shifts, 139–141

Stark spectroscopy
  in far-infrared lasers, 129–213
  high resolution by, 201–204
  investigation of pump saturation by, 196–200
  laser gain curves in electric field, 139–154
  of polar molecules, 133–138

Static electric field, theory of molecules in, 54

Static Induced Transistors (SIT)
  characteristics of, 264
  frequency of operation of, 263

Submillimeter lasers, Raman transitions in, 15–18

Superheterodyne principle, in FIR detection, 43–44

T

TEA lasers, use in FIR studies, 34–35

Transferred Lamb dip spectroscopy
  application of, 191–192, 209
  Stark effect in, 188–192

TUNNETT diodes, 215–266
  design of, 238–242
  doping profile of, 238
  efficiency of, 230–231, 239–240
  electric field profile of, 238
  future of, 260–264
  materials for, comparison of, 243–244
  noise in, 258–260
  oscillation characteristics of, 257
  output power of, 260
  preparation of, 244–246
  temperature dependence of tunnel injection of, 229–230
  time constant in tunnel junction of, 228–229
  transit time effect of carriers on, 232–237

Two-photon pumping, use in FIR studies, 36–37

V

Varactor diode, millimeter wave generation by, 3

W

Wave equation, application to DFB gas laser, dispersion, 308–309

Waveguiding, in distributed feedback gas lasers, 306–307
Wavelength, measurements of, 44–49

Z

Zeeman spectra of Rydberg states, 24